DARK ENERGY
Solved by Big B

- shows what Dark Energy r
- Solves problems of the big bang
- Solves other vexed questions in cosmology

Peer reviewed and published by Pearson (1997), this theory predicted the accelerating nature of the expanding universe BEFORE its discovery in 1998! This means it offers a unique achievement not paralleled anywhere else worldwide.

Most of the Six ideas for new research described could win anybody having maths or physics qualifications a NOBEL PRIZE! **(See page 271)**
Our hope is that this should encourage mathematicians and computer specialists to continue where our study ends.

This book is divided into TWO PARTS

PART I of this book is explanatory with no maths and is suitable for the non-scientist. However, it is also needed as an introduction for the mathematicians. It shows why the big bang theory at least needs to be supplemented by a theory showing how energy can arise from the void. Prediction of the acceleration is part of the solution.
PART II, the major section, provides full mathematical detail. The calculus required is little higher than about school A level standard.
(See page **282** for a **SHORT LIST OF ACHIEVEMENTS**)

The names of some physicists are included in 'Acknowledgements'. Although the assistance given by all of them was real and valuable, help, except for two of them, was not intended. One was the **Nobel Laureate Brian Josephson** of Cambridge University. His reasons, summarised in 'Acknowledgements', are similar to those of the others but all had to admit they were unable to actually find any fault in the logic presented. This is why we say their involvement provides us with credibility.

Furthermore, physicist Stephen Adler (2004) followed our lead, first published by the Russian Academy of Sciences in 1994, in having the quantum realm emerge from an ultimate level of reality.

CONTENTS LIST: PART I see page VIII: FIG.s page IX .
CONTENTS LIST: PART II starts on page 78, FIG.s page 82

Published by www.Lulu.com

Author: Ronald Denzil Pearson

Seventh edition
Date: November 2010

ISBN 978-1-4466-3145-4

Title: DARK ENERGY MYSTERY
Solved by Big Breed Theory

- shows what Dark Energy really is
- Solves problems of the big bang
- Solves other vexed questions in cosmology

PART I is explanatory with no maths
PART II provides full mathematical detail

Includes Many black on white illustrations.

Two plates should have been in colour to provide the meaning required. Unfortunately this proved impractical. They can be viewed in colour, however, by internet download of our book:

The Big Breed Theory of Creation of the Universe

Access details given at the end of this book

Source File: DarkEnergyBigBreedD161110BookEd7ISBN

FOREWORD
Physics is now in crisis!

By their own admission theoretical physicists have lost their way! We show the reason is loss of expertise in Classical Physics, which now resides in a mathematically based allied discipline. This book has been written mainly to help put theoreticians back on track.

You don't believe us? Then read what was reported from a major conference:

That a crisis exists was admitted in reports from the 23rd Solvay Conference on Physics at Brussels that occurred early December 2005. *New Scientist* (10th Dec. 2005) reports the anxiety shown by Nobel Laureate, David Gross. He compared the position with that of 1911 when radioactivity baffled everybody and says,

"We are missing perhaps something as profound as we were back then".

The editorial is headed *"Ideas needed"*, and extracts from it say: *"Physics greatest endeavour has ground to a halt. We are in a period of utter confusion."*

"General relativity is incompatible with quantum theory. Since the 1960's theorists have struggled to solve this problem, so far to no avail. And the trouble is we have nothing to put in relativity's place."

"String Theory cannot explain the accelerating expansion of the universe or tell us where space and time come from".

The New Scientist report suggested that a consensus of opinion at the conference held that string theory had failed. It went on to say that many of the greatest minds in physics were there at last week's conference but that none had an answer.

The latest admission of failure came from top physicist Lee Smolin (2007) who states in his book that his generation has failed by providing nothing new of any value since 1980.
An extract from a website says much the same:

"Physicist Sir Roger Penrose says physics is wrong, from string theory to quantum mechanics."
Then: "Penrose did not flinch from questioning the central tenets of modern physics, including string theory and quantum mechanics. Physicists will never come to grips with the grand theories of the universe, Penrose holds, until they see past the blinding distractions of today's half-baked theories to the deepest layer of the reality in which we live." (Which is what our book does) See:

http://discovermagazine.com/2009/sep/06-discover-interview-roger-penrose-says-physics-is-wrong-string-theory-quantum-mechanics

The reader is likely to be surprised to hear that the author of this book had been expecting this to happen since 1987. What follows next gives the reason.

A letter to this author, dated August 2nd 1987, had come from **top physicist Professor J.P. Vigier of Paris**. He was also the gravitational consultant for the prestigious journal 'Physics Letters A' to which a critique had been submitted.
An extract from the letter went:
"Your letter now seems to be correct as far as our referee's and myself can see……I feel indeed that supporters of the big bang theory (of which I am not) should discuss the energy problem at creation time (if there is one) and your contribution should not be ignored. (You can utilise this statement if you want) and be published. I hope you will succeed in this.
Yours sincerely,
J. Vigier

This letter refers to the critique of an article published in 1984 by Professor E. Tryon (1984). The critique had been submitted to one scientific journal after another during three years. All others had rejected the critique as invalid mostly on grounds it was established physics. In APPENDIX I of PART I the flaw is explained together with others. Also demonstrated is that over-dependence on mathematical rules, without checks by 'conceptual logic' can lead theoreticians onto false tracks.

Another cause is the way students are taught to accept internal contradiction as "Non-intuitive aspects the physicist has to learn to live with". We say they should be told theories having such features can only be regarded as provisional until a better theory appears. Theorists are all masters of mathematics but somehow are making logical errors that are obvious to others outside their discipline. The engineering discipline is mathematically based but is taught in a different way. In the area of overlap with physics we seem to have retained expertise that physics has lost. This loss is probably due to the search for greater mathematical sophistication and the huge intellectual demands required for comprehending Einstein's relativity and quantum theories. Nobody can be a master of everything. Therefore if only the physicists would listen to our critiques it is very likely that we could provide the help they require.

One of our books, *CREATION SOLVED?*, details the way the concept of 'gravitational potential energy' is being misused so throwing theorists off track and the same book details flaws in the logic of the big bang theory. This shows why it makes two hopelessly wrong predictions.

Following the non-mathematical PART I:

APPENDIX I gives details of the way Tryon reaches a wrong conclusion by misusing gravitational potential energy.

APPENDIX II shows how the logic of the 'inflation' period of the big bang theory was flawed:

APPENDIX III shows why an alternative to Einstein's relativity theories had to be found.

APPENDIX IV summarises the mechanics used to replace relativity and describes how it provides quantum gravity. Full mathematical detail of the theory is provided in our book, *QUANTUM GRAVITY via an Exact Classical Mechanics*.

After REFERENCES details of our book series are provided

Reference is frequently made to 'we' and the reader may wonder to whom reference is made. The list of acknowledgements gives many of the growing number of supporters who accept the theory as valid and have helped in some way to assist in its promotion. These are the collective 'we' to whom reference is made.

When a date in brackets follows a name such as, Tryon (1984), this signifies more details in surname alphabetical order will be found in **REFERENCES**

<u>For scientists only</u>: please first read APPENDIX I on page 62

This advice is given because we have discovered that people with a scientific training confuse the 'negative energy', used in this book, with potential energy and then fail to understand the text.

See page VIII for list of contents & page IX for list of figures

Acknowledgements

Physicists

The author is indebted to a number of physicists who have provided the critiques that added credibility, improved rigour and reduced the possibility of misunderstandings in the text.

The Nobel laureate of Cambridge, Professor Brian Josephson FRS admitted in 2006 that he found no fault in logic, though making it clear he hoped a flaw would be found (My theory competes with his, "Mind/Matter Unification Project), The Cambridge Professor of Ocean Physics, Peter Wadhams gave support to part of the theory during the Jeff Rense Radio Show in 2001. Professor J.P. Vigier (Paris) in 1987 supported the first critique and tried to help its publication. Dr. S. Nicholls organised the first lecture at Leeds.

The late Dr. John Day made most useful mathematical contributions. After reading my first book of 1990 he produced a simpler solution to one of the problems and showed that some of Einstein's tensor maths could be transferred to the new ECM approach. This added extra value and John's contributions appear in both books *CREATION SOLVED?* and *QUANTUM GRAVITY.*

Roger Anderton, provided useful critiques but Dr W. Hurlbut and two quantum physicists under the pseudonyms RudolphRigger and NormallyNormal provided other valued critiques that helped show where revision was needed (they were not attempting to be supportive).

Engineers

The inventor of the caesium beam atomic clock, Dr Louis Essen FRS arranged for me to go as his deputy to a Russian conference in 1991 to present ECM theory. Dr G.B.R. Feilden FRS, designer of the first industrial gas turbine made great efforts to have the theory properly assessed by physicists over several years. Professor J.H.Horlock, Vice President of the Open University, also tried to assist in this way in 1985. Andrew Wyon, Managing director of Seward Wyon has given very strong support and encouragement from 1990 onwards. He also made efforts to have the work evaluated by Durham University and tried to help launch an experiment in Earth orbit. Dr. Harold Aspden sent copies of his papers. Al Kelly provided useful data during E-mail correspondence and his book provided extra information. Ron Staniforth provided expertise in the design of accurate quartz clocks that will be useful in design of a new instrument that needs to be flown in orbit. This could create a paradigm-shift. Mr Vince F Wise, Electrical/Electronic/Medical Engineer did a magnificent and voluntary proof reading job. Alan Middleton donated several books giving support by providing data supporting the theories .

Geologist Rory Macquisten has provided constant encouragement and assistance by proof reading, literature searches, the creation of the www.pearsonianspace.com website and in other ways, particularly in regard to thinking of novel ways for marketing the three books of which this is the first to emerge. It was his proposal and help that resulted in publication through the Lulu internet system. We have regularly met for long walks with our dogs to discuss problems of how to communicate the contents of these books knowing that, from experience, established publication routes are not available.

Industrial Chemist: Keith Hudson has provided support and, during a ramble, made a suggestion that led to the jump and dwell model of particle motion that is a key feature of the theory. He also sent a letter of protest to the Bath Royal Literature and Science Institution. (BRLSI) concerning the way a leading member, Victor Suchar, had deliberately attempted to discredit a lecture I was presenting, about the creation problem, by breaking in and taking out 20 minutes without waiting for the discussion.

Computer Expert: Tony Agar has provided essential assistance. Without this input these books would never have been completed.

Marketing Consultant: Roger Baty provided essential expertise in helping with the presentation of the first book, *Intelligence Behind the Universe!* Printed in 1990. This book is now out of print but Amazon is now selling used copies as collector's items at prices exceeding £160.

Non-Science Based Supporters

Michael Roll has given constant support since we met in 1988 in approaching publishers and sending them copies of my 1990 book free of charge and some pamphlets designed to communicate the theory that appears in a popularisation of our series. This also provides essential data in support of the theory. Paul Read created Mike's website without charge to help in this communication. There are many other helpers but they concern only the popularisation and so cannot be listed.

My daughter Sarah has assisted in preparing a DVD that forms an extra part of the series.

Finally my wife Margaret, who died in 2009, provided the background support that is so essential and to whom I am greatly indebted.

<u>For scientists only</u>: please first read APPENDIX I on page 61

This advice is given because we have discovered that people with a scientific training confuse the 'negative energy', used in this book, with potential energy and then fail to understand the text.

CONTENTS LIST FOR PART II see page 78

LIST OF CONTENTS PART I ONLY

LIST OF FIGURES

PART I

THE BIG BREED

Creates all the energy needed to make a nested set of universes

General introduction for both technical and non-technical readers

PART I is a non-mathematical description accessible to readers who are unfamiliar with cosmology and quantum theory and have no knowledge of the calculus. It provides an essential introduction to all those who wish to understand how the BIG BREED THEORY is able to solve some crucial problems that still remain vexed questions.

The MATHEMATICALLY MINDED will find this section an essential springboard before studying the mathematical detail covered in PART II.

CHAPTER 1

INTRODUCTION
The BIG BANG Theory & why DARK ENERGY appeared

A short history of this subject shows how the big bang originated and highlights its shortcomings. The idea that Dark Energy must exist with mysterious repulsive power arose as an expedient: a wrong prediction had to be corrected. This is why an alternative theory is required from which Dark Energy and correct predictions emerge naturally - as will become clear as the story unfolds.

The Roman Catholic priest, Georges Henri Lemaître first suggested the big bang theory in 1927. He proposed that the universe could have been deliberately created as a massive explosion between 10 and 20 billion years ago. It would still, he said, be in a state of continuous expansion. Then since all the energy needed to build the universe would have appeared in a blinding flash no more was created from that time onwards. This concept fitted in with his religious belief and so provided an attempt to reconcile his faith with science.

Then in 1929 the famous astronomer, Edwin Hubble, discovered that light from distant galaxies was 'red shifted'. Visible light exists in a spectrum as electromagnetic waves having a mixture of wavelengths. The shortest wavelengths appear to us as blue light whilst red light has the longest. When a source of light, such as a galaxy, is moving away the wavelength is stretched by the speed of recession and so appears redder than the light as it was emitted. So Hubble deduced the 'Hubble Law' that says distant galaxies are receding at speeds proportional to their distance from us. This provided good confirmation of the priest's idea.

During this time the physicist, Fred Hoyle, proposed a different concept: continuous creation. He said the universe never had a beginning and would never end. It would always look the same at all times since continuous creation was going on everywhere in apparently empty space. Somehow atoms were being spontaneously created to form a very rarefied cloud of gas. This was pushing the galaxies apart. As separations increased new galaxies would form to fill the gaps. Then observers in any epoch would always see very much the same picture.

Controversy ensued and arguments became heated and quite nasty. The young theoretician Stephen Hawking pointed out at a lecture given by Hoyle to the Royal Society in 1964, that his 'Steady State' theory would go 'divergent'. This meant the theory was flawed and soon the concept became discredited. Hoyle angrily dismissed the alternative as the 'big bang' and this name stuck.

The blinding flash of light, caused by creation, would remain trapped in space, according to theorists, and so ought to be still observable. It ought to appear as a faint glow, diffused over the entire sky, with wavelengths stretched out into the microwave band.

Then two people made an accidental discovery. They were Penzias and Wilson working on a large horn for the Bell telephone company. They were troubled by an unexpected radio interference coming from the entire sky. It was in the microwave band.

Theorists soon seized on this as confirmation of the big bang theory. Also by this time it had been linked to quantum theory.

Quantum theory describes the strange behaviour of the minute particles of which atoms are composed. Theorists insist that an all-pervading background medium exists known as the 'quantum vacuum'. Now a vacuum is defined as the absence of atoms but this does not mean that a void of just nothing exists. Instead so-called 'virtual particles' keep popping out of nowhere to vanish again shortly afterwards. These are not considered to be real particles since the latter would have a permanent existence. However, the virtual particles behave like real ones whilst they last and, by striking real ones, produce forces. These are the forces of nature.

There are four forces of nature. One is the 'strong nuclear force' of very short range that binds the component parts of the atomic nucleus together. Another force of short range is associated with nuclear fission. Then an electric force causes very lightweight 'electrons' to be confined to regions of space near atomic nuclei to form atoms. But when electrons are moving they create a magnetic force and so this is now combined with the electric force to be considered as a single long-range force of 'electromagnetism'. Then finally there is a force of attraction between all kinds of 'sub-atomic particles that is very weak compared to all the other forces but is also of long-range. This is the force of gravity.

The virtual particles form a seething mass of 'quantum fluctuations' – random changes of density. The larger the fluctuation the shorter the time they last. Although each virtual particle has a short life the lifetimes of many such particles overlap. This means that the energy from which they are made add up and so space, alias the quantum vacuum, has in fact a very high energy content in every cubic metre: its 'energy density'.

The big bang theory has it that one of these fluctuations was enormous and started to decay immediately. However, before collapse was complete something else was triggered – 'inflation'.

Dr. Alan Guth, in 1980, proposed the theory of inflation. Space had to expand and, since it had a high energy density, a huge input of energy had to be continually supplied. The problem was that this had to be supplied from the void of nothing. The logic led directly to a major false prediction to which we will come later.

This is where all the trouble began and provides the reason for writing this book. Our research had been triggered by Professor Tryon (1984) who wrote an article concerning 'potential energy' that is mentioned in this book in 'Foreword'. This had prompted the submission of a critique that had been rejected by all assessors except Professor J. Vigier of the University of Paris.

The flaws are described in APPENDIX. I

Then the writer and physicist, Professor Paul Davies, had been approached to ask how this could affect the big bang theory with which it was associated. He sent the mathematical basis of Guth's theory of inflation to this author in 1987. This logic had seemed totally at variance with both the mechanics and thermodynamics of convention. Several flaws seemed to combine to render the logic unacceptable. In consequence Davies's letter triggered the search for a more plausible theory. This is the subject of this book and, hopefully, it will intrigue the reader as the story unfolds.

We have not finished with the big bang yet however. The inflation had to last for only a minute fraction of a second and then switch off. The trouble was – Guth's theory had provided no effective means for doing this. They tried having inflation occur in a 'false vacuum', postulated as being many times more energetic than the real vacuum to which reversion occurred when inflation ended. This reduced errors by a many orders of magnitude. However, the same logic prevailed and so the theory still predicted an on going inflation albeit of reduced magnitude.

This magnitude was, however, absolutely huge and predicted a rate of expansion, which according to the string theorist, Brian Greene (1999), is 10^{120} times greater than astronomers can possibly allow. This is a huge number that can be best appreciated by comparison with the total number of atoms of the entire observable universe. This is about 10^{81}. So the 'error' dwarfs to insignificance the number of atoms in the universe – some error! It is known as the 'Problem of the Cosmological Constant' (Not to be confused with an error of the same name that Einstein said was his greatest mistake)

However, theorists pressed on, leaving this difficulty for resolution at some future date. After inflation a cloud of gas existed expanding in all directions, initially at speeds close to that of light. All four forces of nature existed initially as a single 'super-force' but as densities reduced the individual forces split away, gravity first, followed by the other three. Now a gas ball of ever growing size existed propelled by its own inertia. The expansion was, however, slowing always due to the mutual attraction of gravity. The gas cloud thinned in density as it grew but mutual gravity caused non-uniformities in density to grow. These became clouds of gas condensing later to form the stars and planets we observe.

Then the astronomer Schwarzchild, B.(1998) published observations of remote exploding stars called 'supernovae'. Very large stars burn up their nuclear fuel very much faster than smaller stars. On exhaustion they then explode giving off radiant heat and light equal to that of an entire galaxy consisting of billions of stars. Supernovae only last about twenty days but are invaluable to astronomers because they provide the equivalent of a 'standard candle'. There are several different types but each has a known radiant energy value.

However, the bombshell to all cosmologists was the news, resulting from these astronomical observations, showing that the expansion of the universe was speeding up!

It was not slowing at all as theorists had predicted.

In order to patch up the big bang theory and make it fit the new facts the concept of 'Dark Energy' was soon invented. This is imagined as some mysterious substance pervading all of space and having the property of lying dormant for billions of years. At some point its latent power was switched on to overcome gravitational attraction at great distances of separation, so causing all the galaxies to start accelerating away from one another.

It will be for the reader to judge this proposition after the text of this book has been digested. We will show that Pearson (1994) had published the Big Breed theory in Russia. This had resolved the creation switch-off problem of the big bang theory. Inherent in the solution was the prediction of the acceleration not yet discovered!

There is more to perplex the reader. Since Hubble's time

astronomers have had access to ever-improving instruments to help their observations. In particular the magnificent 'Hubble telescope', still orbiting the Earth, has supplied information that earlier astronomers could only have dream about. Hubble's law has been refined so that cosmologists are now confident of extrapolating backwards in time, using this law, to determine the age of the universe. They tell us, with great confidence, that the big bang happened 13.7 billion years ago.

Not everybody, however, is happy with this figure. Astronomers and some cosmologists complain that their computer studies suggest this does not allow enough time for galaxies to form. Worse still, they say, some stars seem to be older than the universe!

The reader can expect some surprises as this book unfolds and resolutions to all these problems ultimately appear.

Theorists never did solve their major problem. They now say the energies needed to allow space to expand are supplied via 'wormholes in space-time' from some higher dimension. Is this an admission of defeat? Does it mean that they now accept the required energy has always existed without a beginning?

This is only a brief outline of the complete big bang theory but is an adequate summary for our purpose. We will revert to the original concept of creation from the void. The void is defined as nothing: a state of zero energy. We argue that only nothing can have existed forever without a beginning: since nothing cannot even exist!

We shall see that a solution to the problem of the cosmological constant appears from simple logic. This will provide the creation switch-off means the big bang theory lacks. It is not a total switch–off, however. A minute fraction of net creation remains controlled by the accelerating expansion that is produced. An accelerating expansion is predicted! So a solution to the second problem will appear as consequence of the first.

Dark energy was never needed since this solution was published in Russia (Pearson 1994) and the peer-reviewed journal, 'Frontier Perspectives' in 1997 (Pearson (1997)) – before the accelerating nature of the expansion was discovered in 1998.

In the next chapter we will prepare the ground needed before we can home onto a solution that satisfies the criterion of creation of the universe ex-nihilo – from nothing.

CHAPTER 2

SETTING THE STAGE FOR A SOLUTION

Before we can begin to try and solve our problem we need more information. First let us see what our senses tell us. Then we need a brief look at the quantum world to see the difference. Most important we need to study the philosophical implications of the weirdness that this uncovers. We need to see what quantum physicists have discovered and tell us about this. Then we have to decide what it all means.

2.1 The World we observe with our senses

What we see, hear and feel by our senses convinces us that our world is real and solid. At our level of common experience things move in predictable ways. We throw a ball and it follows a predictable trajectory. It arcs upwards and then falls in a smooth curve. This is the reality we experience known as the 'macroscopic level'. We can weigh the ball on scales and measure its 'mass' as so many kilograms. We can put several balls on a billiard table and hit one with a cue so that it strikes another ball. If the collision is head on, and both balls have the same mass, then we see the driven ball stop dead and the originally stationary ball move away at the same speed as the first ball had originally As Galileo was the first to shush out, a moving ball has 'momentum': its mass multiplied by its velocity. The first ball loses all its momentum but the second carries the same amount away. Scientists say that 'momentum has been conserved' across the collision event because no momentum has been lost. It has simply been transferred from one object to another. The reader may have noticed we used the term 'velocity' instead of 'speed'. There is a difference. One can drive a car round a 180 degree bend at a constant speed. However, the velocity will have been continually changing until finally it is moving in the opposite direction. Velocity is speed and direction combined.

The conservation of momentum applies to any kind of collision: the billiard balls could be struck with glancing blows and even could be of different sizes. They could be made of plasticine and stick together: but still momentum is conserved.

But to make the ball speed up from rest (to be accelerated) a force of some kind has to be applied. This force moves with the object for some distance, although this is not obvious from the collisions we have imagined. A better example to make this clear is the firing of an arrow from a bow. The bowstring has to push against the arrow for quite some distance in order to give it adequate speed and momentum. During this period of acceleration energy is transferred to the ball or arrow. Such objects have gained 'kinetic energy': the energy of motion. And the energy put into accelerating the object is exactly equal to the kinetic energy imparted. So another conservation law is involved. These are examples of the 'conservation of energy'.

The Big Breed theory is entirely dependent on these two laws, the conservation of momentum and the conservation of energy. So both laws need to be firmly understood

But now another amazing finding by physicists gives a whole new twist to the meaning of energy. Einstein has been (wrongly) given all the credit for the discovery of a famous equation:

$$E = mc^2$$

E represents energy, which for any object that is not moving, is not the energy of motion but is the equivalent of its mass. Symbol m is the mass of the object concerned and, for a stationary object is its 'rest mass'. Finally c is the speed of light, which is huge. In fact it is 300,000 kilometres per second. It only takes about a second for light to travel all the way to the moon. The energy E now means 'rest energy'. Any object is actually MADE out of rest energy when it is standing still! But with the speed of light squared we see that the energy equivalent of mass is absolutely huge. In fact only a gram or two of mass needed to be converted to energy to destroy the entire city of Hiroshima by the atomic bomb.

So the moving ball we threw earlier has not only kinetic energy but rest energy as well and the two add up. The total is only significantly different from the rest value, however, when speeds far beyond normal experience are involved. This is all so-called 'classical mechanics' building on the kind first identified by Sir Isaac Newton 300 years ago.

But now let us go on another imaginary journey. We thump a table and the brain records a pain transmitted from the fist. This suggests to us that wood is a solid substance since the fist cannot penetrate. Look at the wood through a microscope and we see it is full of fibres and holes. Go to basic organic chemistry and what we find is that the wood is made of extraordinarily complex molecules. Each is built from atoms of mostly carbon, hydrogen, and oxygen but entwined in the assemblies are smaller amounts of phosphorus, nitrogen and trace elements. All play an essential part in enabling the wood to grow and exist.

But now go deeper still, beyond the resolution of even the electron microscope. We are now entering the quantum world that quantum physicists have penetrated, starting with a major discovery in 1900 made by the famous physicist Max Planck. He discovered that even energy couldn't be a continuous substance: it has to come in discreet chunks he called 'quanta'. But now we are entering a new world altogether.

2.2 The Quantum World

Soon experiments showed atoms to be made of two kinds of electrically charged particle. All atoms had a central nucleus of minute size that had a positive electric charge. Around them moved, at incredible speeds, particles of about 1800 times smaller mass: the electrons. These were negatively charged. Opposite electric charges attract one another and so keep the electrons orbiting the nucleus like planets going round the sun. At least that was the first interpretation known as the Rutherford-Bohr atom in which the simplest atom, hydrogen, consisted of a single 'proton' as its nucleus, orbited by a single electron. All other atoms have larger numbers of both and also add in some uncharged particles, of almost the same size and mass, called 'neutrons' making up their nuclei. These protons and neutrons all stick together as if glued to one another by the strong nuclear force of extremely short range so these 'nucleons' form minute balls of matter. But even protons and neutrons are composites. They contain even smaller entities called 'quarks' bound by 'gluons'.

The amazing thing was the sheer emptiness of the atom. Scaled up, the space occupied by the nucleus and electrons combined is no greater, relatively, than that occupied by the sun and planets as compared to the volume of space in the solar system! The atom is almost entirely empty space! It seems a far cry from the solid wood we hit with our fist. The reason the wood seemed so solid is that the fist is made of atoms as well as the wood. The electrons round the atoms making up the fist repelled those of the wood - even at great distance as compared with the size of the electrons. Our senses can deceive. All is not what it seems to be from observation by our unaided senses.

2.3 The two-slit experiment

A controversy arose during Newton's time in regard to the nature of light. Even though Newton had made a major advance, in finding that white light from the sun is a mixture of colours, he asserted that light was carried by 'corpuscles': light was a stream of particles fired like bullets from a machine gun. However others, like Hook and Huyghens, disputed this. They said light moved as waves: like ripples on a pond. So who was right? Because of Newton's stature people did not question his corpuscular explanation for the nature of light for many decades. Then an eye surgeon, **Thomas Young**, (1773-1829) performed an ingenious new experiment. He used a mirror outside a darkened room to reflect sunlight through a small hole in the blind. He allowed a narrow beam of this light to fall onto a pair of closely spaced very narrow slits cut in a very thin sheet. He had placed a screen at a considerable distance behind them to display the resulting pattern.

If light has a wave nature, then according to reasoning by Huyghens, if it passes through a very narrow slit it should spread out by 'diffraction'. So if two very narrow slits are arranged close together, reasoned Young, the emerging waves should spread into one another to form an 'interference pattern'. Such a pattern is observed when two pebbles are thrown together into a pond. Ripples spread out in concentric rings until they cross into each other. In that region choppy water is seen and that is an interference pattern.

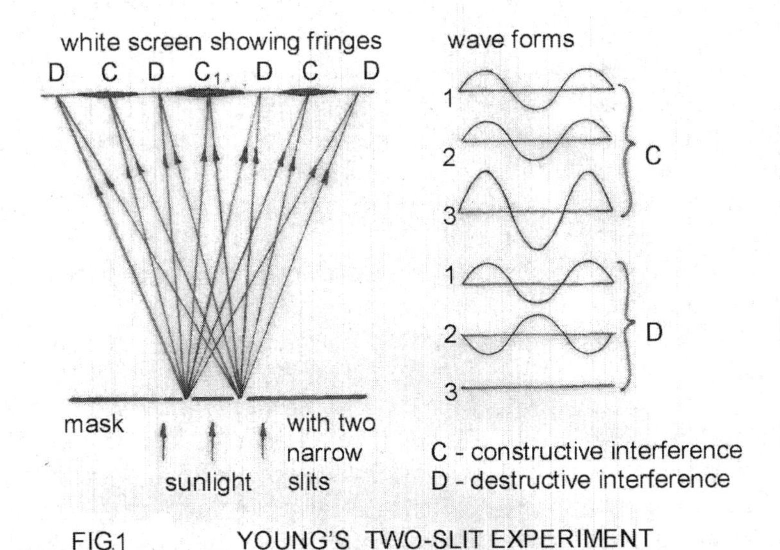

FIG.1 YOUNG'S TWO-SLIT EXPERIMENT

If light behaved as a stream of particles, then two bright smudges merging into one fuzzy band would be expected to appear on the screen. If they were waves then the screen would show an interference pattern. This would appear as a pattern of many fuzzy lines, parallel to the direction of the two slits. These fuzzy 'fringes' would fade with distance from the centreline.

All waves have this nature whether they are light waves, sound waves or ripples on water and can be readily explained by reference to FIG.1. This shows the two-slit apparatus in plan on the left and on the right waveforms are illustrated. The upper one, labelled C, shows two sets of waves, 1 and 2 'in phase' meaning the peaks of both coincide and so do the troughs. They have equal amplitude, which for water ripples means half the difference in height between crest and trough. They add up to double the amplitude in a manner known as 'constructive interference', shown by group C at 3.

In the lower figure, labelled D, the waves are totally out of phase at 1 and 2, representing 'destructive interference'. Now when they are added the amplitude is zero as shown by group D at 3.

The plan view of Young's apparatus, on the left in FIG.1, shows rays of light emanating from the two slits. The central rays arriving at C_1 from each slit are of equal length and so both crests and troughs from both rays arrive together and produce constructive interference. The amplitude is greatest here. With light this means a bright fringe is produced. On either side of this bright fringe the rays from each slit, arriving at D, have different lengths and when this difference is equal to half a wavelength of the light total destructive interference occurs and a dark fringe results. With greater distance from the centre the path length differences are even greater but now equal a whole wavelength. So again there is constructive interference as shown at C and consequently a pair of bright fringes appear one either side of the centreline. For the outer pair shown arriving at D path differences are further increased to once again produce destructive interference. The pattern repeats with brightness fading until nothing is visible.

Young observed exactly what he anticipated: an interference pattern appeared on the screen. So light had proved to be of a wave nature and Newton had been proved wrong – or so it seemed at that time.

However, after Max Planck had deduced that energy must come in discreet chunks this suggested light must also be a stream of particles. Young's two-slit experiment was refined by replacing the screen by very sensitive photo-detectors. These are electronic devices that amplify any light received by many times.

They kept turning down the intensity of light so that if it came in chunks, that Einstein later called 'photons', then they would come ultimately one at a time. Then only one could go through one slit and no interference pattern could then result. After a long time many photons would pass one at a time so that a pattern would build, speckle by speckle, and no interference pattern could be expected to arise. Instead, from the foregoing logic, a single smudge was expected to appear.

To the astonishment and dismay of the entire physics community the 'impossible' interference pattern resulted! It did not make sense. How could a single photon, going through only one slit, produce an interference pattern by itself?

However, the experiment did prove one thing – Newton had been right all along: light did come in the form of a stream of particles. But Huyghens and Young were right as well. Light had both a particle and a wave nature! It was astonishing.

The single photon, for example, had to exist as a wave in order to go through both slits at once to produce an interference pattern, and yet it had to be a particle to make a tiny flash on a screen or trigger a photo detector

More was to come. Prince Louis de Broglie suggested all sub-atomic particles should behave similarly and all would exhibit this wave-particle duality. The two-slit experiment was explored again using electrons, then protons and finally other particles. De Broglie's speculation was confirmed. There was a difference, however, between light and matter. Light waves seemed more real, being the electromagnetic waves first predicted theoretically by Clerk Maxwell. Matter waves seemed more abstract. Nevertheless everything in the quantum world behaved both as particles and waves. This caused a lot of head scratching since the finding posed some intractable philosophical difficulties.

Clearly quantum theory operates on a mechanics that is very weird indeed. Its sub-atomic particles, such as protons and electrons, appear to have a dual nature in that they sometimes manifest as waves, filling a large volume of space, whilst at other times bouncing off one another like billiard balls – so manifesting a particle nature. Particles in transit, as described by Feynman (1985) in a very readable book meant for the non-mathematician, have the potential to dance about over vast distances before arrival at a given location, even though they appear to have travelled in a straight line. He is describing what is known as 'quantum field theory' that does not specifically refer to waves but the end result is much the same.

During transit particles appear to exist in every possible state, such as spinning in opposite directions simultaneously whilst also behaving as waves. This unfathomable combination only appears to collapse into the reality of particles spinning in just one direction on hitting a screen or measuring instrument. But the instruments are made of atoms obeying quantum rules as well. And so do our brains that finally register and analyse the information.

When these extraordinary discoveries were made quantum

physicists were totally amazed and baffled. It just did not seem to make sense.

One said, “This is more like Eastern mysticism than Western science”.

The famous physicist and one of the pioneers of quantum theory was Neils Bohr. He organised a conference to try and make sense of all this weirdness. This ended with what is known as the, ‘Copenhagen interpretation’. This states that particles only exist in some kind of limbo state in the form of waves until observed. Only then do they collapse into the reality of particles.

This meant that consciousness participates in the creation of matter. It is a startling conclusion that many scientists would not accept. Indeed the established view across all scientific disciplines, to this day, insists that consciousness is mere brain function. Then the mind and consciousness must cease to exist when the brain dies. If the Copenhagen interpretation is correct, and mind is merely brain function, then how could matter form before brains existed?

To get around this problem an idea first floated in 1957 by a student called Hugh Everett has now become a mainstream choice. This is known as the ‘Many Worlds’ interpretation. This has it that when an electron has a choice of two paths, such as when there are two slits in a mask and it is equally likely to go through either, then a ghost electron goes through the other slit. Both take their ‘wave functions’ with them. This allows an interference pattern to form as required by quantum theory to dictate where the electrons will go.

To provide the ghost electron our entire universe splits in two to make a replica of itself. Then since this is going on everywhere in each universe both have to keep splitting countless billions of times a second. So an infinite number of parallel universes exist all occupying the same space and splitting to make new ones all the time.

Is this more reasonable than the Copenhagen interpretation; taking into account that no mechanism for the splitting or the energy needs required are ever considered?

Is it possible that something better than either will turn up as our investigation proceeds?

This author considers a more rational interpretation of quantum weirdness is that mind is not mere brain function. Mind must exist at an ultimate level of reality that is truly real. Mind had to exist long before the universe was created. Only then can the Copenhagen interpretation become paradox-free with absurdity made obsolete.

What this implies is that the quantum level, and therefore the macroscopic level emerging from it, represents a form of virtual reality rather than as something truly real. On this basis true reality needs to exist at a deeper level that is not directly accessible to our senses or instruments. Physicists say there must be a fluid-like background medium they call, 'The quantum vacuum', on which the quantum level rests. Since this exhibits the same unreality it cannot be that ultimate level. It has to be regarded as emerging from the ultimate level of what we will call 'i-ther' (since it differs from the old and long discredited 'ether'). The i-ther must somehow create the quantum level by generating the mathematics needed for its organisation.

If this seems an unlikely proposition it needs to be compared with that of Everett that seems to us to be far more outrageous. Furthermore physicist Stephen Adler (2004) followed the lead summarised above in having the quantum level emerge from a deeper level based on a modified form of 'classical mechanics'. This is exactly what was proposed in our first solution published in Russia by Pearson (1994). This book developed from that point.

On this basis our universe of matter must be made from the same energies as that from which the i-ther is composed.

It also follows that if the i-ther were based on wave mechanics, like the quantum level, then it too would need an even deeper level for its organisation – and so on ad infinitum. Such an infinite regression is never acceptable; the buck has to stop somewhere. Consequently a different mechanics must operate at i-ther level.

The mechanics at i-ther level have to relate to true reality. Now the conservation laws of energy and momentum, that initially arose from Newton's classical mechanics, apply equally at the quantum level. Then since all levels are made from the same basic stuff it is reasonable to conclude that these laws also apply at the ultimate level. The simplest possibility to explore is therefore one in which Euclidean geometry and universal time form the basis of mechanics at the ultimate level. We shall consider it reasonable to assume a mechanics similar to Newton's to apply at i-ther level.

To provide credibility for this assumption the resulting predictions need to penetrate the quantum level and match observations made at the macroscopic level. In 1987 when, this research was initiated, this requirement seemed to pose very formidable difficulties.

This was because, unfortunately, Newton's mechanics is not exact and Einstein's general relativity theory, though seeming exact, could not be used. The reasons are covered in *CREATION SOLVED?* and summarised in APPENDIX III of this book, where it is shown that there are internal contradictions and Einstein's theories are incompatible with the existence of anything like the i-ther. Furthermore his concept of curved space-time could not describe the quantum vacuum.

An exact classical mechanics, alias ECM theory, had therefore to be derived, as a first step, in order to solve the creation problem. This mechanics is compatible with the quantum vacuum. As shown in 'Quantum Gravity via an Exact Classical Mechanics', giving mathematical detail, the new mechanics is free from internal contradiction and incompatibility with quantum theory and yet matches all the experimental data just as well as Einstein's general relativity. It also provides a solution to the vexed question of providing a satisfactory solution for quantum gravity. Another attractive feature is the ability of ECM theory to show why the force of gravity is so weak as compared to the other three forces of nature. That this represents a considerable advance is demonstrated by the fact that no other theory of gravity has yet appeared with this potential.

Since ECM is equally applicable at i-ther level, it seems reasonable to regard this as the mechanics required for solving the creation problem. However, if this is disputed all that needs to be said here is that only a minor part of ECM theory is required for application at i-ther level. The ultimate particles called 'primaries', of which the i-ther consists, need to consist of two kinds of energy. The energy of a primary standing still is made of 'rest energy'. To this is added the energy of motion called 'kinetic energy' These two add up to yield 'sum energy'. This sum energy is conserved during collisions. Momentum is based on the mass equivalent of sum energy multiplied by the primary's velocity. And then there have to be two opposite kinds of primary as will be shown later.

The ECM approach adopted is based in the simplest possible way – as briefly described in APPENDIX IV. No resort is made to higher dimensions or sophistication other than several modifications to Newton's original mechanics. The latter is only applicable to speeds low as compared with light and in weak gravitational fields. The revisions extend applicability to eliminate such restriction.

As we shall see this mechanics permits of a solution, we call the 'Big Breed' theory. This depends on the existence of the i-ther that spontaneously develops a chaotic structure fed continuously by massive amounts of energy. These are exactly the conditions that chaos mathematicians have discovered produce incredible self-organisation, as lucidly described by John Gribbin (2005). The structure has similarities to a neural network. G.E. Hinton (1992) shows how artificial neural nets possess learning and memory capability. It does not seem unreasonable, therefore, to speculate that the i-ther could have the potential of evolving the required intelligence.

Such a speculation will no doubt be hotly contested and may indeed be ultimately proved invalid. However, for the main theme of this book this is not a crucial matter. Regardless of the i-ther being able to self-organise, the main importance of the Big Breed theory is its success in solving the problems that still invalidate the big bang theory as first formulated – it cannot describe creation from the void – something, as the reader will discover, is achieved by the Big Breed theory. Another feature in support of the approach is the ability of the structured i-ther to provide an explanation of another vexed question – non-locality.

Non-locality is another extraordinary feature thrown up from quantum theory. Sub-atomic particles can become 'entangled' so that they can affect each other instantly regardless of huge separating distance. Nobody has yet published an explanation for this phenomenon. However, the i-theric structure has already provided an explanation, as described in Chapter 5 and in greater detail in *CREATION SOLVED?*

Regardless of the intelligence issue, however, since the quantum level is so weird and unreal it will be regarded as energy that is organised mathematically by the i-ther. This ultimate level generates waves that are used like numbers, in order to produce what appears to us as matter, together with its controlling forces.

The re-introduction of macroscopic mechanics at i-ther level has energy, momentum and spinning motion smoothly varying. Quantum effects, limiting energies to discrete values, can safely be considered as absent. Also no electromagnetic or the strong and weak nuclear forces will exist at the ultimate level of i-ther. Indeed the only forces to be considered are the transients acting during the collisions of the ultimate particles, the 'primaries'. The only laws of physics existing will be the conservation laws of energy and momentum acting on primaries: the only real particles that exist.

Some readers of our other books have said we are encroaching into theological territory by trying to describe God. A few physicists say quantum weirdness demands the existence of transcendental creative force existing outside space and time. Two holding the latter view have dismissed the Big Breed theory on these grounds.

Our answer is that it is not possible to disprove the existence of either an unknowable God or anything transcendental outside space and time. But neither can they be proved to exist. So the choice is best left to the individual. The i-ther needs to exist to solve the problem of the cosmological constant. Some may choose to consider no extra God or equivalent is required since from a scientific viewpoint the i-ther seems to make any extras redundant. Others can take the alternative option but no controversy remains since the i-ther needs to exist for either choice.

2.4 Comparison with the Establishment Approach

If this proposal seems unreasonable then the reader should compare it with the established approach that also aims to achieve a solution for existence. Michio Kaku (2009) often appears on television to describe this approach and should be studied in order to compare with thc proposals made in this book.

Because the big bang theory is regarded as established, with gravity as part of the original super-force, it has to be regarded as comparable in strength to the other forces, such as the strong nuclear force. Unfortunately, gravity is known to be incredibly weak when compared to nuclear or electromagnetic forces. So their proposition is that the force of gravity is just as strong as the others but leaks out into a 'hyperspace' consisting of all those parallel universes proposed by Everett's Many Worlds interpretation and other similar speculations requiring infinite numbers of universes. Indeed Michio Kaku posits that there are four different varieties each involving infinite numbers. Also how does a <u>force</u> leak?

In the Big Breed theory there is no super-force from which the four forces of nature split as in big bang theory. Indeed only gravity now appears as a force that is possibly truly real and its relative weakness as compared with the others is fully explained by a combination of ECM theory and Big Breed theory. No speculation concerning hyperspace or leaking gravity is required.

Then again, since the big bang now allows the inflow of energy through wormholes to enable space to expand, the energy needed had to exist without a beginning. So other speculations have our universe expanding and then falling back to form a 'big crunch'. Our universe is annihilated. But then another universe rises, like Phoenix from the ashes, to bounce back again. This has been going on always and will continue forever.

Another speculation going strong posits the 'crashing brane'. Huge membranes exist in some higher dimension in almost parallel planes but are not quite flat. They approach each other at high speed and crash into each other. Because they have bumpy surfaces they hit each other in multiple places and each of these creates a separate universe. How can these branes (short for membranes) be imagined as existing in almost parallel planes when the curved higher dimensions in which they exist are themselves impossible to imagine? Then again: where did the branes come from?

Furthermore because problems seem so intractable, theorists

have felt forced to accept more and more sophistication resulting in the 'string theories' requiring seven extra dimensions added to the three of space and one of time to which we naturally relate. These are not to be confused with the concept of parallel universes existing in the same space as our own. These extra dimensions are so highly curved they form little balls or odd shapes. Sub-atomic particles are represented as vibrating strings existing in these higher dimensions. Different particles are represented by different resonant frequencies of vibration of such strings. The main target is to find a unified theory for the four forces of nature and all the sub-atomic particles.

However, three of these forces are already explained in a very satisfactory way by quantum field theory. This is most lucidly described and explained by Richard Feynman (1985) in his popular book, 'QED'. No higher dimensions are used or required. So only gravity remains unresolved. The ECM mechanics was first published in Russia (Pearson (1991) and provides a quantum compatible theory of gravity without need for higher dimensions. So the need for such sophistication needs to be questioned.

And Brian Greene (1999), a string theory enthusiast, admits in his book on page 225 that string theory cannot solve the problem of the cosmological constant!

What I thought I would find in Greene's book was an explanation of the need for the higher dimensions that string theory requires. I had expected the reason to be an extension of that given by Einstein for the force of gravity. This is a geometrical reason. Objects travel in straight lines in curved space-time and so seem to move in curves to us from our perspective in Euclidean space. He says no actual force is involved. So I expected these higher dimensions forming little balls, or other strange shapes, to be attempts to simulate the far stronger nuclear and electromagnetic forces.

No mention of anything like this appeared. Instead the reason was probability. Greene says that probabilities predicted when limited to three dimensions of space became infinite when they could only lie between zero and one. Using higher dimensions resolved this issue.

The reader will be left to judge whether this seems an adequate reason for abandoning less sophisticated solutions.

Could a simple approach eliminate all this confusion? If you think it possible then please read on.

In the next chapter we consider a simple experiment that provides all the understanding required of the concepts called the "conservation of momentum" and the "conservation of energy".

CHAPTER 3

ENERGY AND MOMENTUM

Before the Big Breed theory can be described we need to consider, in more detail, the way both momentum and energy are conserved when large objects collide. This will prepare us for what happens when minute particles of i-ther breed by collision

3.1 Introducing the Creation Logic

First we need to amplify our understanding of the conservation of momentum to build on what was presented in Chapter 2. It will be remembered that the momentum of an object is given by its mass multiplied by its velocity and that velocity is speed with direction specified. In fact the deeper understanding required is best achieved if the reader carries out a simple experiment using a pair of pendulums. This will provide an accurate 'momentum balance'. There is nothing as good as doing experiments to fix the understanding of mechanics in one's mind. However, just imagining an experiment is the next best thing.

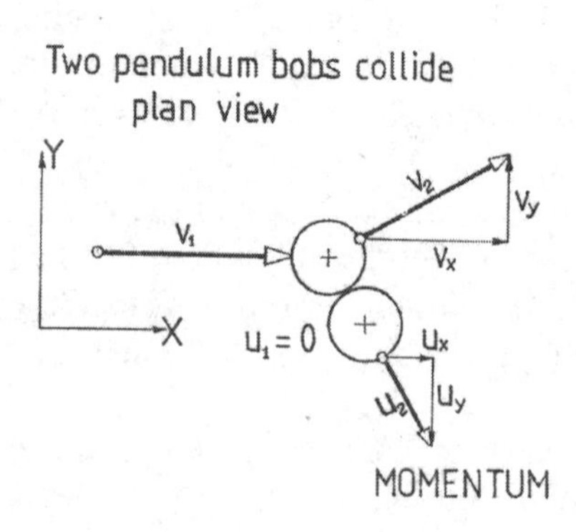

FIG 2 MOMENTUM CONSERVATION BY COLLIDING BALLS

The simple apparatus needed consists of two identical high bounce balls each fixed to a length of thread. The threads attach at their upper ends to a wooden block by sticky tape. It will be found that the threads will be able to slide through the tape to make adjustment of their lengths quite easy. The pendulums are made over 2 metres long to hang from a wooden bar attached to the top of a door and arranged so that bobs, the balls suggested, just touch each other when hanging still. They are also arranged to just clear a flat board placed on the floor.

One bob is moved a horizontal distance of not more than about

50 centimetres. The position of the centre of the bob is marked vertically below it on the flat board. The bob is then released so it swings to hit the other bob - not head on. Both bobs bounce away from each other and a mark is placed to show the maximum position the centre of each reaches before swinging back. Two people are needed for this task: one to mark the maximum 'throw' of each bob.

FIG.2 shows a plan view of such an experiment. The X direction is the direction of travel of the bob that was released. This is the 'driver' that has velocity $\boldsymbol{v_1}$ on impact. The bob hit is the 'driven' that scatters away with velocity $\boldsymbol{u_2}$. The driver scatters away with velocity $\boldsymbol{v_2}$.

Now it can be shown by simple maths that in such cases, where the angle of swing is not excessive, that the maximum speed reached during the swing is directly proportional to the horizontal distance moved. So in the figure the thick arrows show, to scale, the distances of throw of each bob. (The bobs shown as balls can be any size)

This means we can use the horizontal distance measurements as speed measurements to some scale. And the scale does not matter for our purpose. What matters first is to confirm that momentum is conserved in the X direction. We measure the X 'components' of velocity. The driver bounces away with velocity component $\boldsymbol{v_X}$ equivalent to a distance measured in this X direction. The driven bob bounces away with velocity component $\boldsymbol{u_X}$ in the same direction. Add these together. You should find this sum equal to the velocity $\boldsymbol{v_1}$ given by its distance equivalent. (The diagram has been drawn to scale and so can be measured.)

Since both bobs had equal mass what you have done is to confirm, by experiment, the conservation of momentum in the X direction.

But what happens in the Y direction measured perpendicular to the X direction? Now the driver has no velocity and therefore zero momentum in the Y direction and yet both bobs have both velocity and momentum in that direction. Does this mean that momentum is not conserved in the Y direction?

To find out measure both $\boldsymbol{v_Y}$ and $\boldsymbol{u_Y}$. You should find these are equal. However, we should first have defined a positive direction and this can be chosen in the direction of the arrow on the Y symbol, though this choice is purely arbitrary. Now that $\boldsymbol{v_Y}$ has been chosen as positive, then $\boldsymbol{u_Y}$ has to be regarded as negative.

So the negative momentum of the driven bob cancels the positive momentum of the driver. We can write this in symbols:

Using m to symbolise mass, $\boldsymbol{mv_Y + m(-u_Y) = 0}$.

It follows that since the driver also had zero momentum in this direction, momentum has also been conserved in this Y direction.

This also illustrates what scientists mean by the terms ‘positive and ‘negative’. They signify one thing being opposite another. The scientific meaning of the terms ‘positive’ and ‘negative’ has nothing to do with anything being good or bad.

We could have used pendulum bobs of unequal mass and then quite different scattering effects would have been observed. However, on analysis by the method just described it will be found that momentum is always conserved exactly. It is interesting to try this experiment using bobs made of plasticine. They stick together on impact and move away as one. But still momentum is conserved. It is particularly interesting to see what happens when both plasticine bobs are of equal mass and both are dragged away from their rest positions by equal horizontal distances in opposite directions. They are released together and collide head on. Now they both stop dead. However, each collided with equal and opposite momenta and so their sum was zero before collision – exactly the value they had afterwards. So once again this demonstrates the conservation of momentum.

Now, however, each had kinetic energy, due to their speed, before collision but none afterwards. So although momentum was conserved, in this case kinetic energy was not conserved. Does this mean that the law of conservation of energy has been violated?

Actually no. What has happened is that all the kinetic energy lost has been converted to an increase in the random kinetic energy of motion of atoms in the plasticene. This means that both bobs have been heated a little and with very accurate instrumentation a small temperature rise would have been recorded. Energy is always conserved just as momentum is always conserved but energy is always at least partly degraded to a form that is no use to man.

The arrows shown at the ends of the ‘velocity vectors’ all point in the directions of motion. Since momentum is mass multiplied by velocity and mass has no direction, it follows that, for the special case of equal masses, these arrows could equally have represented ‘momentum arrows’.

(Anything that does not have direction specified such as speed, mass or energy is called a ‘scalar’ quantity.)

(Vector quantities all specify direction as well as anything else)

It is important that this experiment be thoroughly understood because now we are going to consider an opposite kind of energy, negative energy. This is required before the Big Breed theory can be described.

3.2 Introducing Negative Mass and Negative Energy

Actually the concepts of negative energy and negative mass are not new. An engineer who then transferred to physics, Paul Dirac, noted that a certain energy equation from Einstein's theory of relativity had a square root. Square roots can be either positive of negative and so he suggested that therefore negative energy states should be possible. However, according to Blanchard (1969), he concluded space was packed with electrons in negative energy states, which is quite wrong. We will look into the question in a way that gives more understanding of what is meant by negative states.

However, before going into this and to give the reader more confidence that negative energies are not just unacceptable fiction, another advocate will now be introduced. This is none other than the famous paraplegic, Professor Stephen Hawking. He uses the concept to predict 'Hawking radiation'. He says space is made from particles of both positive and negative energy that spontaneously appear in pairs from nothing. But then they collide and annihilate each other. We shall see in the next chapter that he has missed something. But at least the idea that negative states can exist is shown to be a scientifically acceptable concept.

We embark on another thought experiment by presenting the plan view of the momentum balance as illustrated in FIG.3 in a novel way. It looks identical to the first presentation given in FIG.2 on first sight. However, consider the direction of arrows marked ***p***.

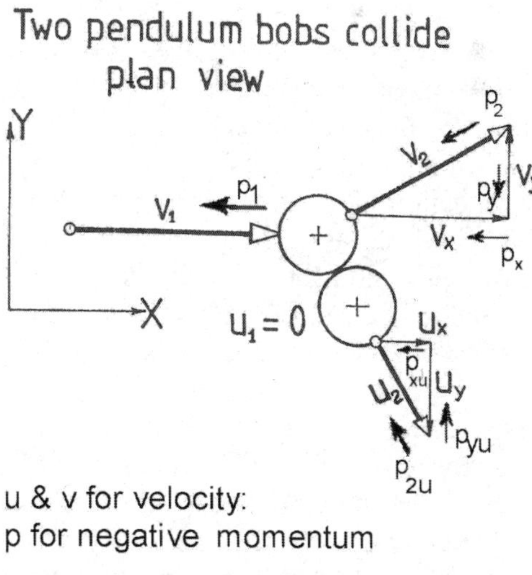

Now these represent momentum and are all pointing in the opposite direction to their corresponding velocity vectors. The question now is, "Can momentum still be conserved?"

Two pendulum bobs collide
plan view

u & v for velocity:
p for negative momentum

FIG. 3 MOMENTUM BALANCE

Carry out the previous thought experiment again with all momentum arrows reversed as shown. It will be found that momentum is conserved, in each case, just as well as when all momenta were taken to be positive. This is because all the opposite directions mutually cancel their effects.

Now since momentum is mass multiplied by velocity, and the latter has not been altered, it follows that we are now considering bobs as made of negative mass. What does this mean?

We consider Newton's three laws of motion, but first a point of interest is worth making. The conservation of momentum can be predicted from Newton's laws. Indeed the simple maths involved is presented in *CREATION SOLVED?*. So the experiment illustrated by FIG.2 actually provides an experimental check of Newton's laws. This illustrates the meaning of the scientific method. A speculation needs to be elevated to a hypothesis by mathematical logic. It does not reach the status of a theory until it passes the test of adequate experimental checks. So the checks we have made support the validity of Newton's three laws that are:

The first is the law of inertia. This just says that any object will keep standing still or maintain its motion in a straight line unless a force of some kind acts upon it. (Fairly obvious to us nowadays)

His **second law** could be re-arranged to state that the acceleration of an object is equal to the 'force of action' acting upon it divided by its mass.

So the meaning of mass being negative is simply that the direction of acceleration is opposite the force of action. (When the balls collided they were given a very high acceleration for a very short period of time but we will not go further into the proof of the connection of our experiment with this second law.)

His third law simply states that when an object is pushed by a force of action then the object doing the pushing feels a force of 'reaction' that is equal but opposite the force of action. (Again fairly obvious to us from experience)

The reader is likely to be bemused by the above interpretation of Newton's second law. How could an object accelerate backwards against the force applied? This seems to be nonsense: the object of negative mass would accelerate into the one pushing it and this is not observed and so makes no sense. It gives this impression, however, because one is naturally thinking of pushing the object with one of positive mass. This is a matter to which we will return. We are not yet ready to consider the interaction of opposite kinds.

If the object doing the pushing had also been of negative mass then its responses would also be reversed and so the response would be exactly as observed for positive masses (as already proved from FIG.3).

However, a negative mass is to be defined as one in which the force of action points opposite the acceleration produced. Then mechanical work, being force times distance moved, will be negative. It would be a braking force for positive mass but will accelerate negative mass. Since the work done on an object, totally free to move, will be entirely transmuted to kinetic energy, it follows that, for negative mass, the associated kinetic energy must also be negative.

If mass is negative then since $E = mc^2$ can still be derived in the same way as it was derived in Chapter 1 of our book *QUANTUM GRAVITY* (and summarised in APPENDIX IV). Then the associated sum energy must also be negative, and this includes the rest energy of the object. In fact all energy forms associated with our universe could have been treated as negative and the maths involved would have shown responses no different from those observed.

Treating everything as negative would be a silly thing to do since it makes the maths harder by needing more attention paid to negative signs, but the point is that in principle this could be done. It means that the entire universe could just as easily be made from negative energy as the positive kind. This is because it is impossible to differentiate between the two on the basis of observed responses.

So contrary to the first impressions most people have, when first introduced to the idea of negative mass, there is nothing strange about it!

What all this really means is that since positive and negative energies are equally probable, it is perfectly reasonable to infer that, at the ultimate level of i-ther, both exist simultaneously as a mixture. As we shall see this allows a solution to most of the major problems that physicists still struggle in vain to resolve.

This is a major philosophical point that needs to be fully grasped. The next step is to realise that the i-ther needs to be made from both positive and negative energies like the Yin and Yang of Chinese philosophy.

With such a mixture and with each existing in balanced amounts the net energy is zero. Since this is equal to that of the void, at last a plausible theory for emergence from nothing becomes possible. For example, at some place there could exist 10 energy units of positive energy and 10 units of negative energy so that a simple sum can be made that can be written:

+10 units -10 units = 0

This represents mutual annihilation. The two components of the

mixture cancel each other out to yield nothing.
Equally, however, spontaneous creation becomes possible as the converse case and can be written:

$$0 \text{ units} = +10 \text{ units} - 10 \text{ units}$$

It follows that the void is unstable and is therefore potentially able to give rise to the universe.

3.3 Primaries Moving at Very High Speeds

A new mechanics had to be devised before a satisfactory solution to our major problem could be found. So far we have only considered motion at low speeds. In Chapter 2 we decided that the new mechanics would need to be applicable at a sub-quantum level of reality that is truly real. This had to penetrate the quantum level and yield equations that matched all the data previously considered as Einstein's unique achievements.

All that matters as far as the Big Breed theory is concerned, however, is that each complementary form of energy provides the substance from which ultimate particles, the 'primaries' are made and each needs to have two components. The components are 'rest energy' and 'kinetic energy'. The reader will remember from Chapter 2 that $E = mc^2$ means that everything can be considered as made from energy. 'Rest energy' is the substance any particle or object is made from when standing still. 'Kinetic energy' is the energy added to create motion of such objects. These two kinds of energy can be added to yield, 'sum energy' (symbol E).

The mass corresponding to sum energy is now considered to be 'inertial mass' and this is now represented by symbol m.

So momentum $p = mv$ now has a more precise definition.
It is important to note that the inertial mass is higher than the mass existing before being accelerated.

So now we need to investigate the details. We need to find how these two opposite kinds of energy will interact. We will see, in the next chapter that the most amazing things will happen.

CHAPTER 4

THE BIG BREED THEORY

At last we are ready to delve into what we hope the reader will consider a satisfactory solution of the creation problem.

In the previous chapter we found that a mixture of particles, to be called primaries, needed to exist at the ultimate level of reality called i-ther in order to permit of creation from the void. Now we need to explore the way creation could happen.

4.1 An accelerating pair

To begin with let us consider a pair connected by some mutually attractive force as shown in FIG.4. If the negative mass leads in the X direction, then the force is pulling back but causes acceleration in the opposite direction: the X direction. The following positive mass feels the same force but oppositely directed. It therefore also acts in the same X direction. Therefore it also accelerates in the same direction as the negative mass. Both accelerate keeping a constant separating distance. Both are gaining energy of their own kind from the void but in equal and opposite amounts.

So this 'accelerating pair' represents pure creation from the void and yet there is no disconservation of energy or momentum. Sums of either property for the pair always remain zero.

If the positions of the masses were reversed then both would slow down: so representing mutual annihilation. What this example shows is that with such complementary energies and masses a paradox-free solution to the creation problem can start to be

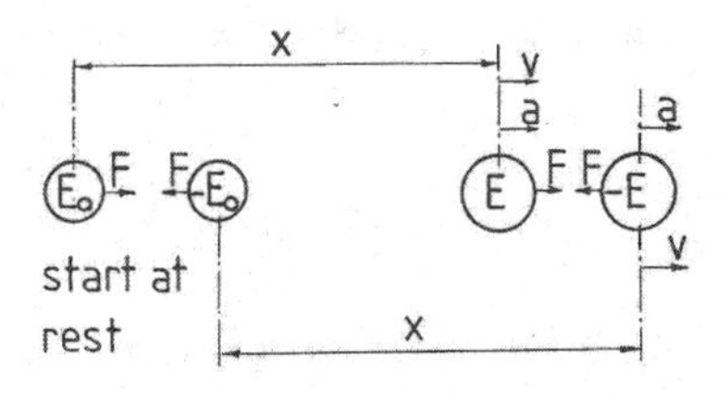

FIG. 4 AN ACCELERATING PAIR

After acceleration "a" from rest to speed "v", energy "F.x" is added to rest energy "E_0" to give the total energy "E". i.e:- $E = E_0 + F.x$

But for the negative member:-

$$-E = -E_0 - F.x$$

So the net energy is always zero!

formulated.

In 1987 an attempt was made to publish this idea as a way in which the universe could have been created: to supply the energy required without involving the unsolved problems incorporated in the inflation phase of the big bang theory. This was rejected by *Nature* and some other journals, without any reason being given.

Then Paul Davies (1989) published a book in which an article written by Clifford Will appears. He illustrates an accelerating pair on page 31. In this case a repulsive force acts with the negative mass following the positive one but the result is the same. So after-all the idea of the accelerating pair cannot be regarded as new.

Will goes on to report that in the 1960's this concept caused considerable controversy over a 15 year period leading to a 'Positive Energy Theorem' that proved negative energy could not exist. However, this rejection was based on the concept of potential energy being negative and causing difficulties. As shown in APPENDIX I and APPENDIX II this difficulty does not exist in the approach we will adopt. So negative states become once again an admissible proposition. In APPENDIX II it is shown that theorists had relied on mathematical rules without checking by the use of 'conceptual logic': the logic of common sense. In consequence a hopelessly wrong conclusion was drawn.

However, accelerating pairs cannot be postulated as a practical solution since no forces of mutual attraction can be expected. However, mixtures of opposite particles, all in motion like the molecules of a gas, will interact by repeatedly colliding with one another. This offers a practical solution worthy of further study.

4.2 Breeding Collisions

We are going to show now that when primaries of opposite kinds of energy collide, they breed like opposite sexes!

If a mixture of particles, to be called 'primaries', are colliding then from the foregoing logic either mutual annihilation or mutual creation could arise without energy conservation being violated. The question arising now is what other factor could determine which case will occur in given circumstances.

Has the reader guessed the answer? If not then let us again consider the conservation of momentum. This also applies to a collision of primaries of opposite kinds of energy. This is best argued by considering 'impulse'. Impulse is defined as force multiplied by the time of action of that force and is readily shown to equal change of momentum. Both interacting primaries are subject to the same duration and the same magnitude of the forces acting. Then since the impulse on each is equal to their change in momentum it follows that momentum must be conserved when opposites collide.

Let us now consider two primaries of opposite energies in head–on collision as shown at **A** in FIG.5. The positive mass is moving left to right at velocity v_1. The negative mass moving right to left at velocity $\boldsymbol{u_1}$ has its momentum arrow $\boldsymbol{p_-}$ pointing left to right and this is the same direction as the momentum arrow of the positive mass $\boldsymbol{p_+}$: they add up.

There is no way even any partial mutual annihilation could occur since the momentum of both would also reduce. Then momentum would not be conserved. Indeed neither primary could change its energy at all. So they are forced to go right through each other to emerge unchanged. Temporary mutual annihilation followed by reconstitution has to happen as they go.

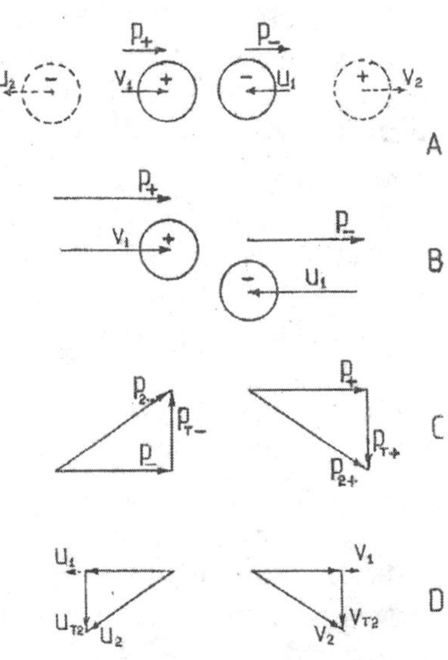

FIG. 5 BREEDING COLLISIONS
v for velocity +ve. energy
u for velocity -ve energy

Next consider the scattering collision shown at **B**. The lines of action are now offset so that each has an extra impulse added that need to be equal and opposite each other. Then each gains extra momentum. If the collision of a pair is considered in which a component of the added momentum is parallel to the incident lines of motion then it is easy to show that one primary will gain, numerically, more energy than the other. It follows that the momentum added by scattering has to be in a direction perpendicular to the incident relative velocity.

Just as at **A** no change in momentum in the original direction can occur for either primary and so the addition of the scattering momentum vectors increases the momentum of each collision partner.

The transverse momenta imparted need to cancel to zero to satisfy momentum conservation in the transverse direction.

However, each adds momentum vectors of its own kind to its original momentum as shown at **C**. The momentum arrow $\boldsymbol{p_{T+}}$ is added for the positive primary and $\boldsymbol{p_{T-}}$ for negative one. This is not an arithmetic addition: it is called a 'vectorial' addition. This is because the momentum vectors that are being added have different directions from that of the incident momenta. The tail of one arrow is added to head of the other to leave resultant momentum vectors $\boldsymbol{p}_{2+}$ and $\boldsymbol{p}_{2-}$ in FIG.5 at **C**.

Then at **D** both primaries deflect in the same transverse direction since u_{T2} is oppositely directed to $\boldsymbol{p}_{T-}$.

Although energy and momentum are very different, since energy represents the amount of substance present and momentum controls the dynamics of that substance, when the momentum of an object is increased, speeds increase and so energy is increased as well. So the need to conserve momentum has forced a balanced energy gain to occur for both collision partners. If added together these gains would of course sum to zero so that energy is conserved. However, we have now found a way for two energies of the universe to self-create from the nothingness of the void without involving false logic and without internal contradiction.

Of course most primaries converge from some angle θ ranging from 0 to 180 degrees and the latter situation is the one so far described. However, by using the standard practice of adding an appropriate velocity to the whole field of interest all possibilities can be made to represent exactly what has been portrayed. The only difference is that the relative velocity of the two primaries represented is now smaller than the velocity with which they converged.

Furthermore hardly any primaries hit head-on or make glancing collisions. Most hit between 1/3 and 2/3 of the way between these limits and scattering occurs in every direction. A careful statistical analysis involving triple integration has shown that, although some collisions will actually lose energy, the gains more than compensate. Indeed if primaries move at not more than about 10% of their ultimate speed then the maths shows that the average gain is 20% of the initial kinetic energy of the participants added numerically (ignoring the negative sign). The gain ratio falls to about 11% when speeds have their ultimate value. Spinning motions have also been studied and found to add almost the same net energy gains.

These gains make the primaries move at ever faster speeds. However, spinning motion acts like rest mass and the tendency of equilibrium to arise between linear and kinetic energies provides a limit on the maximum speeds reached. This limitation has pinpointed the average speed of primaries to be about 70 percent of their ultimate speed: the speed they would have if travelling at the i-theric equivalent of our speed of light.

However, primaries are not matter and so are not limited to our speed of light. To fix their average speed resort to an experiment had to be made. This was the astronomical observation of close binary neutron stars made by Hulse and Taylor (1975) as described by Will (1988). This showed an energy loss due to the radiation of gravity waves that Einstein assumed to move at the speed of light. ECM theory shows these must also be pressure waves propagating through the fluid component of i-ther. This requires the average speed of primaries to be 1.464 times our speed of light.

It follows that, on average when opposites collide, both gain energy of their own kind and, since the energy density of primaries will not change, size will increase bump by bump. There has to be a limiting size. For example raindrops can grow up to a critical size, then they split into smaller drops. Similarly, primaries will grow until they split. In this way a collision breeding process will develop. This is the basis of the Big Breed theory.

From collision rate analysis based on the research of Jeans (1887) the calculated breeding rate worked out at about 10^{43} times that required for the expanding universe. This improves on the 10^{120} factor of the big bang and now brings the problem within handling range. Furthermore mutual annihilation can now arise to cancel the excess in our theory to provide a complete solution.

4.3 Mutual annihilation cancels most of the creation

Unlike the big bang theory the solution to the problem, of switching off creation, lies within the breeding concept itself. We now consider what will happen after some critical number of primaries is packed into a unit volume of the i-ther. The latter is unstable as will now be shown, and can give rise to mutual annihilation.

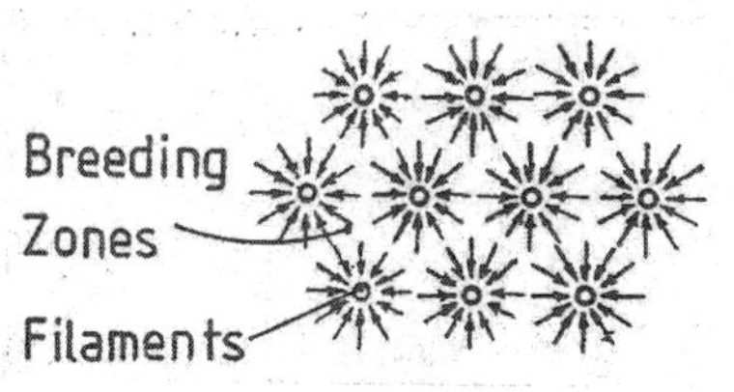

FIG.6 ANNIHILATION IN CELLS

The mixture now acts as a fluid. All systems in nature tend to seek the lowest possible energy state. Such a state of lowest energy arises when flow patterns develop in which primaries converge from all possible directions to a point or line. Add up all the radially inward momenta of positive primaries and the result is zero. This is also true for radially imploding negative primaries. Now, however, with both kinds present mutual annihilation can occur. Indeed the need to conserve momentum now forces this.

Consequently flow cells, minute even compared to the size of atomic nuclei, and as illustrated in FIG.6, spontaneously self-organise with flows to their centres where they mutually annihilate.

A solid core of annihilation forms at the centre of each flow cell. The cores grow until primaries arriving at their outer surfaces are equal to the number squeezing each other out of existence as they move toward the centre points. Cores can be blob-like or filament-like.

In FIG.7 a tangle of filament-like cores form a network that appears to have the potential for further evolution by the self-organising power of chaos. Theorists have shown that chaotic systems, supplied continually from some source of power, spontaneously self organise in ways that could never have been anticipated from the simple initial assumptions. The book by physicist, John Gribbin (2005) is well worthy of study. He shows how mathematicians have been amazed at the way purely randomised conditions show incredible degrees of self-organisation when fed by energy of some kind. And these cores are fed with huge amounts of energy that arise from the breeding zones around them. Creation will not be uniform but will have a random buffeting nature that will continually vibrate the cores and sometimes break them. They can reorganise in an infinite number of ways. By further development of this theory it seems reasonable to suggest the source of quantum waves may well be found. This could provide a worthy project for the programmer of a super-computer since it has Nobel Laureate potential.

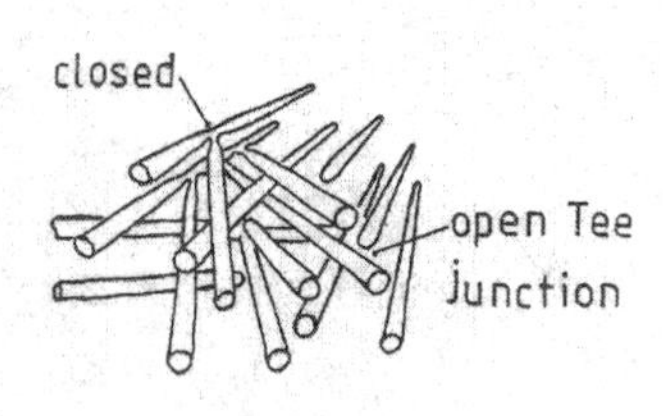

FIG. 7 ANNIHILATION CORES AS A TANGLE OF FILAMENTS

If the i-ther reaches a very high density, so that primaries fill all available space, then any further increase will cause them to squeeze further gains out of existence. Actually the state of equilibrium will occur when primaries still have some space in which to move. We call this the i-theric liquidus state in which annihilation exactly balances creation. Below this density a small net creation will obtain that sets itself to provide an accelerating growth. The greater the acceleration the greater the pressure difference from the origin point of i-ther to its growing edge. But the higher the pressure the lower is the net creation rate. It follows that the rate of creation and the acceleration produced act on each other to form a self-governing system.

For example, a growing sphere of i-ther can be imagined to be like an onion: as a nested set of thin spherical shells each being accelerated radially outward. Each shell needs a pressure drop from its inside to its outside radius in order to produce the acceleration. It follows that pressure and therefore also energy density of i-ther will be highest at the central origin point, reducing with radius increase until the ever-growing outer edge of the i-ther is reached. It then follows that, with a theory able to quantify the acceleration it has become possible, for the first time, to calculate the size of the universe!

In this way the basis of a solution to the problem of the cosmological constant has now been described and, as spin-off an accelerating expansion is predicted of, potentially, the right order of magnitude. The Dark Energy mystery has therefore been at least partly resolved at this stage.

The initial inflation of the big bang generated all the energy required to build our universe in a minute fraction of a second. For the Big Breed an initial inflation produced only a minute amount of energy. After that a small net creation continued everywhere and will continue to do so indefinitely. Nothing can stop this everlasting accelerating growth due to continuous net creation. So the reason for our universe being so huge has appeared as spin-off: it has to grow forever – unless it hits another one of comparable size that has been growing from a different location.

This prediction was made in 1992 as a result of the first private publication of this creation theory in *Origin of Mind.* A publication by the Russian Academy of Sciences followed (Pearson (1994)) and another in Frontier Perspectives (Pearson (1997)) It was not until Schwarzchild (1988) published astronomical data concerning remote supernovae that cosmologists accepted that the expansion of the universe was speeding up: not slowing down as they had expected. Consequently resort is now made to the concept of 'Dark Energy' with mysterious powers of repulsion at great distance to explain the observations.

The reader is asked to consider this proposition and compare with the theory just described. Has the i-ther model already explained the nature of Dark Energy? Furthermore the prediction arose from the solution to the problem of the cosmological constant for which no other solution has yet been advanced.

4.4 Other Factors

The breeding concept described in §4.2 assumed rest energies to remain constant during collision, the only changes being of kinetic energy. In the complete analysis it was found possible for some collisions to produce a net annihilation of rest energy to cancel part of the breeding going on. In fact this could cancel nearly all of the breeding energy gains.

However, primaries can spin as well as travel in straight lines. Some spinning motions can be increased by the collision whilst others are reduced. Analysis showed that increased spin dominated massively to increase overall energy gains. This swamped all possible annihilation of the kind previously mentioned. So there is no doubt that all cells of i-ther will contain breeding zones surrounding their annihilation cores.

It needs to be understood that the spin of primaries will not have the same meaning as that of the quantum spin The latter is restricted to finite values such as ½ or 1 but at i-ther level classical mechanics has returned. Now primaries can spin like a ball with any value of surface speed. Since real particles are being considered their spinning motion represents kinetic energy.

Now spin kinetic energy will appear the same as the rest energy of an object. Imagine, for example, two objects connected by an attractive force of some kind so that they form a spinning dumbbell. From a distance they would appear as a single stationary particle having only rest energy. A particle having rest energy can never reach a linear speed equal to the ultimate value.

This means that primaries, even if they had zero rest energy, cannot travel at their ultimate speed when they are spinning. Indeed the complete analysis suggests they travel at about 70 percent of ultimate speed, and that ultimate speed is 1.96 times the speed of light for reasons that will be considered in PART II.

4.5 The Limiting Conditions for Creation and Annihilation

How did the first primaries emerge so that they could start to breed? This is a vexed question that several people have posed. The answer is that the void is unstable because of its potential to create, simultaneously, particles of opposite kinds of energy. Conditions needed for instability to arise will now be explained.

For spontaneous creation to occur from the void, it is necessary to satisfy, simultaneously, the laws of conservation of both energy and momentum. If this can be arranged then the void can be considered unstable so that pure creation can happen, though this would need to occur at a very low rate everywhere. At some place, by pure chance, sufficient numbers of primaries would need to concentrate. Then collision breeding could be initiated.

In FIG.8 in Case **A** a pair consisting of equal and opposite masses emerge together from the same point at a narrow diverging angle causing them to separate. Energy is conserved and so is momentum p_X measured in the X direction, which is the median of the two trajectories. Unfortunately momentum is not conserved in the Y direction that is perpendicular to this median, since both momentum components p_Y point in the same Y direction.. If the angle of divergence is made zero to eliminate the difficulty, then the pair is unable to separate and so no creation can occur.

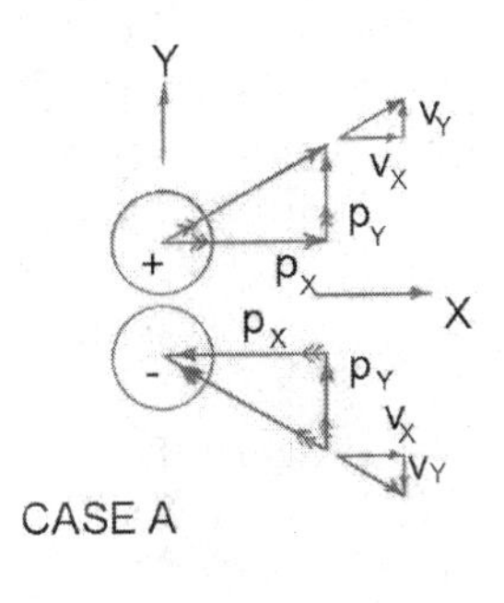

A solution is illustrated at Case **B**. This illustrates the lower limit of possible numbers allowing spontaneous creation from the void. This number is four. All have to explode into being from the same point. Two have to be of positive inertial mass and need to emerge with equal and opposite momenta p_X. Therefore they move off in exactly opposite directions at speeds v_X. This satisfies the conservation of momentum for the positive primaries. The other two need to be of negative mass having a total inertial mass numerically equal to that of the positive inertial masses. In this way energy conservation is satisfied. The negative primaries also need to move off in exactly opposite directions at speeds v_Y to satisfy the conservation of momentum having opposite directions p_Y.

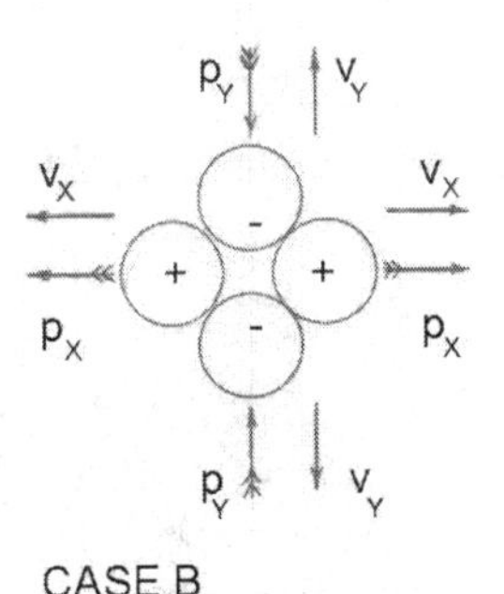

FIG.8 SPONTANEOUS CREATION OF PRIMARIES

It is not necessary for the directions of motion of opposites to be mutually perpendicular as shown: almost any angle will suffice.

A second condition for creation, of either the spontaneous kind or for collision breeding, is that motion should occur in the direction of increasing separation of the primaries. This will be defined as 'explosive' with velocity vectors diverging away from one another.

When velocity vectors are reversed, then conditions for mutual annihilation will exist. This will be considered as 'implosive': the mirror image of explosion. In this case total annihilation can occur. However, there is nothing to stop pure creation following when only four primaries are involved. It follows that more than four need to implode together in order to create the conditions favouring annihilation and the consequent formation of cores. This is why annihilation cores cannot arise until some critical number density has been reached.

4.6 DISCUSSION and CONCLUSION to CHAPTER 4

Primaries made of two opposite kinds of energy have been shown to gain energy of their own kind as they collide in twos. Numbers increase as they split after growing to a limiting size. This is the breeding process. When large numbers collide together from all possible angles, however, then mutual annihilation results. The combination produces steadily accelerating growth of the i-ther.

In the big bang theory space and matter all arose at the same instant with all forces of nature tangled in the mix. Everything arose together with matter and the forces acting on it considered fully real. Our solution is different.

Primaries are only the precursors of matter. These and the transient collision forces are the only things we have considered to be fully real. The creation of matter and the forces of nature have not yet been considered. This will be left to Chapter 6.

4.6.1 How Hawking went wrong

To digress we are now in the position to criticise 'Hawking Radiation'. This paraplegic is a totally amazing person in that to achieve anything at all in his condition is not short of miraculous. However, this does not exempt him from the critique that others have to accept since the progress of physics needs to be considered.

This theorist describes a quantum vacuum consisting of pairs of virtual particles that emerge spontaneously from the void. This is permitted, he considers, since one of each pair is made of positive energy with the other made of negative energy. This allows energy to be conserved. Then they collide again in pairs to annihilate each other. When close to a black hole the negative primary is sucked inside and the positive one is released to give the radiation that the theory predicts.

What Hawking has not considered is the need to conserve momentum as well as energy. As we have shown, by considering FIG.8, pairs cannot emerge from the void. Neither can they cancel by pair collision – as explained with reference to FIG.5.

Furthermore it is the positive primaries that would be sucked in: not the negative ones, so that the black hole would grow instead of being eroded as Hawking describes.

Before the creation of matter is considered, let us first amplify our explanation of the way the problem of the cosmological constant is resolved. This is considered in the next chapter.

CHAPTER 5

THE COSMOLOGICAL CONSTANT SOLVED

In the previous chapter the creation of the ultimate reality of i-ther, as a sub-quantum medium, was described by the Big Breed theory. Now we look at the way this solves major problems of the BigBang.

5.1 The Problem of the Cosmological Constant solved!

A solution to the problem of the cosmological constant that still eludes the big bang theory has now appeared. All the energy required to build our universe has emerged naturally as a result of the mutual creation and annihilation processes inherent in a mix of positive and negative primaries (and it needs to be realised at this stage that these have nothing whatever to do with positive or negative electric charge). In the theory being advanced electric charge and the forces of electromagnetism do not exist at the level of i-ther and neither do nuclear forces. These forces only appear at the quantum level of reality, simultaneously with their associated sub-atomic particles. But matter has not yet appeared! The creation of matter is dealt with in the next chapter.

It is also interesting to note that the basic structure of this theory was first published in 1992 in the pamphlet, *Origin of Mind*. At that time the prediction that the universe had to exist in a state of ever-accelerating expansion seemed an embarrassment since all cosmologists were saying that the expansion must forever slow down. However, in 1998 observations of the exploding stars called 'Type 1a Supernovae' reported by Schwarzchild, B. (1998) showed the expansion is speeding up.

Cosmologists all admit this came as an unwelcome surprise

PLATE I Frames Selected from Video Describing Creation

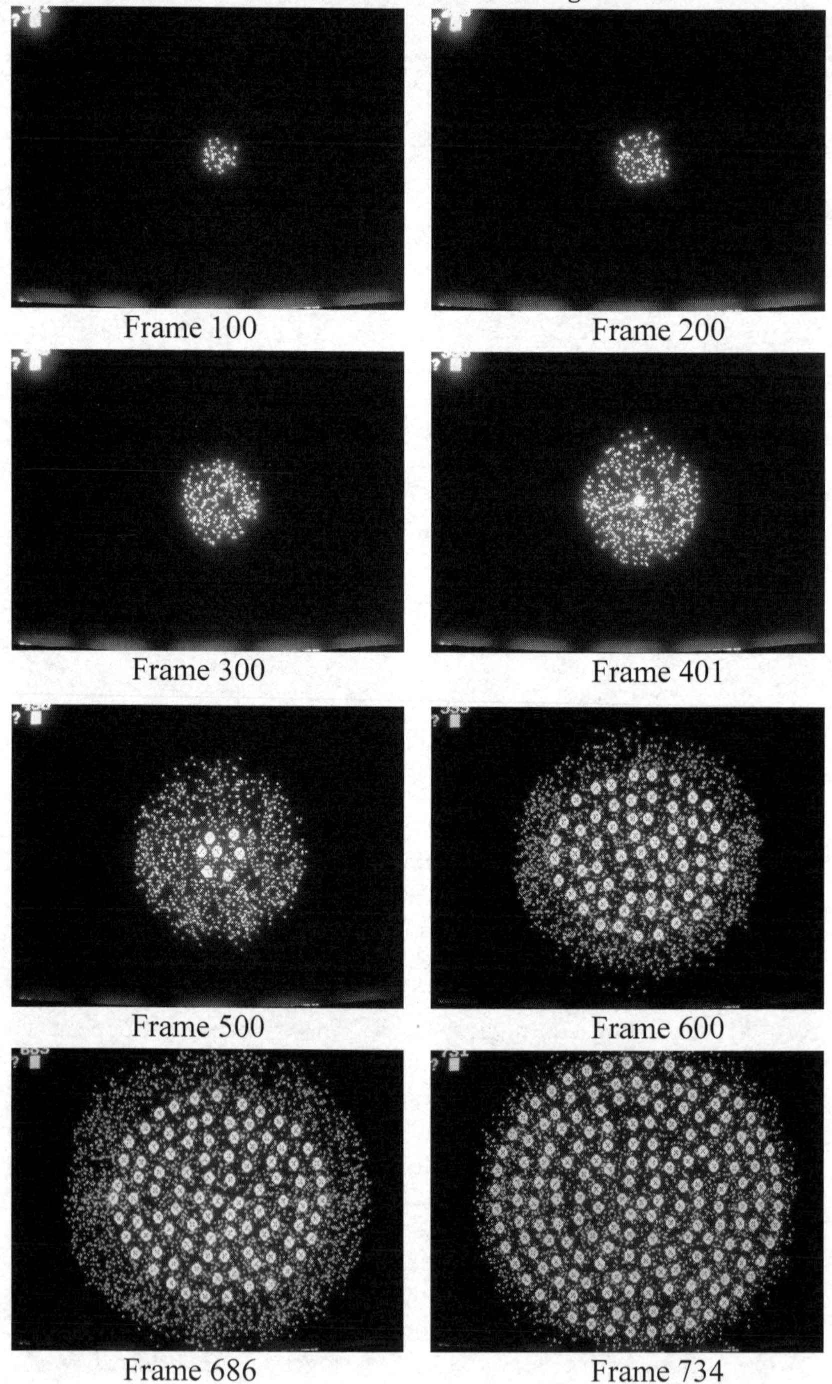

Frame 100 Frame 200

Frame 300 Frame 401

Frame 500 Frame 600

Frame 686 Frame 734

and caused them to make a rapid revision of their contention. Then they invented a mysterious substance they called ‘Dark Energy’ with unexplained and, using their word, ‘mysterious’ powers of repulsion to overcome the attraction of gravity at great distance.

I ask the reader to consider this: “What is the earthly use of adding such refinement to a theory, as an add hoc feature to match observation, when its major false prediction of an unacceptably high rate of expansion remains totally unresolved?”

Cosmologists have now resorted to assuming all the energy, which is required to allow space of the universe to expand, to be delivered through wormholes. One objection to this is that it requires all the energy needed to have existed without a beginning. Another is the location of that energy. To just assume it hides in some higher dimension does not seem very satisfactory. And again wormholes arise from the black holes suggested by Einstein’s theory of general relativity. The mechanics needed for the solution described in this book does not allow wormholes to exist. The ECM theory describes something similar to black holes. However, the speed of light only falls to zero at zero radius and this is never quite reached since an ultimate finite density is involved: that of primaries. And the force of gravity saturates.

The ‘singularity’ at the centre of Einstein’s black hole, has matter existing at infinite density and zero size. This is a feature all physicists admit is impossible.

The big bang theory has to resort to Dark Energy to provide an explanation for the accelerating expansion of the universe

The big breed theory, described in the previous chapter, provided a solution to the main problem and then the acceleration appeared naturally as a prediction: no resort to mysterious Dark Energy with mysterious powers of repulsion at long range was ever needed! This all arose because a creation switch-off means appeared naturally. The way the switch works should grab the readers’ interest.

The sequence of frames shown on the opposite page is taken from a video illustrating the way in which ‘inflation’ now appears. It will be remembered that, during the inflation phase of the big bang theory, all the energy eventually to become the universe was created in a split second and remained constant ever afterwards.

In the Big Breed theory only a minute fraction of this energy is created during inflation that is illustrated in the sequence. After the initial burst, due to the fortuitous appearance of some breeding groups, collision breeding produces exponential growth.

PLATE 2 Frame enlarged to show random bunching.

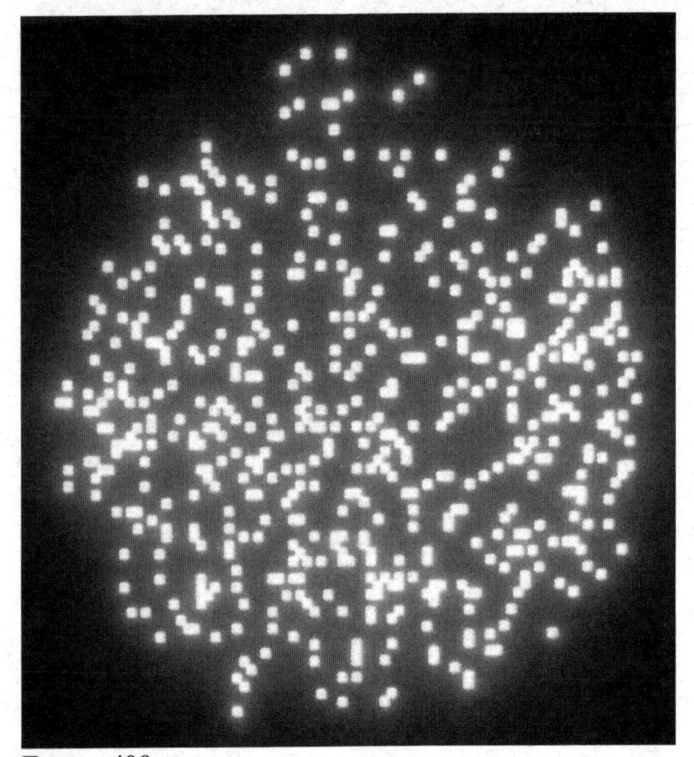

Frame 400

The primaries form into a seething mass. Unlike the big bang, as shown in PLATE I creation is brought to a stop at frame 734. At least it appears as a complete stop on the relevant time scale of milliseconds. Measured from the initial burst, this initial exponential growth is illustrated by frames 100 to 400.

What is illustrated is a thin slice cut through the diameter of a growing sphere of primaries. It is therefore a two-dimensional representation of three-dimensional reality.

Also cores are minute. They are small even compared to the components of atomic nuclei since they have to build the sub-atomic particles of matter. The huge acceleration produced requires the development of pressure gradients. The highest pressure is at the origin point and pressures fall toward the outer boundary. The number of primaries per unit volume is their 'number density'. This is forced to follow a similar pattern. It follows that a critical density, at which instability occurs, is first reached at the central point. Frame 401 shows a white blob, representing the first annihilation core, suddenly appearing.

Then as densities grow at all radii the zone of instability grows. Successive rings of annihilation cores then appear. Six more have appeared by frame 500 but this only stops growth inside itself. The cores are jumping about due to buffeting caused by random surges in the motion of primaries. The outer annulus is still growing. However, as shown by frames 600 and 686 the radial growth of the rings of cores exceeds that of the growing edge. Eventually growth is brought to a stop at frame 734 with cores close to that edge.

However, on cosmological time scales measured in millions or billions of years a minute rate of growth continues. Annihilation can only equal creation at the 'i-theric liquidus' density when primaries have as little room to move, relatively speaking, as the molecules of water. This is not reached since confinement is by acceleration. The greater the acceleration of each spherical shell of the i-ther and the greater the total diameter of the growing ball, the greater is the pressure needed. So a balance is automatically formed that adjusts the rate of creation to the size of the i-theric ball. Creation can never stop and the i-ther can never cease to grow.

What is being represented is the first stage of self-organisation by chaos limited by a two-dimensional representation. In three dimensions the cores can be either blob like or can form long filaments. These can tangle and join up with blobs to form a complex structure. It is the hope of this author that enthusiasm will be generated in people having access to a supercomputer. The self-organisation of randomness fed by energy is a phenomenon described by the mathematics of chaos theory – as lucidly described by John Gribbin (2005). Astonishing organisation has been demonstrated that cannot be explained from considering the simple structures that originally existed.

The Big Breed theory only explains the creation of the i-ther, which is the origin of the energies needed for building the universe of matter. The i-ther had to evolve a conscious intelligence before it could create the mathematics needed to create the quantum world of mathematically organised energy. This cannot reasonably be discounted since, as already described in Chapter 2, the weirdness of the quantum world suggests it is not fully real but exists more in the nature of a virtual reality.

Of course this is painted as a moving picture on the i-ther using itself as a three dimensional screen and using quantum waves as its brush strokes. Consequently this gives the impression of being totally real, especially since our bodies are included as part of the picture.

Then the electromagnetic and nuclear forces, as described more fully in *CREATION SOLVED?* and *INTELLIGENCE BEHIND THE UNIVERS III*, also exist as mathematically contrived illusions. In those books and given mathematical support in *QUANTUM GRAVITY via an Exact Classical Mechanics* it is shown how only gravity could exist as the only truly real force.

5.2 A curious Observation on Randomness

PLATE II shown on page 38, together with one at frame 200, is given in full colour in our companion volume *THE BIG BREED THEORY of Creation of the Universe*. This can be downloaded from the internet. Indeed that introductory volume is printed in full colour throughout. In this technical book, however, all figures are in black and white. Consequently the random nature of the motions of primaries is illustrated but no distinction can be seen between primaries made of either positive or negative energy. However, the front cover provides an illustration in colour.

On the cover illustration positive primaries are shown in magenta, whilst negative primaries are shown in cyan.

Programming of the 'VideoCr9' (in GWBASIC) described in the previous section led to an observation not previously anticipated. To illustrate this observation, an enlargement of frame 400 is shown as PLATE II. The latter is the frame just before the first core appeared. Only the randomness of motion is portrayed due to lack of colour. The reader is therefore requested to turn to the front cover, which shows the primaries in colour with annihilation cores shown as white blobs.

In the computer code, as new primaries appear they emerge as pairs. Each pair consists of a positive primary (magenta) and a negative primary (cyan) starting from the same point. After that both undergo random motion to produce a seething mass. Now it is well known that randomness does not correspond with a uniform distribution. Instead considerable bunching always arises. This bunching is illustrated in all the others and the front cover.

What was not anticipated was that the two kinds would bunch in different places as these frames show.

The significance for the Big Breed theory is that it had been assumed that statistics would demand that half of all collisions would be between primaries of like kind and half between opposites. Now it is realised that this would apply only when distribution of position is uniform. What the non-coincidence of bunching shows is that the probability of collisions between opposites will be far smaller than had been expected.

This only reduces the gross creation rate, which is unimportant since annihilation will adjust to match. The importance of the finding is that as far as mass motions are concerned each phase of the mixture will couple mostly to its own and not its partner. Each is accelerated by its own pressure gradients with little cancellation due to the opposing momentum of the opposite phase.

5.3 How our new theory answers other vexed questions

In fact our new theory answers some other vexed questions. One of the questions raised in Chapter 10 of *INTELLIGENCE BEHIND THE UNIVERSE II* was: "Why make the universe so huge?" The answer is now obvious, "Because it has to grow forever and always at an accelerating rate: it can never stop growing!"

Then again: "Has the universe always existed?" No. Primaries had to be emerging from the void at a slow rate for a considerable time and had a life span so they vanished back to nothing again spontaneously. Only when by chance a bunch did emerge, that had time to trigger the breeding process, did a growth point form. This allowed breeding collisions to take off.

"Will the universe ever end?" No. It has to go on growing forever. Only if the acceleration were to be eliminated could the universe end. Without acceleration pressures and densities would fall to zero and then the whole of creation would vanish back into the void from which it emerged. Then nothing would again exist.

However, such a scenario is impossible. It has turned out that the void of pure nothingness is an unstable state able to give rise to something – the i-ther. This will keep growing, and the central pressure will keep growing until the liquidus state is reached. Only there will growth stop altogether. A spherical patch of i-ther, devoid of any structure, will grow rapidly but always a wide annular region outside the liquidus will exist that keeps growing and will retain its self-organising structure. The i-theric liquidus state is the only true state of equilibrium towards which everything tends to settle.

The way the universe of matter could arise from the i-ther has not yet been addressed. The clue was given in Chapter 2 where it was argued that the weirdness of the quantum level of reality needed a conscious observer to "collapse the wave function" according to the Copenhagen interpretation of 'wave-particle duality'. This accepted that some form of conscious intelligence was involved in the organisation of matter.

And since that consciousness had to precede the creation of matter it could not be mere brain function. There might be a 'brain mind' and a 'brain consciousness' that vanish on death. However, this could not be the creative force of matter because a consciousness that preceded matter was required. This demanded the existence of an ultimate level of reality that was truly real and had the potential, within its structure, of evolving conscious intelligence – hence an intelligent ether (i-ther) had to exist.

This suggests that abstract waves are being used like numbers to organise what seem to us as sub-atomic particles. Our world now appears more as a three dimensional virtual reality than as something truly real. All that exists seems now to be a seething mass of primaries and the solid structures they spontaneously form. Michael Roll, the director of the 'Campaign for Philosophical Freedom' is fond of saying, "Reality lies in the invisible."

This is all explained in more detail in *CREATION SOLVED?* where it is argued that in the fullness of time and the organising power of chaos the i-ther could ultimately evolve this conscious intelligence.

This could happen because the huge rate of creation going on in the outer regions of cells, together with annihilation at their centres provides the essentials of a power supply. Then as described in the very readable book by Gribbin (2005) when any randomly formed arrangement is supplied with power, amazing degrees of self-organisation occur. The filaments and blobs of the i-ther look like the neural networks of our brains and Hinton (1992) has shown how artificial neural networks have memory and learning ability.

Most of this picture was presented, in summary form, in 1993 at the Sir Isaac Newton Scientific Conference in St. Petersburg. It was selected for publication by the 'Russian Academy of Sciences' and published in the proceedings, Pearson (1994). It has also appeared in an American scientific journal, Pearson (1997) and in an Indian scientific journal, Pearson (2005).

5.3.1 Non-Locality Provided with an Explanation

Non-locality is the apparent connection of quantum particles separated by huge distances. The i-theric filaments are composites of positive and negative mass that sum to zero or almost so. It they are absolute solids then they also have a high Young's modulus. These factors combine to provide a near infinite speed of propagation of longitudinal waves that could transfer information almost instantly from one point in the universe to any other.

In the next chapter the creation of the matter to which we relate will be considered in more detail.

CHAPTER 6

CREATION OF OUR UNIVERSE

In Chapter 4 the creation of the ultimate reality of i-ther, as a sub-quantum medium, was described by the Big Breed theory. Then in Chapter 5 it was suggested that the i-ther had the potential for self-organisation that could have led to its evolution into a conscious entity. However, no reference has yet been made to matter. In this chapter the way matter could have arisen is explored.

6.1 The Creation of Our Universe

Only the creation of i-ther with its seething mass of primaries and annihilation cores has so far been considered. However, the possibility of evolution of i-ther to an intelligent entity (intelligent ether) has been argued in Chapter 5. Only when consciousness had also appeared, within the i-ther, could our matter be produced.

Our speculation is that sub-atomic particles, together with their associated nuclear and electromagnetic forces, were programmed in a similar way to our use of computers. For some reason as yet unexplained, abstract waves were used instead of numbers in the computational procedure. In this way the weird nature of wave-particle duality is incorporated. The 'wave-particles' would be produced as computational sub-routines containing replication features like those of the computer viruses by which we are plagued. Unlike computer viruses, however, replication cut off routines would be incorporated. These would ensure cessation of replication after a chosen number of generations.

The result would be an exponential growth of the precursors of matter that would suddenly cut off after a preset time. In this way creation of the universe of matter could now occur in a manner very similar to the inflation Guth has described to provide the big bang theory. The difference, of course is that now there is no problem of a cosmological constant needing resolution because a creation switch-off means has been incorporated.

Indeed there are now effectively two big bangs. The first initiates growth of the i-ther that then evolves consciousness. The second is a deliberate creation produced by that consciousness. And both have the means for appropriately switching off creation.

No energy needs feeding through wormholes in space-time to allow expansion to continue. All the energy required has been provided by the ever-growing i-ther.

The flash of light that has now become the cosmic microwave background (CMB) is explained by existing theory but applied to the abstract system of matter that has now been deduced. The only differences from the established approach are the absence of the force of gravity as part of a super-force. Gravity needs a separate explanation that is covered in the next section.

It could well be that as the i-ther continues to expand more galaxies will be required to fill the gaps. Therefore new explosive creations may need to be programmed to occur as time goes on. Indeed, these may already have been observed as quasars but this is only a speculation. However, it is not impossible that continuous creation could be superimposed in a manner providing a revision of Hoyle's steady state hypothesis. The main revisions are the finite age and size of the universe together with the causes of creation together with incorporation of their switch-off mechanisms.

We now have a theory that fits in with the Copenhagen interpretation of wave-particle duality that demands the participation of an observer. Mind, however, now appears as real and immortal, due to the self-organising structure of i-ther from which everything we see emerges.

6.2 What causes the Force of Gravity?

Only gravity remains unexplained. In our approach matter and the other three forces are abstract instead of being real. They are part of the way the i-ther creates the illusion of matter being real. The i-ther is now imagined as a three-dimensional computing system. All computers need a power supply for their operation.
Energy is therefore sucked in from immense distance via the filaments of i-ther and is deposited where matter is organised.

This excess energy produces an excess energy density where matter exists. Then because the i-ther is also effectively a porous solid, due to its cores of annihilation being embedded in a 'gas' of breeding primaries, the excess energy leaks away.

It leaks by viscous flow and simple analysis given in *QUANTUM GRAVITY* via etc. shows this produces an inverse square law of pressure gradients surrounding all material objects. This yields exactly the correct form of pressure variation as required for producing the gravitational force as one of buoyancy type and matching the ECM equations.

However, in order to produce a force of attraction, the net pressure of i-ther needs to be negative. Now it is the random motion of the atoms or molecules comprising a gas that cause its pressure. So for the net pressure of i-ther to be negative requires the kinetic energy of the negative primaries to exceed that of positive primaries. Then in order to restore overall energy balance to zero the rest energy of positive primaries has to exceed that of negative primaries. Asymmetry is needed to explain gravity.

The net pressure gradients are very small in the solar system. When analysed mathematically as in *QUANTUM GRAVITY* the reason for gravity being apparently weak as compared with the other three forces is explained. About the correct magnitude for the force of gravity is the result.

We consider this adds considerable credibility to the complete Big Breed theory since no other theory of gravity has yet appeared that shows why the force of gravity is relatively weak.

Indeed the standard approach says gravity is about as strong as the other forces at creation time. Then resort is made to this force leaking into a hyperspace consisting of an infinite number of parallel universes. How can such a speculation have any hope of yielding a comparable achievement?

Another difference is that, in the standard approach, all forces are assumed to be fully real. In the Big Breed approach only gravity seemed at first to be truly real with the others being abstract simulations. However, the force of electromagnetism could be arranged to have an optimum value based on gravity as its controlling factor. So a form of unification is still possible.

And so it seemed up to August 2010. In PART II Chapter 12 we find that even gravity has to be considered an abstract force!

Clearly such paradigm-shifting concepts are likely to have some effect on cosmology. At present astronomical data is interpreted according to Einstein's theory of special relativity according to the cosmologist Wright (2005). This does not accept that light propagates through any medium. The Big Breed theory demands the existence of a medium and so some difference of interpretation is to be expected. For example astronomers say existing big bang theory does not allow sufficient time for galaxies to form and that some stars are older than the estimated age of the universe. These enigmas are considered in the next chapter.

CHAPTER 7

HOW WILL COSMOLOGY BE AFFECTED?

The short answer to this question is, "Quite a lot". Several factors that have never been considered in the literature combine to render quite different interpretations of data produced by astronomy.

7.1 We are observing the entire universe

Most theoreticians say the observable universe is likely to be only a fraction of a much larger entity so that the impression we have of being at its centre is an illusion. General relativity is applied and from this some conclude that, "The universe is finite yet unbounded". Space is curved, so they say, and closes back on itself.

This interpretation cannot be maintained when the assumption of Euclidean geometry is adopted. Now the universe appears as a huge ball with a finite edge or boundary. Furthermore, as we shall see, the speed of recession at the edge, as observed from light taking about ten billion years to reach us, cannot be much more than half the speed of light. This means that we should be able to see right to that edge.

Another factor is that a background radiation has been studied that is considered to be evidence of the flash of light caused by the big bang. Astronomers, using the Hubble space telescope, have discovered that we are moving at a speed of 400 kilometres per second relative to this radiation. Therefore with the speed of light being 300,000 kilometres per second, and the edge moving at half this speed, it seemed clear in 2007 that we are moving at only 0.27 percent of the speed of the edge, as it was about 10 billion years ago. However, it is now realised that this cannot be used as evidence supporting our proximity to the origin point since the CMB would tend to propagate relative to the same local space as the observer.

However, since the universe seems symmetrical and we can see to its edge we conclude that our distance from its central origin point, though finite, is small enough to be ignored for the following analysis. This appears to be at variance with the established view.

7.2 Difficulties in Communication with Cosmologists

The cosmologist E. Wright (2005) has described how he analyses astronomical data and provides a wealth of data regarding the distances and red shifts of remote type 1A supernovae. These are exploding stars of accurately known light intensity so that by measuring the flash of light they emit, their distances away from us can be estimated to fair accuracy. Unfortunately he then renders the 'luminosity distance' data useless to us by processing it through the eyes of special relativity. The next section shows why this is the case. He has been approached by several physicists, as well as us, to provide some unprocessed distances. He has always replied but, instead of giving any unprocessed distances, he always supplied a mass of other data instead that did not help.

The only useful data provided was the statement that the greatest supernova red shift ever observed is 1.755. (What this means will soon become clear.)

Other difficulties in communication have also been encountered. Repeated attempts were made to publish the new theory of mechanics, described in APPENDIX III, to a large number of journals during the period 1988 to 1991. All submissions included the maths and showed good agreement with observation. Every one rejected the paper without even attempting to show any fault existed. These are still considered to be the unique achievements of Einstein's relativity theories. Furthermore none even mentioned its excellent agreement with observation or the way it solved the problem of incompatibility with quantum theory. Most rejection letters simply made statements, as for example, "Einstein's theories of relativity have withstood the test of time so no alternative is required". Yet all physicists admit quantum theory and Einstein's theories are incompatible and the whole point of the papers submitted was that a new mechanics was being presented that eliminated this difficulty. Then they admitted in 2005, as quoted in Foreword, that a replacement for Einstein's theory is needed but nobody has any idea of how this can be achieved!

Is it unreasonable to suggest that difficulties in communication could be causing a waste of public funds?

7.3 Light must speed up as it approaches Earth

Cosmologists, like E. Wright (2005), interpret astronomical data through the eyes of Einstein's theory of special relativity. If used directly that theory predicts huge and wrong mass increases for remote galaxies. However, no such direct application is adopted. Instead a whole string of observers is imagined passing the information about the light emitted from a remote galaxy from one imaginary observer to the next. This theory ignores the existence of any background medium, despite the fact that quantum physics demands the existence of an all-pervading 'quantum vacuum'. The latter simply means that atoms are absent. However, a seething mass of so-called 'virtual particles' is accepted to exist. This acts like a huge gas cloud extending to the very edge of the universe. Is it reasonable, therefore, to ignore the existence of any background medium?

Other cosmologists use general relativity in which wavelengths are stretched by expanding curved space-time. This produces the red-shift with all galaxies stationary relative to local space-time. This approach is consistent with light propagating through expanding space and so has similarities to the analysis to be presented next. Unfortunately general relativity cannot be adopted when the expansion is caused by a net creation going on everywhere. This is because theorists say the Hubble constant had to be much higher in the past and our solution gives $H_O = C_{NA}/3$ where C_{NA} is the net creation rate that is almost constant through time, so H_O does not change very much. In consequence neither special nor general relativity are applicable to space expanding due to any net creation. Therefore a new approach had to be adopted based on our *QUANTUM GRAVITATION via Exact Classical Mechanics*.

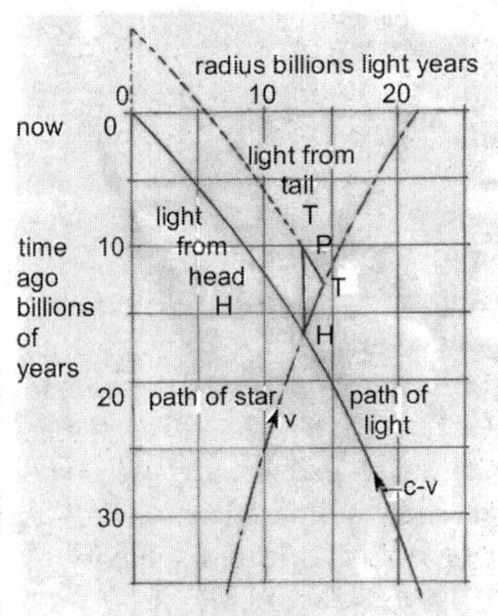

FIG.9 STARLIGHT SPEEDING UP AS IT APPROACHES THE CENTRE OF THE UNIVERSE

Ho=30km/s/Mpc: Z=1.75: v/c=0.4 at star

Summarised in Appendix III, this mechanics is really part of the Big Breed theory.

It demands that any mass increase is deduced from speeds measured from the local background: called space among other names. Then since the galaxies are receding at speeds proportional to their distance from us, as first discovered in 1929 by the famous astronomer, Hubble, their local space recedes with them. Light now travels to us in straight lines from great distances. Since we are almost centrally placed, light comes inwards to us following almost radial paths.

Since space is receding the light starts out from a base that is receding. As it comes radially inwards it has to move against the motion of space. It is like trying to row a boat against the current in a river. The boat's speed as viewed from the bank is less than the speed relative to the water by the flow speed of the current.

Light, however, as it approaches meets space that is not moving as fast as it was at greater distance from Earth. If the light is propagating at a constant speed through space, then it follows that it must speed up as it approaches Earth. The path taken is illustrated in FIG.9 that plots the time in billions of years ago against radial distance in billions of light years. The curvature of the line marked 'path of light' illustrates this speeding up. The plot is to scale but is based on the speed of light relative to space, given symbol c, being constant. The curve marked 'path of star' shows how the expansion is speeding up.

Now momentum has to be conserved as the photons of light speed-up and this means that their energy continually reduces. Since the wavelength of light varies inversely as the photon's energy, it follows that an increase of the wavelength of light results. This has not been previously recognised owing to the acceptance of special relativity: a theory that does not recognise any background medium.

It is the increase in the wavelength of light that is known as the 'red shift'. Blue light has higher energy per photon and a shorter wavelength than red light. Hence, as the energy of light from a star reduces, its wavelength increases. The light shifts towards the red end of its spectrum

Most of the red shift is presently assumed to be due to the so-called 'Doppler shift', somewhat modified by special relativity. As waves are emitted from an object, like a train whistle moving away from us, the wavelengths are stretched due to the speed of the train carrying the whistle and so it sounds lower pitched than would be the case if the whistle were stationary. The same effect stretches the wavelengths of light.

This is also illustrated in FIG.9 by the triangle marked HTP. However, instead of using waves, the flash of a supernova is represented, whose duration has been exaggerated thousands of times. Real supernovae only last about 20 days.

A post is imagined placed at the head of the flash marked H and this post remains at a fixed distance from Earth. So as the star moves, emitting light, space flows past the post moving with the star to the point T. This is the tail of the flash where the light cuts off. Light from this tail has to travel against the flow of space until it gets back to the post at P. The light only took the time H to T to be emitted but to get back to the post the time required is that from H to P, which is longer. Waves are stretched in the ratio of time H to P divided by the time H to T. This process will be called the 'Doppler shift' although it is not identical to the standard definition.

But the light still has to move to Earth against the flow. So this argument shows that the Doppler shift and that due to light speeding up are multiplied rather than added together.

The result is that much lower speeds of recession are now

deduced from the same red-shift data. A fuller explanation is given in Chapter 7 of *CREATION SOLVED?* And full mathematical detail is provided in PART II of this book.

7.4 The Speed of Light increases with distance

Einstein's theories contain the assumption of the speed of light being a universal constant with time treated as a variable. The mechanics needed for the Big Breed theory is based on universal time and then it transpires that the speed of light must increase with altitude, as explained in Appendix III. This shows the speed of light must increase as the pressure of the quantum vacuum reduces. As will presently be shown, this pressure reduces with radial distance.

To produce an accelerating expansion of i-ther, pressure gradients have to develop. Imagine i-ther as a huge onion built in concentric layers. To produce the acceleration of any layer the pressure inside has to be higher than that opposing from outside. This means i-theric pressure has to reduce with distance from the origin point. It therefore follows that the speed of light must increase as distance from the centre of the universe increases. This tends to counter the speeding up considered in the previous section but there is still a red shift caused by the speeding up for a reason derived in PART II Chapter 9 of this book.

7.5 Photons gain energy as they move in towards us

The theory of gravity included in the new mechanics is based on pressure gradients of the i-ther producing a force like the buoyancy force causing a balloon to rise in the atmosphere. As explained in Appendix III, however, it pushes down instead of up. The pressure gradients considered in the previous section therefore have a gravity-like effect and cause the photons of light to gain energy as they move toward our astronomers. This counters part of the effect of light speeding up as it approaches as described in §7.3.

So the analysis of red shift data now needs to take into account four factors, the Doppler shift, the effect of light speeding up as it approaches, the speed of light relative to space increasing with distance and the energy gains of photons as they fall to the origin.

7.6 How I-theric Pressures Change with Time and Distance

Before red shift data can be analysed, to take account of the factors described in sections 7.3 to 7.5, the way the i-ther evolves with time needs to be found. This part of the research program took many years to complete. It turned out to be the most difficult part to derive. This had to start from the way the rate of net creation varies with i-theric pressure and one early relation chosen is illustrated in FIG.10. During the inflation period, as annihilation cores just fill all available space, the pressure jumps almost instantly from zero to the shock front pressure P_{SH}.

So $P/P_{SH} = 1$ at this point as shown in FIG.10. As expansion proceeds this pressure jump occurs to mark the growing spherical edge of the i-ther.

Since the i-ther has to be accelerated by the net creation going on everywhere, pressures have to be higher at the central origin point than at the edge. This is required to provide pressure gradients that will drive the acceleration of each small volume of i-ther. The increased pressures were originally thought to be the cause of a fall in the net creation rate as shown in FIG.10 by the reduction of creation rate as ratio P/P_{SH} increases.

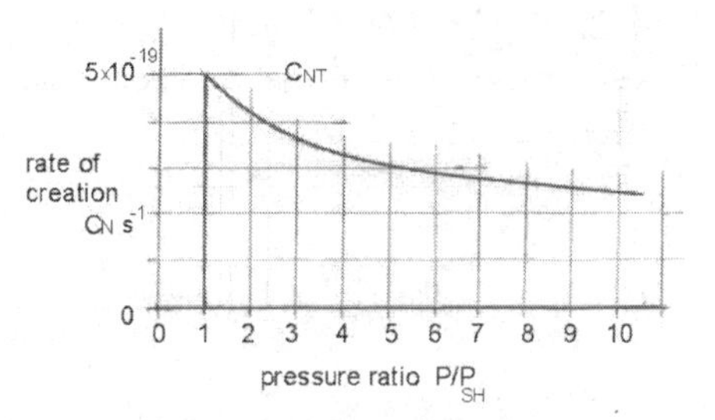

FIG.10 CREATION RATE FALLING WITH INCREASE OF PRESSURE

Plotted for $C_N = C_{NT}(P/P_{SH})^{-0.3}\ s^{-1}$

Only a small part of the creation/pressure profile is shown and ultimately this was found to be quite wrong. The correct relation is derived in PART II Chapter 10. However, the results are reproduced in the following text and figures so that they can be compared with the more accurate evaluations. What is surprising is that the false start reported here gives most of the features of the later derivation.

In FIG.11 the first stage of an attempted solution is illustrated by pressure/radius and speed/radius curves. This is plotted for the present era (now). It is as if some form of information carrier existed, which could move at infinite speed. Then the entire universe could be observed without looking back in time. Pressure at the central origin point, where the radius is zero, is P_0. Pressure at any other radius, given in billions of light

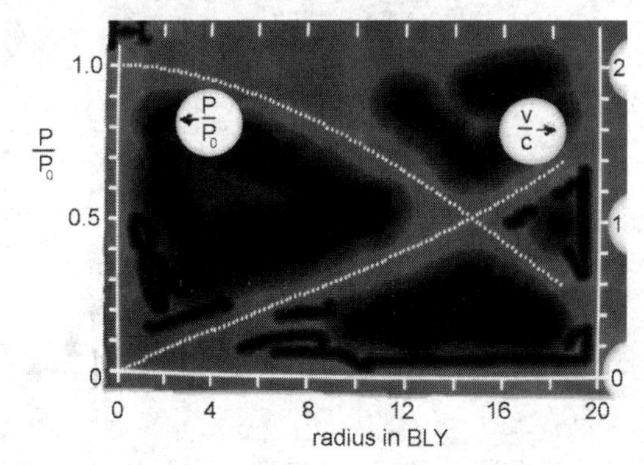

FIG.11 PRESSURE & VELOCITY PROFILES NOW

years (BLY), is plotted as the proportion P/P_0. As previously explained the way pressures fall with radius provides the driving force needed to accelerate every part of i-ther. By drawing a tangent to the pressure curve at any point, its slope there is determined. This slope is the pressure gradient at that point.

The accelerating motion produced is shown by the velocity curve marked v/c. It is the velocity v divided by the speed of light c.

Since cosmologists refused to divulge the 'luminosity distance' of the most remote supernova yet observed, an estimate had to be made in a different way. The latest way is described in Chapter 10 in PART II of this book. Briefly the rate of net creation sets the acceleration rate and the latter determines the radius of the i-ther to its shock-fronted edge. These three factors are therefore interrelated and permit the size, for the first time ever, to be determined from theory.

This computation returned a radius for the shock-fronted edge of i-ther to be 18.4 billion light years. This is less than the value given by E. Wright for the distance of the most remote supernova. He gives a value of between 9.3 and 13.7 Gpc, which converts to between 33 and 45 billion light years. His estimate, however, is based on processing via Einstein's relativity that is now to be regarded as providing an inapplicable and obsolete methodology.

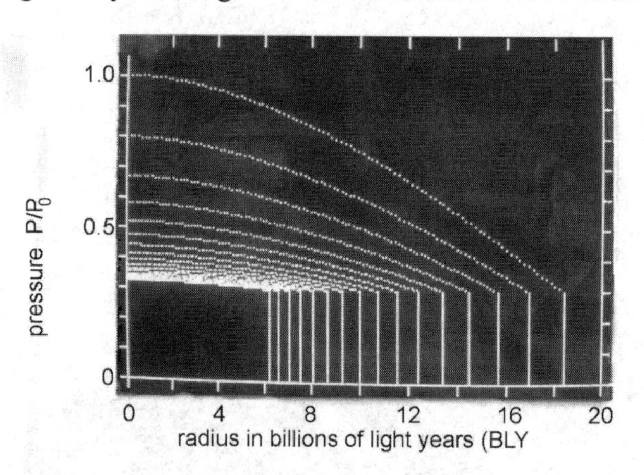

FIG. 12 PRESSURE/RADIUS PROFILES AT I BILLION YEAR (BY) INTERVALS TO 15 BY AGO

The next step is found by finding the acceleration at every one of the 100 points representing each of the two curves shown in FIG.11. Then together with the velocity at each point the place where each point would be a billion years ago is determined. This means a statement of the radius of each point at that time in the past, together with its associated new pressure and velocity. The computation, takes into account a net creation rate varying with pressure. The computation is repeated 15 times to take us back in time to 15 billion years ago. This number of time steps is limited by the memory capacity of the rather ancient computer code that has been used. This is GWBASIC. The computing code is "Un290609".

The result is depicted in FIG.12.

It will be noticed that the shock fronts have all been added. These are the vertical lines marking the edge of each profile.

Clearly the pressure profiles flatten rapidly as we go back in time. From the viewpoint of the evolving i-ther the pressure gradients start off flat at origin time, many billions of years ago beyond our 15 time steps. Then the plot shows pressures at the central point start to rise at an ever-increasing rate.

When the pressure profiles are flat calculation is very simple making it possible to determine the radius of the edge of the universe for any time in the past greater than 15 billion years. This calculation shows that if the i-ther started from a radius of one kilometre its age would be 340 billion years. However, this takes no account of the speed of the shock front exceeding that of primaries. In PART II Chapter 10 such an extra speed is shown to reduce this age drastically to not more than about 60 billion years.

The point is that plenty of time is available for the i-ther to evolve intelligence and consciousness by the organising power of chaos.

In FIG.13 path lines have been added to the plot given in FIG.12. These are obtained by joining every fifth point in each profile array to the corresponding points in all other arrays. They show the paths taken by primaries of i-ther as it grows in both radius and pressure.

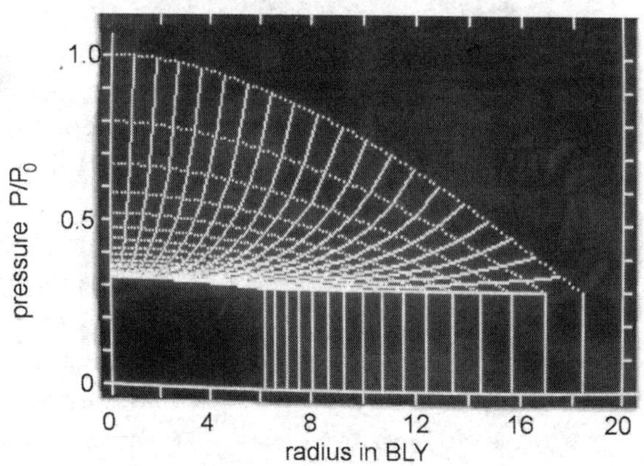

FIG.13 PRESSURE/RADIUS PROFILES WITH PATH LINES ADDED

It is now assumed that the computing ability of the i-ther has produced the big bang that created matter and its associated light propagating routine. The information contained in FIG.13 can now be used to determine how light would travel from near the edge of the universe to its centre point. As described in sections 7.3 and 7.4 the path traversed depends on the way the speed of light varies with pressure and on the velocity of recession v of i-ther

With this information the path of light can be traced back in time from the central; origin point until the edge of the universe is reached. To present the results of this computation it is necessary to first plot the information contained in FIG.13 as time in billions of years against radius in billions of light years. This is presented as FIG.14. The light path is also plotted. Almost a straight line results. The edge is not quite reached but this is due to the coarse nature of the time steps. (The unfortunate choice of scales is due to having to use the same ones as for the other plots owing to shortage of memory.)

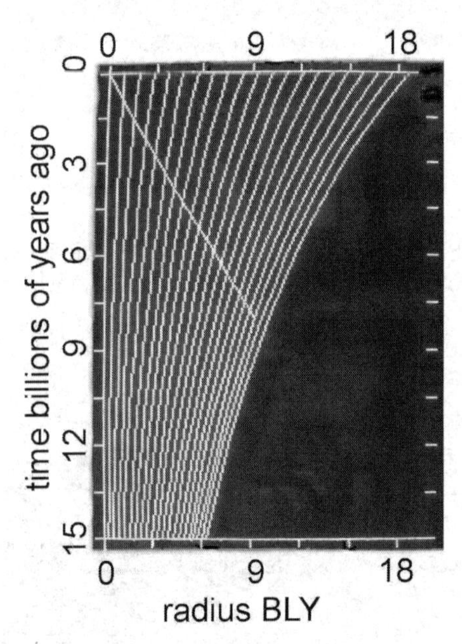

FIG.14 TIME/RADIUS PLOT SHOWING LIGHT PATH

Clearly the effects of light having to speed up due to recession velocity, as explained in §7.3 and illustrated in FIG.9, are almost cancelled by the speed of light reducing as pressure increases during the approach.

This plot suggests that light from a star at the edge of the universe took about 9 billion years to reach us. This does not take us back to the big bang, however. The star had to reach this position before it emitted the light that we observed. The computations gave the speed of recession at 90% of the distance to the edge as 0.6 times the speed of light as measured at Earth. Immediately after the big bang matter would be travelling nearly as fast as light. An approximate calculation suggests the average speed to the point where light that could reach Earth was emitted would be about 0.7 times our light speed. Therefore to reach 90% of the way to the edge it would take about 9/0.7 = 13 billion years. Hence according to this calculation the big bang took place 9+13=22 billion years ago. Fortunately this is longer than the 13.7 billion years given by cosmologists.

Astronomers say 13.7 billion years is not long enough to allow galaxies to form and that some stars seem older than the universe. The increased age should help remove this difficulty.

The same data can now be used to compute red shifts as explained in sections 7.3 to 7.5. These are plotted in FIG. 15. The symbol cosmologists use for red shift is Z and this is related to the increase in the wavelength of light. The relation is defined as

$$Z = \frac{\text{wavelength observed at Earth}}{\text{wavelength emitted at star}} - 1$$

So $Z = 0$ at the star.

According to E.Wright (2005) the greatest red shift of any supernova yet observed is 1.755. So the wavelength received from this star has been stretched 2.755 times!

The Doppler shift is given by Z_D. The total for the Doppler shift and light speeding up is the highest curve Z. So the Doppler shift makes a relatively small contribution. The observed red shift is Z_G, shown by small circles The difference between Z and Z_G is due to light gaining energy by falling into regions of higher pressure according to the new theory of gravity that is based on its being a force of negative buoyancy. (Explained in Appendix III) In PART II Chapter 12 this blue-shift is disallowed from the new reasoning that develops.

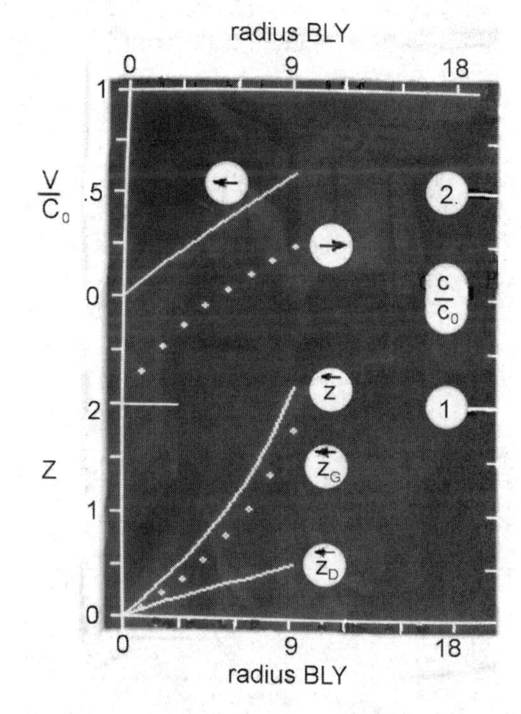

FIG.15 RED SHIFTS RECESSION SPEED AND THE SPEED OF LIGHT

FIG.15 Also shows the recession velocity measured along the path light takes to reach us. This is ratio v/c_0 where v is the recession velocity and c_0 is the speed of light at the centre of the universe. The latter is 300,000 kilometres per second.

The value of v/c_0 is 0.59 at 90% of the radius to the edge, which is 10.1 billion light years in this plot.

The way the speed of light increases with radius, also measured along the light path, is shown by the row of small circles. It is expressed as the ratio c/c_0 where c is the speed of light at radius r along that path. It is what is defined as the 'observed speed of light' that is given the symbol c_T in *QUANTUM GRAVITY*.

The value of c/c_0 is 1.74 at 90% of the radius to the edge.

This set of results can be compared with those finally derived following a major breakthrough that occurred on 1st July 2010. This exciting development is described in Chapter 10 of PART II.

7.7 Hubble's Law and Acceleration Theoretically Derived!

Hubble's law connects the speed of recession of remote galaxies to their distance from us. So far this law has been considered by all cosmologists as empirical: meaning obtained only from experimental observation. The Big Breed theory, however, allows of a theoretical derivation as will now be shown.

So that readers having little knowledge of mathematics will not be put off I will try to present this in an easily understandable way. If you can understand compound interest on a bank account then this should be easy to follow.

We start by considering a volume V of i-ther that increases by a small volume denoted δV in a short time denoted δt due to a net rate of volume creation C_{NA} that is constant for all time and space.

These symbols can be arranged to form a creation equation:

$$\delta V = C_{NA} \times \delta t \times V \qquad [1]$$

This means that after the time δt has elapsed the volume has increased by amount δV that is directly proportional to the total volume V. When a further time δt has elapsed the volume started from is $V + \delta V$ and so the volume increase is now larger than it was in the first time interval. So at every time step the increase is greater than in the previous time step. This is identical with compound interest but in mathematics is known as an 'exponential' rate of growth. And using calculus this is carried out in a way that avoids all that laborious arithmetic. Now the increments δV and δt are considered vanishingly small, to be written dV and dt, so that with ordinary arithmetic an infinite number of calculations would be needed. Equation [1] can be divided by V on both sides to yield (with $\times$ signs ignored as per convention):

$$\frac{dV}{V} = C_{NA}\, dt \qquad [2]$$

But if a sphere of radius r increases by dr its volume $r^3 4\pi/3$ changes by $4\pi r^2 dr$ (adding an extra outer layer of thickness dr).

Then dividing the volume of the outer layer by the original volume yields:

$$\frac{dV}{V} = 3\frac{dr}{r} \qquad [3]$$

The term 'integration' means adding up all those little steps that we could have done by simple arithmetic and is indicated by the long S, which is the integration symbol shown in the next equation: Substituting for dV/V from [3] into [2], dividing by 3 throughout, and then presenting in integral form yields:

$$\int_{r1} \frac{dr}{r} = \frac{C_{NA}}{3}\int_{0} dt$$

And with 'exp' meaning 'exponential' for which Naperian log tables can be used to look up answers this integrates to yield:

$$r = r_1 \exp\left(\frac{C_{NA}}{3}t\right) \qquad [4]$$

This is probably mystifying but it means $r_1 \times 2.718$ *(to the power of* $(C_{NA}\, t/3)$*)* and this number called $e = 2.718$ is very special. To throw more light on this our companion volume *CREATION SOLVED?* provides a simple explanation in its appendix. The magic of this number now allows velocity to be determined. Equation [4] relates distance to time and so the velocity v of a particle is also embedded but needs to be extracted. Velocity is a short distance divided by the short time needed to cover that distance. It is the slope of a graph of r against t. Or in mathematical terms: Velocity $v = dr/dt$. This means [4] can be 'differentiated' using a simple maths rule, allowed only by the magic number e, to yield:

$$v = r_1 \frac{C_{NA}}{3} \exp\left(\frac{C_{NA}}{3}t\right) \qquad [5]$$

Substituting from equation [4] in [5] then yields the result:

$$v = \frac{C_{NA}}{3} r \qquad [6]$$

Hubble's Law is $v = H_0 r$ and so the foregoing has derived this law in which the 'Hubble constant' H_0 is related to the net creation rate by:

$$C_{NA}/3 = H_0 \qquad [7]$$

Since acceleration $a_{cc} = dv/dt$, differentiating [5] yields:

$$a_{CC} = \frac{dv}{dt} = r_1 \frac{C_{NA}}{3}\frac{C_{NA}}{3} \exp\left(\frac{C_{NA}}{3}t\right) \qquad [8]$$

after substitution from [4] into [8] and using [7] the result is:

$$a_{cc} = H_0^2 r \quad [9]$$

Hence it appears that when continuous net creation is involved Hubble's law is predicted with the Hubble constant invariant with time as well as space.

This simple theory has also predicted that the universe must exist in a state of ever-accelerating expansion! This had predicted the acceleration by 1992: something not discovered by astronomers until 1998 and a matter that took all cosmologists by complete surprise. There was never any need to invent 'Dark Energy'.

Unfortunately the effect of pressure on creation rate has been ignored and several years of fruitless effort resulted in attempts to provide a fully satisfactory solution. Only on 1st July 2010 did 'Eureka Day' dawn when all was resolved – and this proved the pressure effect cancelled out. The preliminary results given here had turned out to be accurate to within mostly ±1%.

7.8 Discussion and Conclusion to Chapter 7

The results of computer code "Un290609" have been described. This shows how the pressures of the i-ther grow very slowly from an initial burst about 340 billion years ago (revised in PART II Chapter 10) until around 15 billion years ago. From that time to the present pressures at the centre of the universe increase dramatically at an ever-accelerating rate and also a rapid accelerating growth of the universe occurs due to the universal net creation of i-ther.

However, this PART I was prepared about 2007 before new insights in 2010. In Chapter 13 of PART II the big bang is put back even more than 22 to 31 billion years ago and the present radius of the i-ther, measured to its shock-fronted edge, is now estimated as 25 billion light years. We are situated close to, but not quite at the centre of the universe and are now able to see right to that edge.

The increased age of the universe from the presently accepted value of 13.7 billion years should help resolve two difficulties that astronomers complain about. They say there is not enough time for galaxies to form and that some stars seem older than the universe.

As we shall see in PART II Chapter 10 the creation profile given in FIG.10 is modified somewhat. Furthermore the pressure at the shock-front was only a guess. It was not until 1st July 2010 that the major breakthrough occurred. I think the reader will be intrigued to see the difference this false start makes to the final result.

CHAPTER 8

DISCUSSION AND CONCLUSION

Now we come to the summing up – to see what all this means

In Chapter 1 the big bang theory was described showing how theorists consider the universe of matter, together with all four forces of nature and the surrounding space, to have exploded into being 13.7 billion years ago. Space is not synonymous with the void of nothing from which all was originally assumed to emerge. Space consisted of a seething mass of 'virtual particles' that together contained enormous energy. Due to this, and what we show later in APPENDIX II to be false logic, the result was the prediction of a rate of expansion of the universe that is many trillions of times too high.

This result was due to the assumptions and logic used making it impossible for the huge creation represented by 'inflation' to be effectively shut off. Cosmologists finally abandoned the initial attempt to have everything emerge from a zero energy state and reverted to accepting all the energy needed had to have existed without a beginning. This energy was assumed supplied from some hidden dimension through wormholes in space-time.

In Chapter 2 we described the world of common experience: the macroscopic level of reality. Then we focused attention on the world of the atom, described by quantum theory. This was so weird and unreal that the Copenhagen interpretation had it that apparent reality only appeared when a conscious entity observed sub-atomic particles. This meant that mind was involved in the creation of matter. It followed that minds had to exist prior to the creation of matter. So, instead of being truly real, as had been assumed in the big bang theory, matter now had to be considered as more like a three dimensional semi-virtual reality. It was real energy but mathematically organised by waves generated by the deeper and ultimate level of reality we have called i-ther.

In Chapter 3 simple experiments were described involving just a pair of easily made pendulums. These showed the meaning of what is called, 'The conservation of momentum', in such a way as to be easily understood by the non-scientist.

Then the logic was extended to show how a complementary

and opposite kind of energy could exist, to be called, 'negative energy' that was associated with negative mass. It was shown that if the entire universe were made from negative energy then it would behave in a manner indistinguishable from the positive kind.

This meant negative energy and mass are equally likely to exist at some sub-quantum level of reality.

The **BIG BREED** theory could now be explained and this is covered in Chapter 4. All the required preliminary information had been supplied in the previous chapters.

The responses of an assumed mixture of primary particles, to be called 'primaries', were now explored. The primaries had to be made of two opposite yet complementary kinds of energy: positive and negative. This permitted their spontaneous emergence from the void or conversely their mutual annihilation back to the void. The law of conservation of momentum also had to be satisfied and this determined which of these options would occur.

The astounding and valuable discovery emerged that when opposites collided in twos, then each increased in energy of its own kind. This meant creation from the void of pure nothingness could occur in a totally different way from that of the big bang theory.

This was the origin of the Big Breed theory. Collision by collision primaries gained energy and mass until they exceeded a critical size and then broke up. The net effect was a violent explosive creation rather similar to the 'inflation' stage on which the big bang depended. However, unlike the big bang approach, a means for effectively switching off most of the creation appeared. This happened when numbers per unit volume exceeded a critical value. The same law of momentum conservation now dictated mutual annihilation. The flow field self-organised into minute flow cells each containing a solid annihilation core at its centre. Primaries of both kinds converged to every core and mutually annihilated each other inside these cores.

Chapter 5 considered this solution to what is known as, 'The problem of the cosmological constant'. No other solution appears to have been published. What is disappointing is that although first published in Russia in 1994 followed by a publication in an American peer-reviewed scientific journal in 1997. Nobody appears to have taken any notice of this. It is unfortunate because an inherent feature of the theory demanded that the universe must exist in a state of ever-accelerating expansion of about the right value.

The acceleration was discovered in 1998 taking all cosmologists by surprise. They promptly invented the idea of 'Dark Energy' pervading the entirety of space. It was speculated that this had some mysterious kind of repulsive power at great distance that overwhelmed gravity. We say this speculation was never needed since the acceleration had already been accounted for by the Big Breed theory. It was an essential feature of a solution to the problem of creation from the void that established theorists had been unable to discover. Instead huge efforts are being made to loft hugely expensive space probes by NASA to search only for clues.

So we say that the i-ther emerging from the void **IS** that Dark Energy for which cosmologists still search. It is for the reader to decide whether or not this is an acceptable statement.

Finally in Chapter 7 the implications for cosmology are described. It is shown that the entire universe should be accessible to astronomers. Then since it looks the same in all directions this meant we must be close to the central origin point of the universe. Astronomers find light from distant galaxies is 'red shifted', meaning its wavelength is stretched mainly due to recession speed and explained by the 'Doppler effect'.

However, the Big Breed theory threw up an additional cause for this red shift. Since space is expanding with the universe of matter, it followed that light from a distant galaxy had to start off travelling against a moving medium. Its speed of approach was therefore reduced. It followed that this light had to speed up as it approached. It lost some energy due to having to gain speed and this meant an additional increase in the red shift.

Plots of the final solution were provided showing how the i-ther has to exist with higher pressures at the centre than at the edge in order to drive the accelerating expansion. The spherical ever-growing edge is like a shock front. This corresponds to an initial inflation during which annihilation cores develop in ever growing rings until they catch up with the edge and cut off nearly all the creation going on.

The tangled structure of annihilation cores as filaments and blobs, is continually supplied with power caused by the creation going on around them. As chaos mathematicians have discovered, these are the conditions in which spontaneous organisation invariably occurs. So the final conclusion is that the i-ther has the potential of self-organisation to produce the waves on which quantum theory depends. The universe now needs to be considered as a three dimensional semi-virtual reality. It is comprised of real energy but this is organised by quantum waves. The sub-quantum level of i-ther has to be regarded as the only true reality. It has huge computing power but no control of its own ever-accelerating growth. This is why the universe is so huge.

APPENDICES

What these appendices give are examples of 'conceptual logic'. This is a non-mathematical way of working out what is expected to happen using logic based on common sense. Only when this appears satisfactory is the mathematical analysis worth attempting. This means the assumptions do not lead to internal contradiction and appear to have the potential of leading to a mathematically based theory that should match known data. Nobody should let mathematics lead the way.

At least this is a view from an engineer, though it does not seem to be shared by physicists and cosmologists. However, APPENDIX I demonstrates how blind acceptance of mathematical rules can lead to a false picture and throw the mathematically based theorist onto a false track

PART I of this book, which we hope you have read so far, is in itself an example of conceptual logic in action.

APPENDIX I – the critique of Tryon's article

The critique was of an article published in the 8th March issue of *New Scientist* in 1984. The article, by Professor Edward Tryon, was entitled, *What Made the World?*

This said that the force of gravity, which attracts every bit of matter in the universe to every other bit of matter, had 'potential energy'. This is energy due to its position. As an object is lifted from a central position in the universe its potential energy increases as it tries to pull back to that centre due to the force of gravity.

It is customary, Tryon stated, to measure all potential energy from an infinite distance so that fixes the datum, the zero from which measurements are made. Therefore, he argued, gravitational potential energy, at any finite distance, is less than at the datum: less than zero and therefore has to be regarded as negative. It so happens that when the potential energy of all matter in the universe is added up we find it is about equal to the energy from which matter is made. Therefore, he concluded, the universe could have been created ex-nihilo, meaning, from nothing. He explained that the 'negative gravitational potential energy' would cancel that of matter.

It had seemed obvious that, using the Newtonian mechanics, on which Tryon had based his article; matter would still exist when raised to the zero datum at infinity. Clearly something was wrong with Tryon's logic. Conceptual analysis soon pinpointed the error.

It is permissible to choose one's datum anywhere that is most convenient for ordinary calculations, and infinity turns out to be the most convenient. This is because it eliminates a term in an equation. Arbitrary choice is permitted since the 'constant of integration' cancels out when only changes in altitude are being considered. However, when creation is being considered greater care needs to be exercised.

The only place where the energy of a particle has not yet been affected by gravity is the place where it has just been created: before it has moved. This position is best defined as 'absolute datum' for potential energy, and needs to be adopted. Tryon had failed to realise that creation is a special case for which the choice of datum cannot be fixed arbitrarily since in this case no cancellation of the constant of integration can occur. For the origin point of the universe as datum the potential energy becomes positive and cannot balance the energy of matter in any way.

This error in logic turned out to be more serious still. It was not just due to Tryon. Davies (1989) includes a chapter by Clifford. Will who writes on page 32 that since gravitational potential energy is a binding force they call it 'gravitational binding energy'. Since this is negative they conclude it reduces the mass of an assembly of objects as they fall together and combine. In an extreme case, he says, this might cause the mass to become negative. He goes on to say that, in the mid 60's, this resulted in the 'Positive Energy Theorem' that forbade such an eventuality. It created a controversy that lasted 15 years. This suggests a lack of basic understanding since they reach the opposite conclusion to Tryon who is also in error.

One needs to look at the problem a totally different way. Suppose an electron is situated on a negatively charged point and is freed by field emission to be accelerated. Gravity is to be ignored. It is accelerated by the electric field and work is done upon it that is converted to kinetic energy. Then since $E = mc^2$ it gains mass. But electrical potential energy is lost. If the previous argument, concerning gravity, had been applied the electron would have lost mass due to reduction of electrical potential energy.

The mathematicians, theorising about gravitational potential energy, had used a totally false form of logic by simply treating negative signs by mathematical convention! This is why engineers who have been trained differently need to be incorporated in teams of mathematicians and theoretical physicists.

More needs to be said about the example of the accelerated electron. Actually on impact with a collecting surface, all the extra mass would be converted to heat that, after being radiated away, would restore the electron's original rest mass. The kinetic energy gained has come from the electric field and was then dissipated.

A somewhat different situation arises during fission of an unstable radioactive atom. Some rest energy of the atom will be converted to kinetic energy by the electric force of repulsion between the positive electric charges of the fission products as they fly apart. Then this kinetic energy is lost by collision with other atoms. Consequently the total rest energy of fission products will be less than the original atom and so will weigh less.

This argument seems to conflict with the conclusion reached for our previous accelerated electron. Agreement is restored, however, if the electric field has in turn been supplied with its energy by the emitting surface. So the latter has lost the energy that became the kinetic energy of the electron.

What is also extraordinary and alarming is that the idea of negative gravitational energy being able to cancel part of matter is still accepted as established physics! If only assessors had listened to the critiques made by this author, between 1984 and 1987, this particular mass misconception would have been eliminated.

(Physicists call, 'mass energy' what we term 'rest energy', the change in terminology is made to avoid confusion in the new mechanics that is described in APPENDIX IV)

APPENDIX II

FLAWS IN THE BIG BANG THEORY

Professor Vigier had failed to obtain publication in *Physics Letters A,* for which he was the gravitational consultant, despite his stated wish that this valid critique of Tryon's article, should be published. Consequently this author sent a letter to the physicist and writer, Professor Paul Davies, in 1987 asking how this could affect the big bang theory.

Davies replied by sending a draft version of Dr Alan Guth's theory of 'Inflation' that was intended for a new book. This was published later: Davies (1989).

The theory began by showing why a background substance called the 'quantum vacuum' alias 'space' contained a large amount of energy in every cubic metre: its 'energy density'. Since both matter and space had to emerge from the void of pure nothingness, having zero energy, this high energy density was cancelled by an 'intrinsic negative pressure of the vacuum'. It was something fully accepted by established physics but arose without any causative means. The vacuum of positive energy would have produced a positive pressure. *To try and balance the energy density by a pressure arising from nowhere was clearly a flawed concept.*

This negative pressure was so huge that, substituted in one of Einstein's equations, it overwhelmed the energy density of space causing the latter to become negative.

An energy density cannot become negative after first being defined as positive.

Used in Newton's equation of gravity this negative energy density reversed its force to one of repulsion. This was the causative agent for producing the massive 'inflation' that is the basic creative explosion of the big bang.

So an anti-gravity force was being postulated as the driving means. The negative density acting on negative density would produce cancelling effects, as shown in Chapter 3. So responses would not be affected and gravity would still remain attractive.

Furthermore Newton's theory, as quoted in the article, is only applicable for speeds that are negligible compared to that of light when speeds close to light were being shown to apply. So having used one of Einstein's equations already it introduced inconsistency to revert to the old Newtonian. Why make such a switch? One reason is not hard to find. Einstein's general relativity is based on curved space-time that does not admit that any real force of gravity exists. Consequently general relativity could not be switched to produce the driving force inflation required.

Furthermore the negative pressure would have created implosion, instead of an explosion, but this is ignored altogether.
This flawed maths is reproduced in *CREATION SOLVED?*

An inadequate understanding of basic mechanics did not appear to have been noticed by all assessors concerned.

A critique pointing out the errors in conceptual logic was returned to Davies. This must have caused some embarrassment since no further response was received. However, the critique must have had some effect since the final book, Guth (1989) pp 57-59, contained a highly revised version. This left out some important equations and added complexity to others. It was now made almost impossible to follow but scrutiny shows the logical errors are still in place.

A plot of the inflation had been included, p.36, showing that this occurred at constant density and pressure. *This is also impossible since to produce the huge accelerations involved would have required the development of enormous pressure gradients.*

Not surprisingly a huge false prediction of a universe expanding at a rate 10^{120} times greater than is remotely possible was returned. Top physicist Stephen Weinberg (1989) wrote that this represented a major crisis for physics.

These examples further highlight the need for some people of an engineering background to be recognised as having the expertise that needs to be restored to its source.

The gist of the original (not an exact copy) and its critique are provided in Chapter 10 of the book *CREATION SOLVED ?*

APPENDIX III

What is wrong with Einstein's relativity theories?

A brief look back at 'Foreword' should be sufficient to convince readers that all is not well with Einstein's theories. Physicists all admit now that they are incompatible with quantum theory. Even Stephen Hawking (1988,1996) says in his popularisation, on about page 11, that one of these theories must be wrong.

It is not difficult to see which is wrong: relativity theories contain internal contradiction - as will be explained!

Einstein first developed his theory of 'special relativity' in 1905 that was restricted to the motions of objects travelling at constant speeds. He postulated that the speed of light was a universal constant and was the same for all observers. This applied whatever speed they had relative to others. Any of these observers, Einstein admitted, could equally claim to be standing still with everything else in motion, except for co-moving objects.

For motions at low speeds such assumptions are admissible. However, as Einstein showed by simple mathematics, when speeds become significant as compared to that of light a mass increase will occur. Indeed his famous equation $E=mc^2$ shows this must be so. The energy of an object is E, m is its mass and c is the speed of light. An object in motion has kinetic energy that adds to its rest energy and so E and also m are greater the greater the relative speed.

Now any observer is an object and so two identical observers in relative motion each see the other as more massive than themselves. Does this count as an inconsistency?

Relativists refuse to concede that it does. I have tangled repeatedly with physicists on this issue but none accept there is anything wrong. So another example has been tried.

Two spaceships S and A originally see each other as standing still. S remains so but A accelerates away to some high speed and then the acceleration cuts off to cruise at constant speed.

S looks at A and sees a mass increase as is to be expected.

A looks back at S and sees S as having this mass increase whilst remaining unchanged himself. Yet S has not changed!

Again relativists all refuse to accept there is anything inconsistent here. One argued that the mass increase of S seen by A is the correct value as seen by him. He soon admitted they consider the mass increase as 'effective'. This means they regard the increase as an illusion.

Light can be shown experimentally to produce 'radiation pressure' when it falls upon a surface. This demands a loss of momentum by the photons that are the carriers of light. This means they have some kind of mass since their momentum is mc.

Again physicists' say the photon has zero mass. It only has 'effective mass' to explain radiation pressure. They are saying that the mass is zero for any other calculation. Is this consistent?

In high-energy particle accelerator research anything travelling close to the speed of light has to be considered as having momentum based on a real mass increase. So relativists choose to consider mass increase as real or illusory depending on circumstances. According to one relativist the appropriate choice is taken from a list to ensure that correct interpretations of the maths are made. He said this disposed of any contradiction.

Or is the difficulty simply being swept under the carpet?

Another physicist contended that the difficulties of mass increase were merely 'counter intuitive' and that physicists had learned to live with counter intuitive concepts. It was, he insisted, unsophisticated thinking to simply rely on the logic of commonsense.

So I said, "But suppose a test was made by sticking to the rules of common sense with any kind of contradiction disallowed and then producing a theory matching observation just as well. Would that not be preferable to a theory that had to mask its internal contradiction under the cover of it being counter intuitive?"

His answer was that this was impossible since no such theory had been achieved. If it had it would be common knowledge by now.

My stance is that ECM provides such an alternative and that when a discipline like physics has hit the buffers a paradigm-shift is needed. A fresh start from the place where things started to go wrong is really what is required. Unfortunately all attempts to publish the solution between 1987 and 1990 were rejected, mostly on grounds of relativity having stood the test of time. Not a single assessor even mentioned the success of the theory in paralleling all the achievements that are still considered unique to Einstein.

Then again associated with relative speed is 'time dilation'. Clocks tick more slowly when having a high linear speed according to the observer.

Herbert Dingle (1972) taught relativity for years before the penny dropped. Then he says two observers in relative motion each have an identical clock. Each sees the other's clock as running slow. Dingle points out that this inconsistency invalidates the theory.

He was virtually made an outcast for such heresy!

Having failed to convince physicists that mass increase invalidates special relativity I resorted to what seems to me a critique that cannot be sidestepped. Each observer is in very high-speed relative motion to the other and they each carry a pair of negatively charged objects. Now it is known that electrons in motion produce magnetic fields around themselves causing them to attract each other if they move side by side. Indeed this is the cause of the 'pinch effect' made use of in certain fusion experiments.

So according to special relativity an observer moving with the charged objects experiences no such force but the force is observed by the relatively moving observer.

This critique was dealt with by calling it "nonsense".

So the non-relativist cannot win any argument to convince the relativist that anything is wrong. The difficulty seems to have arisen from the way students are taught when at an impressionable age. Then major intellectual effort is needed to master the difficult concepts and maths associated with the theory and particularly its extension to cover gravity. It is very hard to jettison intellectual capital that has to be acquired by great effort. Is it possible that the form of training used is inadvertently producing a hypnotic effect?

Einstein had, by 1916, developed his 'general relativity'. This was an extension that included acceleration and gravity but still retained the assumptions on which special relativity is based. Now the force of gravity was abolished and made an illusion. Objects moved in straight lines in curved space-time.

Yet theorists still try to achieve their goal of 'quantum gravity' by attempting to match general relativity to quantum theory. They postulate 'gravitons' as the carriers of the force of gravity. My question is, "How do they provide a match this way when, according to Einstein, there is no actual force?"

With these critiques to ponder I leave the reader to consider the validity of Einstein's approach. The alternative that was needed for integration with the Big Breed theory is briefly summarised next.

APPENDIX IV
An Exact Classical Mechanics leading to Quantum Gravity: A brief description of the ECM mechanics adopted

(An example of setting the stage using conceptual logic)

The ***Exact Classical Mechanics*** **ECM** required is described in more detail in the primer ***CREATION SOLVED?*** written for people having little knowledge of physics or mathematics. Its appendices provide details of the Newtonian mechanics which formed the starting point and provide a simple introduction to the calculus needed. Full mathematical detail is provided in ***QUANTUM GRAVITY via an Exact Classical Mechanics*** *(QGECM)*

However, only a small part is needed for application at a sub-quantum level of reality. So a brief summary is all that is needed here.

Since the idea was to test the possibility of the simplest initial assumptions representing reality, Euclidean geometry and universal time were adopted. However, to achieve credibility this mechanics had to penetrate from the ultimate level called i-ther, penetrate the quantum vacuum, which emerges from the i-ther, and then yield predictions matching the achievements of Einstein's theories of relativity. This ruled out an unrevised Newtonian mechanics and Einstein's theories of relativity were inappropriate for reasons given in Appendix III. Further reasons are given under the heading **Local Frames in Cosmology,** that appears later in this appendix.

In the ECM theory the local quantum vacuum (space) forms the reference, called the 'local frame', from which all speeds and associated kinetic energies are measured. Then all observers record an identical mass, for any object, whatever their speed relative to anything else. Light propagates through the medium and so moving observers will not see the speed of light as the same in all directions. The local frame is the centre of the Earth for distances somewhat greater than that of the orbital radius of the moon. Then a transfer to another local frame occurs moving with the sun. It means that a blob of quantum vacuum is carried along surrounding the Earth to move through the blob locked around the sun.

The quantum vacuum is considered to emerge, with fluid-like properties, from the ultimate reality of i-ther. This allows some motion relative to the i-theric structure. On very large cosmological scales, however, such relative motion is ignored, being very small compared to the speed of light.

Photons. For application at the macroscopic level the behaviour of light had to be included. Since light is quantum based it would not exist at i-ther level. The mechanics at i-ther level had to be of a revised Newtonian form. Consequently the propagation of light had to be based on its particle nature: not its wave nature. This meant light being treated as a stream of photons. However, they also carry momentum, as otherwise they could not produce radiation pressure when they hit a surface. The ECM assumptions, unlike relativity, allow the photon mass to be considered real. Since photons have no rest mass, however, they have only 'kinetic mass'. Consequently in ECM theory photons are considered to be made from kinetic energy.

This also means that mass is proportional to energy. Also it means that matter is to be considered as made from the arithmetic sum of its energy when not moving in the quantum vacuum, its 'rest energy' E_0 and its energy of motion or 'kinetic energy' E_K. The combination is to be called 'sum energy', symbol E.

By considering the acceleration of a massive object comprising both rest energy and kinetic energy the constant of proportionality appears together with another equation looking like something straight out of special relativity. It is:

$$E_0 \Big/ \sqrt{1-(v/c)^2} = E = mc^2 \qquad \text{[IV1]}$$

There is a crucial difference: v is the absolute velocity of the object: not the relative velocity. This makes a huge difference yet this equation ensures ECM matches the same experiments as those considered to support only Einstein's special relativity.

Furthermore, in special relativity the speed of light c was assumed to be a universal constant and in consequence time had to be accepted as variable. As will be shown later, since time is now universal it is c that becomes variable.

How Absolute and Relative motions are incorporated

A fuller description is provided in the primer *CREATION SOLVED?* and in *QGEC M.*

Objects moving with different absolute speeds, as defined by measurement from a 'local frame', have relative speeds in relation to one another. Any forces due to collision will depend on their relative kinetic energy and relative momentum. However, the mass used for calculating these properties will be the value measured from the local frame.

Local Frames in Cosmology.

Experiments made before 1900 to measure the speed of the Earth through an imagined stiff medium called 'ether' gave null results. At that time the universe was considered static and so everything had to travel relative to the ether, which provided an absolute frame of reference. Light also propagated through the ether. With such a model the Earth could not possibly be standing still as the experiments seemed to indicate and so a dilemma existed. This was apparently resolved by the creation of Einstein's theories of relativity that operated without any ether being required. These theories had the surprising quality of accurately predicting the outcome of all experiments made for their falsification. However, in every case the observer sat in what is now the 'local frame'.

No solution to the problems of the big bang could arise without having some kind of background medium re-introduced and indeed quantum theory also required one, calling it the 'quantum vacuum'. However, the blob of vacuum locked around the Earth and moving through a larger sun-locked blob mentioned earlier provides a paradox free model fitting in with these null results.

A spherical blob of large size enclosing the Earth would ensure that little refraction of starlight would occur since the blob surface would be almost perpendicular to the beams of light. This is required for consistency with 'stellar aberration' discovered by the astronomer Bradley (as illustrated in *QGECM*)

Absolute speed will change during transition across the boundary of any blob from one local frame to another. Therefore absolute kinetic energy and mass increase will change. However, a cancelling change of rest mass and rest energy will occur so that the sum energy, will not have changed – at least if the boundary is thin enough for any effect caused by gravity to be negligible. Furthermore no change of speed will occur as measured from Earth and so no detection of a boundary could be made, except by the use of an absolute speed-measuring probe carried with the spacecraft. Instruments able to check this conclusion are evaluated in QGECM.

The physicist Selwyn Wright (2008) also publishes a blob model having returned to similar conclusions regarding the flaws in any relativity concept. However he thinks the Earth blob only extends a few hundred kilometres.

This is why I want to find out where the blob boundary is by an experiment in Earth orbit and also discover whether the blob is stagnant or has an internal flow structure.

In my opinion such experiments, made using a Fitzgerald type torsion mounted electrical capacitor, would yield more valuable physics than almost anything else. The experiment would also be very cheap as compared with most other space probe experiments.

The way red shift data need to be analysed for a universe having an expanding background medium moving with its galaxies was covered in Chapter 7 and so will not be repeated here.

The Speed of Light Increases with Altitude. The next step is to consider photons losing energy when they rise in a gravitational field. Their kinetic energy couples with gravity just like matter. This explains the 'gravitational red shift' of light and also demands that a horizontal beam of light is bent by gravity.

The outside of the bend is longer than the inside; so light has to go faster on the outside of a bend in order that it can move perpendicular to its waves. In relativity it is time that goes faster on the outside, but now the same reasoning and with universal time, dictates that it is the speed of light that increases with altitude.

Then $E=mc^2$ no longer means mass and energy are equivalent since c can no longer be regarded as a universal constant. The concepts of potential and total energy now cause confusion when an attempt is made to find whether it is mass or energy that remains constant when an object changes level. The sum of kinetic and potential energies is conventionally called 'total energy' in Newtonian mechanics. Total energy has the convenient property of remaining constant for an object in free rise or free fall.

This concept cannot be incorporated in an exact mechanics, since it prevents equations yielding mass increase as an object falls.

Instead an object rising under its own inertia has to be considered as losing speed due to transferring its kinetic energy to space. Space returns the energy when the object falls back. The interchange of energy with space maintains energy conservation but since the kinetic energy E_K is varying it follows that the energy E being defined as the sum of only the rest energy E_0 and E_K is also now a variable. Remember 'sum energy' is E and is $E = E_0 + E_K$.

The concepts of both potential energy and total energy need to be ignored altogether. These are clearly useful artifices for application in approximate theories like Newtonian mechanics but have no reality. They are inapplicable in an exact classical mechanics.

But $E=mc^2$ no longer means mass and energy are equivalent since if an object is lowered on a cable, for example, the value of c will be reduced. Then either E or m must change for this equation to remain satisfied. So the question is, "Which of these will change?"

By considering lowering an object on a cable, the energy released appears as heat at a brake fitted to a winding drum: none is supplied to the lowered object. So it is deduced that the rest energy E_0 of that object will remain constant. **So E_0 never alters.**

Then the object's mass has to increase to compensate for the change in c. Since clocks tick more slowly when its balance wheel is made more massive it should tick more slowly at a lower level. The maths shows this gives exactly the correct 'gravitational time dilation' except that it is not time dilation any more.

Furthermore as a consequence of the predicted gravitational mass increase with rest energy remaining constant, energy has now to be considered the substance matter is made from: not mass. Mass is now relegated to the property of energy that determines the dynamical behaviour of energy in accord with Newton's second law of motion.

The force of gravity can now be considered. Since photons are to couple with gravity all kinetic energy must couple with gravity just as well as rest energy. Consequently it is the sum energy, the sum of rest and kinetic energy that has to be considered to couple with gravity. It is this sum energy of two separated objects that dictate the force of gravity acting upon them: not what is actually the rest mass as used in Newtonian theory. (it makes a difference). This also allows both light and matter to use the same equation for the force of gravity. For planets orbiting the sun, for which the 'gravitational radius' r_0 =1477 metres, the revised equation for the force of gravity F in Newtons with E in joules and r the orbital radius in metres is:

$$F = r_0 \left(E_0 + E_K\right)/r^2 \quad or\ simply\ \ F = r_0\, E / r^2 \qquad [IV.2]$$

These equations illustrate the way the kinetic energy E_K of objects adds to the gravitational force. These are necessary refinements to Newtonian mechanics that extend applicability to the speed of light and to the most extreme gravity of black holes.

A little conceptual logic shows that with this modification planetary orbits will no longer execute perfect ellipses as the pure inverse square law of Newton's equation of gravity dictates. Starting from the point of closest approach, the 'perihelion' a planet will rise to the point of greatest distance at 'apogee'. Its speed and therefore kinetic energy will reduce and so E will reduce. The force of gravity will therefore reduce more than it would by a perfect inverse square law and logic suggests therefore that a greater angle than 180 degrees would be required to reach apogee.

Twice this angle gain is known as 'precession'. So the new theory ought to predict the precession of the planet Mercury - just as does general relativity. Mathematical evaluation returned a positive verdict but it fell short of Einstein's prediction.

The question then arose, "Could space have a non-uniform density and if so would this produce a more accurate prediction?" The next section provides the clinching factor. **A theory emerges matching the achievements for which Einstein is so famous!**

The non-uniform density of space Finally it seemed possible that if the energy per unit volume of space, its energy density, were non-uniform, it might have the same effect as the 'curved space time' that Einstein had used to explain gravity. A quantum rule was found in a textbook by Novikov (1983) and when used provided the magnitude of the non-uniformity required. Amazingly, together with the changed equation for the force of gravity, the result now gave exactly the same equation for the perihelion advance of Mercury as Einstein's theory of general relativity. The same non-uniformity also doubled the deflection of starlight passing close to the sun.

A final check is the 'Shapiro Time delay' found by bouncing radar beams from Mars and Venus. In ECM this is due to the speed of light reducing as it passes near the sun. The resulting equation matches observation just as well as general relativity.

Mostly identical end equations to those of both special and general relativity appear but have a totally different interpretation. Instead of relativity's 'time dilation', for example, clocks tick more slowly due to mass increase together with the electromagnetic forces inside atoms in linear motion. The speed induced mass increase alone is shown due to the real life increase of cosmic rays.

The combined effect of these revisions of Newtonian mechanics therefore provides a theory matching all the achievements of Einstein's theories of both special and general relativity. However, unlike Einstein's theories, ECM contains no internal contradiction or incompatibility with quantum theory.

Quantum Gravity

When combined with the Big Breed theory a highly satisfactory theory of quantum gravity results that fits all the experimental observations that are still considered the unique achievements of general relativity. The Big Breed theory provides the mechanism by which the i-theric pressure gradients are produced that ECM requires. A force of gravity now appears that is consistent with the other three. Energy is required to organise and maintain matter together with all four forces of nature. This energy is transmitted along the filaments from great distance to be deposited where matter is to exist. The i-ther's filamentous structure also provides the equivalent of a porous solid through which the excess energy, due to this deposition, has to leak back by viscous flow. This produces exactly the inverse square law of pressure gradient that the ECM theory of gravity requires. Then Spaniol and Sutton (1994) showed how 'G' is linked to electromagnetism.

Since the other forces are fully explained by quantum field theory, a **Grand Unification Theory** has also appeared!

Other matters of importance

Since this ECM theory was derived mainly for application at the i-ther level of reality, where light and matter do not exist, the wave nature of light is not considered. When applied at other levels the theory therefore considers only the particle nature of light. However, the wave nature can be recovered by the use of Planck's equation that shows the energy of a photon to be equal to Planck's constant multiplied by the frequency of its associated waves.

A problem arises since Professor Selwyn Wright (2008), has also produced a theory that eliminates the contradiction in Einstein's theory of special relativity mentioned earlier. Wright also adopts the same concept of a common frame of reference needing to be used by all observers. However, his theory is based on the wave nature of light and, during our correspondence; he agreed that either the wave or particle nature ought to yield identical end equations.

The problem was that to match Maxwell's equations for electromagnetic waves all objects and space itself had to contract in the direction of motion with time dilating, according to what are known as the 'Lorentz transforms'. This produces an inconsistency between the two approaches. However, when considering the acceleration of an object, such as in a particle accelerator, when the Lorentz contraction and time dilation are multiplied together the effects cancel to yield the same result as given by zero contraction and no time dilation. So my suggestion was that the Lorentz effect be considered a useful mathematical artifice. This simplifies calculation when waves are involved but has no actual reality.

At i-ther level only the sum energy and inertial mass of primary particles, the 'primaries' need be considered. However, as the full ECM approach shows, the rest energy and kinetic energy components of sum energy are interchangeable and can transmute from one to the other.

Further reading

A fuller introductory description with minimum maths and an explanation of the maths that is used is provided in *CREATION SOLVED?* of our series. Full mathematical detail is given in *QUANTUM GRAVITY via exact classical mechanics QGECM.* The latter contains detailed proposals for some new experiments. Those mounted in space probes are designed to locate the boundary of the Earth blob to measure the size of Earth's local frame.

THE NEXT SECTION, PART II, ENDS WITH THE PROMOTION OF:

FIVE NEW PROPOSALS OF NEW RESEARCH FOR THE DEDICATED COMPUTER PROGRAMMER

THE NATURE OF DARK ENERGY REVEALED

A quick look at §7.5.1 page 184 shows why a net small rate of creation everywhere leads to the theoretical prediction of Hubble's law. The latter shows that the velocity of recession of remote galaxies is directly proportional to their distance from us.

The derivation, however, yields more than this: fundamental to the solution is the accelerating nature of the expansion. One can either deduce that 'Dark Energy' with repulsive force was never required, or the nature of Dark Energy was revealed before the concept was invented!

PART II

OPPOSED ENERGY DYNAMICS OpED

for MATHEMATICAL analysis of

THE BIG BREED THEORY of energy creation

Notes

CREATION SOLVED?, to which frequent reference is made, is a book using minimum maths and giving details of errors made by cosmologists. It also outlines solutions in a way understandable to the non-technical reader.
A reference such as, for example, § 3.6, means Chapter 3 and section 6 of that chapter.

When a date follows an author's name, such as Feynman (1985) it means that a reference to an article or book is given in the list of references.

PART II **CONTENTS**

LIST OF FIGURES AND TABLES

Figure numbers shown as F: TABLES as T

PAGE

THE NATURE OF DARK ENERGY REVEALED!

Hubble's Law Derived from the Big Breed Theory

A quick look at §7.5 page 183 will show why a net small rate of creation everywhere leads to the theoretical prediction of Hubble's law. The latter shows that the velocity of recession of remote galaxies is directly proportional to their distance from us.

The derivation, however, yields more than this: it shows Hubble's law derived from a net creation everywhere, yields an accelerating expansion: the acceleration becomes the square of the Hubble constant times distance.

Fundamental to the solution, therefore, is the accelerating nature of the expansion. Consequently the nature of Dark Energy was revealed, in 1992, before the concept was invented: being the ultimate substance of i-ther on which all else depends!

And six new projects for the mathematician and computer specialist are listed in Chapter 12 CONCLUSIONS pages 271 to 279. Some have NOBEL PRIZE potential.

(We only wish we had our degrees in a subject recognised as being eligible but if we can encourage others this has to be our reward)

CHAPTER 1

SUMMARISING WHY THE BIG BREED THEORY IS REQUIRED

The main points described in PART I are summarised in this chapter before going further. The big bang theory makes two wrong predictions by which it is invalidated. A different philosophy 'opposed energy dynamics' OpED is provided that yields a solution and is covered in mathematical detail in this section.

1.1 Introduction – Why the Big Bang is Invalidated

The theory to be presented in this book is to be regarded as providing the energy needed for creation of the universe and is applicable to the existing big bang theory or any alternative. The present big bang theory is unsatisfactory since it makes some very wrong predictions.

The big bang theory was an attempt to show that the universe could have arisen from the void, defined as nothing and so having zero energy. This is also the starting point of our new Big Breed theory. This seems a justifiable postulate since only nothing could have existed without a beginning since no beginning is needed for nothing! However, the big bang used some very questionable logic that is criticised in Chapter 10 of *CREATION SOLVED?*. Briefly Dr. Alan Guth, the originator of the 'inflation' that generated all this energy, attempted to cancel the energy density of space (this is the energy confined in unit volume of space) by a 'negative pressure of the vacuum' for which no causative agent was provided. In any case negative pressures have the same effect as for a vacuum tube with atmospheric pressure existing externally. If fractured *implosion* occurs: not *explosion*!

Other logical errors are described in APPENDIX II of PART I of this book. The result was that the explosive creation of energy called 'inflation' could not be effectively switched off. This led to a predicted rate of expansion of the universe that, according to Greene (1999), is 10^{120} times greater than astronomers can allow! This is known as the 'Problem of the Cosmological Constant' (quantum version) for which no other solution has been provided than that derived in the following chapters. So Guth started with false concepts and the result was a dramatically wrong prediction for the rate of expansion of the universe.

This should not be confused with a concept of the same name that Einstein added to general relativity. He used it to yield the prediction of a static universe. He had mistakenly sought to correct his theory that showed the universe was expanding: but this has nothing to do with the problem invalidating the big bang.

After leaving their major problem unresolved they went on to predict the expansion of the universe to be forever slowing, due to the mutually attractive effects of gravity on all matter. Late in the 1990's astronomers observed remote type 1a supernovae and discovered the expansion was speeding up! After Schwarzchild (1988) published this finding all cosmologists were taken by surprise. They soon invented 'Dark Energy' having the mysterious property of generating long-range repulsive forces in order to patch up the theory and make it match the data.

As this book unfolds, however, the reader will find that a solution to the major problem, concerning the totally absurd rate of expansion, also predicts the acceleration. No long-range repulsive forces are involved. The end result provides a totally new way of interpreting astronomical data concerning remote galaxies and supernovae. Indeed we predict this will revolutionise cosmology!

Pearson (1994) first published this solution in Russia, so no Dark Energy was ever needed. The only western journal that allowed publication was Frontier Perspectives: Pearson (1997). This was buried in an article concerning consciousness that few people appear to have read. However, even this publication predated the discovery of the accelerating expansion and so adds credibility. Unfortunately the prediction of acceleration was omitted since its inclusion would have jeopardised acceptance. As will be seen, however, this is a fundamental feature of the Big Breed theory of creation.

By now the reader may have appreciated that Guth and his assessors lacked an adequate grounding in the logic required. It had seemed to this author that an inadequate understanding of mechanics prevailed within the discipline of physics and cosmology and that this now resided in the discipline of engineering.

The book series of which this is a part aims to return the lost expertise to its source. In no way does this attempt to discredit physicists, cosmologists or mathematicians whose expertise is greatly admired. The hope is that this contribution will be accepted in the spirit intended: - to help science to recover from the impasse to which many physicists now admit their discipline is stranded.

CHAPTER 2

BASIC EQUATIONS

The creation of energy from the void is considered, caused by the collision of 'primaries' made of opposite though complementary forms of energy. The exact equations of motion, adapted from ECM theory, allow the effects of collision to be explored for any speeds up to the ultimate speeds that would relate to primaries of zero rest mass. Primaries are not of matter and have no electric charge, but exist at the sub-quantum level: the source of all that is.

2.1 INTRODUCTION

Why a Sub-Quantum Medium needs to exist

Chapter 1 began with a short description of the reasons that led to the derivations detailed later in this book. However, it is necessary to amplify the argument somewhat in order to fully justify the methodology that first chapter summarised.

In standard quantum theory the ultimate or base level of reality is provided by virtual particles arising from nothing and vanishing after a short life, as determined by application of the uncertainty principle. Even though they have no permanent energy the finite lifetimes of all sum to give a huge average energy density. This is the source of the 'problem of cosmological constant' introduced in Chapters 1 and 2 of *CREATION SOLVED?* and critically analysed in its Chapter 10. The seething mass of virtual particles forms the quantum vacuum, sometimes considered to form 'quantum foam' near the ultimate Planck scale. There are also undefined 'matter waves' that organise the way sub-atomic particles move. These waves are the basis of the 'wave-mechanics' of quantum theory. Although quantum field theory, as described by Feynman (1985), obtains some solutions without using the wave concept, a very similar mechanism is inherent.

That the quantum vacuum exists is not disputed but, since it has ephemeral and unreal qualities, it will now be regarded as an emergent level. A deeper and fully real ultimate level of reality therefore needs to exist for creation of virtual particles, matter-waves and the information required for their organisation. It was argued in *CREATION SOLVED?* Chapters 3 and 11, that these have to be regarded as abstract mathematically organised constructs. Hence something truly real needs to exist to create those waves and the information required. This ultimate sub-quantum medium, referred to as the 'i-ther', is therefore postulated as the only true reality. The i-ther could not operate on wave mechanics, as does the quantum level, since then an even deeper level would have been required to furnish its waves. An unsatisfactory infinite regression would then be implied.

So with wave mechanics disallowed some alternative needs to be specified for application at the level of i-ther. Since the force of electromagnetism is dependent on the wave mechanics of the quantum level it also cannot exist at i-ther level. So at this ultimate level electric forces are also disallowed. This is argued at greater length in Chapter 11 of the companion volume *Physics Hits the Buffers: Why?* and in Chapter13 of *CREATION SOLVED?*.

A classical mechanics, similar to Einstein's general relativity had therefore to be adopted. Unfortunately, as explained in *CREATION SOLVED?* Chapter 5, this theory is inapplicable at this ultimate level of reality and contains unacceptable internal contradiction. However, an appropriate ECM theory was described in *CREATION SOLVED?* Chapters 6 to 8 in which energy density gradients produced in the background medium had effects almost identical with the curved space-time introduced by Einstein. This is also covered in mathematical detail in another book of our series: *QUANTUM GRAVITY via an Exact Classical Mechanics ECM*

Unlike relativity theory the frame of reference in ECM theory had to be chosen as the 'local frame': not the observer. The same restriction applies to the ultimate particles, to be called, 'primaries' of which the i-ther is assumed to consist. In ECM theory both the rest energy E_0 and kinetic energy E_K of primaries could be considered real. Their arithmetic sum E, obtained by adding E_0 and E_K , becomes the 'sum energy' of the primary, a name chosen since the term 'total energy' had already been defined in the literature as 'potential plus kinetic energy'.

In *CREATION SOLVED?* § 3.6 1 it was also argued that Euclidean geometry, with time universal and not directly associated with space, was the most reasonable assumption from which to start. The way this was shown to match the available data so well gave confidence in this assumption so that no need arose for any modification. These assumptions also matched those of quantum field theory so that, since a satisfactory theory of gravity emerged, this could also be regarded as the long sought theory of 'quantum gravitation'.

The reader may feel this statement needs further justification and so a little more detail will now be provided. Early quantum theorists had found Euclidean geometry to be adequate for modelling the strong and weak nuclear forces and that of electromagnetism. Mainline physics, however, discarded this geometry with the advent of general relativity that adopted 'curved space-time' as its basis. Then in order to formulate a quantum theory of gravity that matched general relativity, higher dimensions, such as those involved in string theories, were introduced and used as basis for all four forces of nature and as a means for predicting the constants of nature and associated particles. Theorists were attempting to find a deeper understanding of nature by the introduction of such higher dimensions.

Such proposals can be incorporated in the present scheme but only as abstractions. Our theory considers there are no real higher dimensions. As described in *CREATION SOLVED?* Chapter 13 the quantum world is most likely to be a contrived virtual reality. So if higher dimensions are found to be absolutely necessary, then they can be incorporated as operators, like imaginary numbers. Imaginary numbers form a component of complex numbers. Complex numbers are ordinary numbers with imaginary ones added. The latter are ordinary numbers multiplied by the square root of minus one. Although √-1 does not exist and so has to be regarded as imaginary, it has important mathematical uses that provide the means of obtaining solutions to certain kinds of problem, particularly in electrical engineering. When the final solution is obtained, however, the imaginary part is truncated since it does not represent anything that really exists. It is suggested here that the higher dimensions of string theory could be retained though used in a way analogous to the way imaginary numbers are used.

None of this complexity need concern us since it will not apply at the ultimate level of reality that this part of the book addresses. Here primary particles, the only truly real particles existing in the universe, will be all that really does exist. These 'primaries', moving randomly at high speeds and in continual collision, are assumed to form a gas-like medium. Hence the primaries had to operate on the revised Newtonian mechanics that ECM theory has provided. Hence the model proposed has a classical mechanics at both the ultimate level of i-ther as well as at the macroscopic level. The quantum level, based on wave mechanics and quantum field theory, is sandwiched between these two classical levels.

2.2 Exact Classical Mechanics at the Sub-Quantum level

The only difference from the derivation given in *CREATION SOLVED?* §12.2 to §12.5 is that the ultimate speed has to be assumed different from that of light. As explained in Chapter 13 of that book, light appears only at the contrived quantum level as a mathematical abstraction whose speed has been chosen in some way. At the ultimate level the primaries are not matter: they are the source of matter. As the derivation develops it will be seen that the ultimate speed of primaries needs to be higher than that of light. It will be given the symbol $\boldsymbol{C_U}$ to differentiate from the $\boldsymbol{c}$ used for light. This does not mean primaries are 'tachyons'. This name is reserved for hypothetical particles of imaginary mass since they depend on $\sqrt{-1}$ having a real counterpart (which is impossible). These still have $\boldsymbol{c}$ as their basis but now this is their lower speed limit: they are not considered to have an upper limit.

Primaries behave like ordinary mass except that they have a far higher ultimate speed limit. Repeating the derivation given in *QUANTUM GRAVITY*....§1.1 a start can now be made. This considered the horizontal acceleration of a massive object but C_U now has to be used in place of c. With $\boldsymbol{v_p}$ as the speed or velocity of primaries the result can be immediately presented for primaries of positive inertial mass m_p as:

$$\frac{E_0}{\sqrt{1 - \left(v_p/C_U\right)^2}} = E = m_p C_U^{\,2} \qquad [2.2.1]$$

Exactly the same expression applies for negative mass except that the suffix $_n$ replaces $_p$. Each primary of either kind, but now without those suffices included, has a real kinetic mass m_K, equivalent to its kinetic energy E_K and, together with its rest mass m_0, yields an inertial mass m. so that $m = m_0 + m_K$, This inertial mass carries momentum $p{=}m\ C_U\ (v/C_U)$. Also as shown by [2.2.1] there are energy equivalents so that the sum energy E of primaries is the arithmetic sum of E_0 and E_K so:

$$E = E_0 + E_K$$

It is now therefore permissible to write the kinetic energy, for either a positive or negative primary, by: the composite equation

$$E_K = E - E_0 = E_0\left\{\left[1 - \left(v/C_U\right)^2\right]^{-1/2} - 1\right\} \qquad [2.2.2]$$

When $v<<C_U$ equation [2.2.2] can be expanded by the binomial theorem to become:

$$E_K = \frac{1}{2}E_0\left(\frac{v}{C_U}\right)^2 + \frac{3}{8}E_0\left(\frac{v}{C_U}\right)^4 + \frac{5}{16}E_0\left(\frac{v}{C_U}\right)^6 + \ldots$$

[2.2.3]

And since from [2.2.1] putting $\boldsymbol{v_p}=0$ yields $E_0 = m_0\,C_U^2$ [2.2.3] can be written in the form:

$$E_K = \tfrac{1}{2}\, m_0\, v^2 + 0.\,375\, m_0\, v^4 + 0.3125\, m_0\, v^6 + \ldots.. \quad [2.2.4]$$

Furthermore [2.2.1] can be rearranged to yield:

$$E^2 = E_0^{\ 2} + E^2 (v_p / C_U)^2$$

Then since $E^2 = m_p^{\ 2} C_U^{\ 4}$ *and* $p_p = m_p v_p$

Where p_p is the momentum of the positive primary and p_n will be that of the negative primary. The latter is defined as pointing in the direction <u>opposite</u> motion in contrast to the momentum direction of the positive primary that points in the direction of motion. The resulting equation becomes:

$$E^2 = E_0^{\ 2} + (P_p C_U)^2 \quad [2.2.5]$$

This is an exact expression needed for the exact derivation used in the first formulations of the OpED theory.

Equations [2.2.1] to [2.2.5] are applicable to both positive and negative mass and energy states as required by the opposed energy dynamics that now follows.

2.3 OPPOSED ENERGY DYNAMICS: OpED

The theory to be developed requires primaries of negative energy to complement those made of positive energy - as already explained in PART I of this book. The interactions of primaries of opposed energies will be defined as 'opposed energy dynamics'.

An engineer turned physicist, Paul Dirac, first proposed the idea of negative energy in the 1930's. According to Blanchard (1969) Dirac took the square root of both sides of equation [2.2.5], though, taken from special relativity, and noticed it had a positive and negative root. Then he said space existed as close-packed electrons in negative energy states. A very energetic photon could be absorbed by one of these to switch it to a positive energy state in order to account for certain observations. This idea turned out to be wrong and so the proposal lapsed. The same objection does not apply for the creation problem and in any case no primary could be switched from a negative state to a positive one.

A better understanding appears by considering Newton's laws of motion. This led to the development of the "Opposed Energy Dynamics" whose derivation is the subject of this section. An introduction was provided in PART I but some amplification of the basic argument seems justified at this point.

In Newton's laws of motion an object free to move is pushed by a 'force of action' and accelerates in the same direction as that force is pointing. When the object of inertial mass m has reached speed v it has gained positive momentum $p = mv$ and positive kinetic energy E_K.

An object of negative energy (or negative mass) will accelerate in the direction opposed to the force of action. It consequently acquires negative momentum and negative E_K. This may appear impossible at first sight but if two objects collide, both made from negative energy, both have their responses reversed. Consequently they both bounce apart in a manner identical with the responses we observe. Our universe could as easily be made from negative energy as the positive kind, since responses would be identical.

The argument showing negative energy primaries need to exist as a balanced mixture to complement primaries made of positive energy was detailed in *PART I* and so will not be repeated further. However, it was shown there that both creation and destruction can occur in such a mixture without any violation of the first law of thermodynamics provided this is revised to read:

'ENERGY CAN ONLY BE CREATED OR DESTROYED IN EQUAL AND OPPOSITE AMOUNTS'

Impulse

We first consider a gas like substance containing either positive or negative primaries. They all travel at high speeds constantly colliding and bouncing away from one another like molecules in a gas, neither gaining nor losing energy on average. In a 50/50 mixture of both kinds, however, such collisions only account for half the total. The other half will be collisions of opposites and now some very strange effects are predicted.

The mechanics is best developed by applying Newton's third law, summarised in *CREATION SOLVED?* Appendix A, which states that for every 'force of action' produced by one object upon another the latter pushes back with an equal but opposite 'force of reaction'. Two objects in collision are subject to equal and opposite average forces F over the same time t and the product Ft is defined as **impulse** $\boldsymbol{i}$. So each colliding object is subject to an equal and opposite impulse. We can write:

$$di = Fdt$$

But force = rate of change of momentum = $d(m\ v)/dt$..

Hence: $$di = [d(m\ v)/dt] \times dt$$

or:

$$di = d(m\,v) \qquad [2.3.1]$$

So the change in impulse is also the change in linear momentum. If two primaries of opposite mass are temporarily connected so that the positive primary is subjected to an impulse, then the negative one will be subjected to an equal and opposite impulse. Hence the law of conservation of momentum is applicable between opposed energy particles just as it applies to those of like energy.

A scattering collision between two primaries of opposite energy can now be considered in which both are travelling in opposite directions but on parallel paths before collision as illustrated by FIG.5 p.27. The positive energy primary moves rightward with velocity v_{p1} with the negative one moving leftward with velocity $\boldsymbol{v}_{n1}$. Changes of momentum components are to be evaluated in the parallel and transverse directions X and Y respectively. The positive X direction is taken as moving from left to right.

If in the X direction the impulse i is positive on the positive particle then after the impulse the X component of velocity is increased from $\boldsymbol{v}_{p1X}$ to $\boldsymbol{v}_{p2X}$ and becomes:

$$v_{p2X} = v_{p1X} + i_X/m_p \qquad [2.3.2]$$

But for the negative primary, suffix n, the impulse is $-i_X$ from Newton's third law and so points in the same direction as velocity v_{n1X}. Since responses are reversed the resulting X velocity component changes to:

$$v_{n2X} = -v_{n1X} + i_X/m_n \qquad [2.3.3]$$

Clearly for the special case in which $m_p = m_n$ any positive net X component of impulse will cause an increase of speed and therefore of energy of the positive primary and a decrease in the corresponding negative values. For head–on collisions that cause no scattering the revised law of energy conservation would be violated. It follows that when pair collisions occur there can be no net X component of impulse. In the initial phase of collision an impulse will arise but an exactly equal compensation will occur as separation occurs. Offset collisions, which produce scattering, will therefore only introduce extra net impulse perpendicular to the line of approach direction: transverse impulse in the Y direction. As will be shown later this conclusion is only slightly changed when the colliding masses are unequal. Then the direction of the scattering impulse will be at a small angle ϕ to this transverse direction.

What is also important to note is that, with momentum defined by arrows, both incident X momentum components point in the same direction and so add up. For the case of zero scattering this means that, for energy to be conserved as well as momentum, no changes in either can arise. The primaries have to pass through each other, first mutually annihilating and then reconstituting.

For primaries approaching along parallel paths that are offset, so that scattering occurs, additional impulse in the *Y* direction is imparted yet the condition for *X* components remains unchanged for pair collisions:

$$v_{pY} = i_Y/m_p \quad \& \quad v_{nY} = -(-i_Y/m_n) \qquad [2.3.4]$$

Both primaries deflect in the same *Y* direction. The extra transverse impulses imparted are equal and opposite, so satisfying Newton's third law and the conservation of transverse momentum. Clearly the extra transverse velocities will add extra energies, so providing the

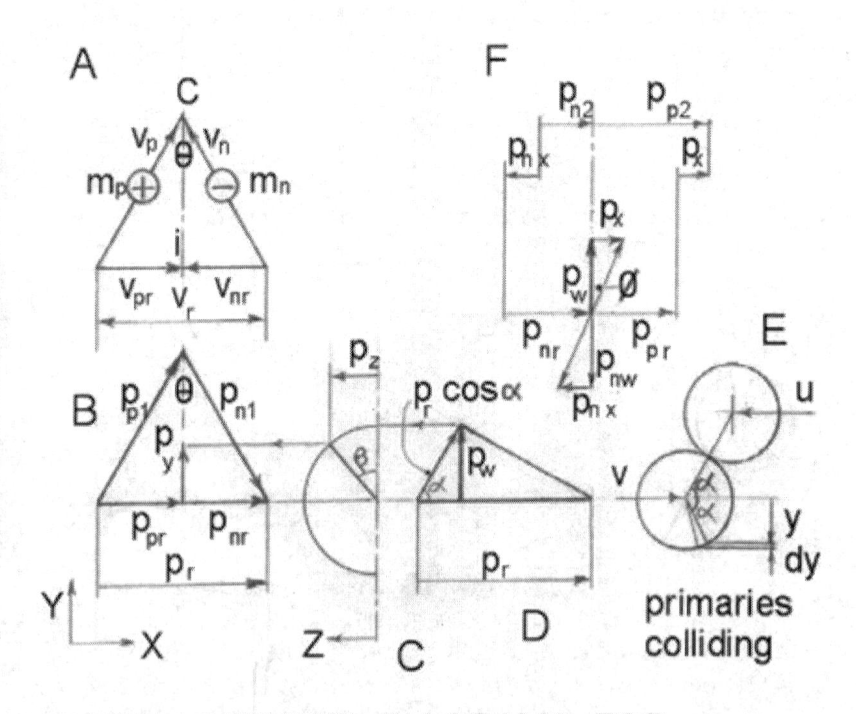

FIG.2.4 VECTOR DIAGRAMS FOR POSITIVE/NEGATIVE MASS COLLISIONS showing scattering in Y & Z directions

basic principle of collision breeding from the void that Big Breed theory of creation depends upon

The historical development of this study will be followed. At first, in 1990, it was thought that primaries would need to be travelling at almost their ultimate speed and that this would correspond with the speed of light. This meant that the derivation had to use the exact equations [2.2.1], [2.2.2] & [2.2.5]. This derivation will now be presented, with spin of primaries ignored, though the effects of such spin will be considered in a later chapter.

2.4 THE FIRST SOLUTION OF 1992

In his first book, Pearson (1990), the idea that primary particles made of both positive and negative energies was proposed as the only reasonable way creation of the universe from the void could be explained. However, not until 1992 was the idea of 'opposed energy dynamics' published as the 'Technical Supplement' to a 70 page book *'Origin of Mind'*. This called these ultimate particles 'cosmons' but the name had to be changed after a presentation in Russia in 1993. That name had already been adopted by a Canadian attendee, Adolphe Martin, but had a different meaning. So now they are to be called 'primaries'.

It was assumed that these travelled at a speed almost equal to that of light, $\boldsymbol{c}$, so the difference could be ignored. Then in FIG.2.4, showing a velocity vector triangle at A, $\boldsymbol{v_p = v_n = c}$. It had not then been realised that the ultimate speed of primaries would need to be much higher than that of light. The latter is the limit for matter. However, since primaries are not matter they do not necessarily have the same limit. For this section the assumption of that early solution will therefore be retained.

The aim was to provide a general equation for the collision between pairs of primaries arriving from all possible angles, which mean the collision angle θ can have any value between 0 and π radians, as measured from an 'absolute' frame of reference to be defined later. Then collision effects would depend only on their relative velocity and momentum. This requires a relative frame of reference to be used by deducting the velocity C to i, which is $\boldsymbol{v_p\cos(\theta/2)}$, from the entire absolute field of interest. Then with the observer at i the two primaries appear to be converging from diametrically opposite directions at relative velocities $\boldsymbol{v_{pr}}$ and $\boldsymbol{v_{nr}}$.

From the viewpoint of either primary the relative velocity of these components adds to the value $\boldsymbol{v_r}$.

Then the primaries would seem to collide either head-on, as a special case, but most would scatter due to lines of motion being offset from that of the special case as shown by the collision of two spherical primaries at **E**. Then a scattering would result. It was expected, from the simple case described on page 27, §4.2 in PART I of this book, that an energy gain would result. However, this was deduced as viewed from this relative frame of reference.

The difficulty presented arises when an absolute frame of reference is used, as demanded by the ECM theory derived in *Quantum Gravity*.... First this absolute frame needed to be defined. If, as expected, primaries could be breeding from the void by collision, then the average motion of all randomly moving primaries, within a large but local region of space, would define that absolute frame.

If this is not clear, then a starting point can be considered in which the minimum number of primaries emerge spontaneously from the void. Clearly the void is unstable since it can give rise to equal quantities of positive and negative energy without violating the conservation laws of energy and momentum. Then if collision breeding takes place and the conservation laws are obeyed at each collision, then the net momentum of all primaries must sum to the same zero value as that of the small starting group. This meant the net momentum of all primaries together would sum to zero and so be equal to that of the void from which they had all emerged.

Then vectors $\boldsymbol{v_p}$ and $\boldsymbol{v_n}$, shown at **A**, represent the absolute velocities of a positive and a negative primary respectively.

From the absolute frame some primaries would gain energy from a scattering collision but others would lose energy by bouncing back with respect to velocity i to C.

Collision probabilities also need to be evaluated. The numbers colliding would be directly proportional to their relative velocity $\boldsymbol{v_r}$ as well as both absolute and relative collision angles. In order to discover whether collision breeding would arise, analysis would need to integrate over the entire possible range of conditions.

FIG.2.4 is a revised version from *Origin of Mind* to form the starting point of the analysis. FIG.2.4A shows a velocity vector diagram for the two primaries of numerically equal mass $\boldsymbol{m_p}$ and $\boldsymbol{m_n}$ colliding at angle θ. The velocities $\boldsymbol{v_p}$ and $\boldsymbol{v_n}$ of the positive and negative primaries respectively, as already stated, were considered to be so close to $\boldsymbol{c}$ that the difference could be ignored. With the observer at i for defining the relative frame each primary has an equal and opposite relative velocity of:

$$v_{pr} = c\sin(\theta/2) = -v_{nr}$$

The corresponding momentum vector triangle is shown at **B**. Then since the primary of negative energy has its momentum arrow pointing opposite motion, both relative momentum vectors p_{pr} and p_{nr} point in the same direction parallel to the relative velocity vector and so add to give a total value p_r. During the collision the two momentum components along the chain dotted line shown at **B** of numerical value **pcos(*θ*/2)** could cancel to leave only p_r and putting $m = m_p$:

$$p_r = 2\, m\, c\, \sin(\theta/2). \qquad [2.4.1]$$

However, the possibility of this cancellation, though recognised, was ignored at the time but is considered later in this chapter.

No change in the total relative momentum can occur, otherwise momentum would not be conserved, as explained in §4.2 page 27, and so the primaries pass through each other, first mutually annihilating and then re-creating as they emerge.

Scattering adds an extra pair of momentum vectors as shown at **F** that are added at angle ϕ to the direction X of relative velocity. Since the component perpendicular to X is shown as p_w this adds an extra component in the X direction of $p_x = p_w \tan\phi$. This adds to p_p so that the exit positive momentum is the larger value p_{p2}.
(Reversing the positions of p_p and p_n as compared with that shown at **B** has no significance since momentum diagrams do not refer to position)

The balancing negative X momentum component p_{nx}, however, deducts from the negative X component p_{nr} to give a smaller exit negative X component p_{n2}.

Now since: $E = mc^2 = (mc)c = pc$

It follows that an energy imbalance is also predicted and this violates energy conservation. Consequently the net impulse due to collision can only be perpendicular to the relative velocity vector.

It follows that the momenta added by scattering in pair collisions are limited to p_w acting perpendicular to the X direction.

The two primaries are shown at **E** colliding at angle α. Another vector diagram is shown a **D** for evaluation of the extra scattering momentum component p_w for the positive primary. The initial impulse acts at angle α that tries to add momentum $p_r\cos\alpha$. (The line of length $p_r\sin\alpha$ is not a momentum vector since it is tangential to the surfaces in contact). However, for reasons given in the previous paragraph the X component is eliminated during partial interpenetration, with partial annihilation followed by reconstitution, to leave the finally added scattering collision vector p_w. Hence the value of p_w becomes:

$$p_w = p_r \cos\alpha \sin\alpha \quad \&$$
$$p_r = 2mv_p \sin(\theta/2) \ \ with\ v_p = c$$
$$Since:\ \cos\alpha\sin\alpha = \sin(2\alpha)/2 \ \ then$$
$$p_W = mc\sin(\theta/2)\sin(2\alpha) \qquad [2.4.2]$$

Scattering makes the problem three dimensional

The *X* direction is an axis of revolution and the primaries shown colliding at **E** can have $\boldsymbol{p_w}$ added in any direction taken about this axis. This is illustrated in FIG.2.4 at **C**. This shows a semicircle of radius $\boldsymbol{p_w}$ drawn in the *Z* direction perpendicular to the *X,Y* plane forming a 'scatter disc'. This shows the scattering vector at an angle β to the direction i to C defining the *Y* direction and has any value from 0 to 2π.

The result projected onto momentum diagram **B** is to produce added momentum components $\boldsymbol{p_y}$ and $\boldsymbol{p_z}$, using [2.4.2], given by:

$$p_y = mc\cos\beta\,\sin(2\alpha)\sin(\theta/2) \quad and$$
$$p_z = mc\sin\beta\,\sin(2\alpha)\sin(\theta/2) \qquad [2.4.3]$$

The four probabilities of collision

The first collision probability, $\boldsymbol{d\beta/\beta}$ is due to scattering in all directions, with equal probability. An element of angle $\boldsymbol{d\beta}$ represents the proportion of primaries deflected to move in direction β out of the total angle 2π. Since the two halves of the scatter disc have symmetry it is only necessary to use one half so β varies from 0 to π. The added momentum components are given by [2.4.3].

This does not affect the total number of collisions but is important for determining energy gains or losses.

The second collision probability, ***dN/N***, is that of collision scattering angle α. Out of a total number ***N*** of negative primaries, approaching the positive primary on a collision course, only the number ***dN*** will collide between collision angle α and $(\alpha + \boldsymbol{d\alpha})$. FIG.2.4E illustrates the problem that assumes a positive spherical primary of radius ***R***. The projected area ***dA*** of a ring element $\boldsymbol{d\alpha}$ will be given by:

$$dA = (2\pi R\sin\alpha)(R\cos\alpha\, d\alpha) \quad \& \quad A = \pi R^2$$

Then since ***dA/A = dN/N***, the 2nd probability becomes:

$$dN/N = \sin(2\alpha)d\alpha \quad \& \quad \int_0^{\pi/2} \sin(2\alpha)d\alpha = 1 \qquad [2.4.4]$$

The third probability, *prob3*, is that the rate of arrival of primaries will be proportional to their relative velocity. Consequently:

$$prob3 \propto \sin(\theta/2) \qquad [2.4.5]$$

The fourth probability $d\eta/\eta$ concerns the number of primaries available for collision $d\eta$ in a range of collision angles from θ to $(\theta + d\theta)$ out of the total number available η.

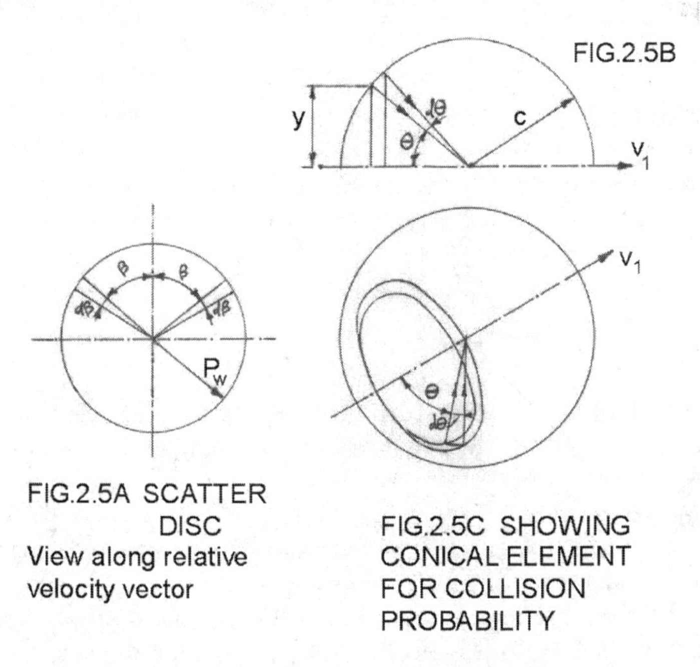

FIG.2.5A SCATTER DISC
View along relative velocity vector

FIG.2.5C SHOWING CONICAL ELEMENT FOR COLLISION PROBABILITY

The problem is illustrated in FIG.2.5B showing a conical element further illustrated in FIG.2.5C. The surface of a sphere of radius c is imagined. Then with the positive primary fixed at its centre the total number of negative primaries available, moving in from all directions, will be proportional to the surface area of that sphere. The sphere has an axis through its centre that is parallel to the absolute velocity of the positive primary. The velocity vector of the negative primary has angle θ from this axis and so of all the numbers available, proportional to the total surface of the sphere, $\eta = 4\pi c^2$, those between a cone of half angle θ and a cone of half angle $(\theta + d\theta)$ will be available for collision. The surface area of this cone element is $(2\pi c \sin\theta)(c d\theta)$ and so the probability is:

$$d\eta/\eta = (\sin\theta \, d\theta)/2 \qquad [2.4.6]$$

The total number of collisions ζ is obtained by integrating the product of the three contributing probabilities. Then with the two involving θ combined the result is the triple integral:

$$\int \frac{d\xi}{\xi} = \int_0^{\pi}\int_0^{\pi/2}\int_0^{\pi} \frac{d\beta}{\pi}\frac{dN}{N}\frac{d\eta}{\eta} = \int_0^{\pi}\int_0^{\pi/2}\int_0^{\pi} \frac{d\beta}{\pi}\sin(2\alpha)d\alpha \sin\left(\frac{\theta}{2}\right)\sin\theta\, d\theta$$

This integral needs to be evaluated to yield $\zeta = 1$. So we write:

$$\sin\theta = 2\sin(\theta/2)\cos(\theta/2) \;\&\; put \quad d\sin(\theta/2) = \tfrac{1}{2}\cos(\theta/2)d\theta$$

Then: $$\int_0^{\pi} 2\sin^2\left(\frac{\theta}{2}\right)\cos\left(\frac{\theta}{2}\right)d\theta = 4\int_0^{\pi}\sin^2\left(\frac{\theta}{2}\right)d\sin\left(\frac{\theta}{2}\right) = \frac{4}{3}$$

Integrating the other terms leaves the result:

$$\frac{d\xi}{\xi} = \frac{d\beta}{\pi}\sin(2\alpha)d\alpha\frac{3}{4}\sin\left(\frac{\theta}{2}\right)\sin\theta\, d\theta \qquad [2.4.7]$$

Then: $$\int_0^{\pi}\int_0^{\pi/2}\int_0^{\pi}\frac{d\xi}{\xi} = 1$$

Now to find the average energy gain, the gain per element needs multiplying by $d\zeta/\zeta$ from the above equation [2.4.7]

The energy gain per element

Only the positive primary will be considered since both collision partners gain energy in equal amounts.

The scatter momentum components p_y and p_z given by [2.4.3] need adding vectorially to the original absolute momentum p_{p1} to give the resultant emergent momentum p_{p2}. This means adding to components $p_{p1}\cos(\theta/2)$ and $p_{p1}\sin(\theta/2)$ as follows:

$$p_{p2}^{\;2} = \left(p_{p1}\cos(\theta/2) + p_{p1}(\sin(\theta/2)\sin(2\alpha)\cos\beta)\right)^2 + \left(p_{p1}\sin(\theta/2)\right)^2 + \left(p_{p1}(\sin(\theta/2)\sin(2\alpha)\sin\beta)\right)^2$$

The first term can be expanded so that the above becomes:

$$\left(\frac{p_{p2}}{p_{p1}}\right)^2 = \cos^2\left(\frac{\theta}{2}\right) + 2\cos\left(\frac{\theta}{2}\right)\sin\left(\frac{\theta}{2}\right)\sin(2\alpha)\cos\beta + \left(\sin\left(\frac{\theta}{2}\right)\sin(2\alpha)\cos\beta\right)^2 + \left(\sin\left(\frac{\theta}{2}\right)\sin(2\alpha)\sin\beta\right)^2 + \sin^2\left(\frac{\theta}{2}\right)$$

Then noting that standard identities are:

$$\sin^2 A + \cos^2 A = 1: \quad \sin^2(\theta/2) = \frac{1-\cos\theta}{2}: \quad and$$

$$\cos(\theta/2)\sin(\theta/2) = (\sin\theta)/2$$

Then since $\boldsymbol{E = pc}$ for this case the equation for energy gain $\boldsymbol{\delta E}$ as compared with initial energy $\boldsymbol{E_+}$ becomes:

$$\frac{\delta E}{E_p} = \sqrt{1 + (1-\cos\theta)\sin^2(2\alpha)/2 + \sin\theta\sin(2\alpha)\cos\beta} - 1$$

[2.4.8]

Then the average energy gain ratio $\Delta E/E_p$ becomes:

$$\frac{\Delta E}{E_p} = \int_0^{\pi}\int_0^{\pi/2}\int_0^{\pi}\frac{\delta E}{E_p}\frac{d\xi}{\xi} \qquad [2.4.9]$$

Where $\boldsymbol{\delta E/E_p}$ is given by equation [2.4.8] and $\boldsymbol{d\zeta/\zeta}$ is given by [2.4.7]. However, [2.4.8] is only one solution and it disallows some annihilation that could take place. The positive and negative momentum components of magnitude $\boldsymbol{\pm p_{p1}\cos(\theta/2)}$ shown in FIG.2.4B are equal and opposite and so could cancel one another. With this term omitted, as represented as $\boldsymbol{p_{p1}\cos(\theta/2)}$ in the above derivation, this annihilation is included. Then the energy "gain" equation becomes:

$$\frac{\delta E}{E_+} = \frac{1}{2}\sqrt{(1-\cos\theta)\left(\frac{3}{2} - \cos(4\alpha)\right)} - 1 \qquad [2.4.10]$$

Results of computation, using, GW BASIC file "PrCL108", for all angles divided into $\boldsymbol{N}$ equal steps, gave the integrated value of $\boldsymbol{d\zeta/\zeta}$ (which should be 1.0) and $\boldsymbol{\delta E/E_{p1}}$ from [2.4.9] and [2.4.10]:

***N* steps**	$\boldsymbol{d\xi/\xi}$	$\boldsymbol{\delta E/E_+}$ **from [2.4.9]**	$\boldsymbol{\delta E/E_+}$ **from [2.4.10]**
18	1.002225	0.1596	-0.2944
72	1.000048	0.1594	-0.2935

Increasing the number of computational steps from 18 to 72 gave higher accuracy but the differences were clearly very small.

The major conclusion to be drawn, however, is that if both cases are equally probable, as is to be expected, then contrary to the view taken in 1992, the mixture would exhibit a net annihilation and rapidly vanish back into the void. Fortunately it will be shown in later chapters that equation [2.4.10] has overestimated the annihilation by neglect of the fact that only partial annihilation can occur when collisions are not exactly head-on when $\propto = 0$. Furthermore energy gains also arise from spinning motions and these two effects ensure that a net creation on average always occurs when primaries collide two at a time.

2.5 Mutual Annihilation in Cylindrical flow Cells

Even a net gain of 0.001% of incident energy per pair collision would be enormous. It would lead to a very rapid increase in both the radius of the growing ball of i-ther and its density. However, when primaries converge from all directions, the net momentum of all primaries is zero, even before collision. This forces the positive and negative energies to mutually annihilate.

Flow cells can form that can ideally be represented as having either spherical or cylindrical symmetry. Of course this ideal cannot be realised since, with an array of cells that need to nest together, cusp-shaped breeding volumes will exist that are being ignored. However, an equivalent radius $\boldsymbol{R_0}$ will be assumed having the same volume as the non-spherical or prismatic shapes that will actually exist. For example, the equivalent cylinder will have $\boldsymbol{R_0}$ equal to 0.9094/2 of the cross-corner dimensions of the hexagonal prism.

Only the cylindrical case will be considered at this stage with cells of diameter $\boldsymbol{2R_0}$ having length $\boldsymbol{L}$ and a core radius $\boldsymbol{r_c}$.

The logic used in 1992 suggested that, to accelerate by a pressure differential, a huge number of primaries forming a fluid, positive and negative momenta would cancel. Then acceleration could occur without any pressure or density change. A creation rate $\boldsymbol{C_R}$ per unit of volume $\boldsymbol{V}$ was assumed so that with $\boldsymbol{\delta V}$ as the increase of volume in time $\boldsymbol{\delta t}$:

$$\delta V/V = C_R\,\delta t \qquad [2.5.1]$$

As the radial imploding flow velocity $\boldsymbol{v_r}$ increases, however, its kinetic energy is derived from the random kinetic energy of the primaries moving at speed $\boldsymbol{c}$ so that the total energy remains constant. Only the remaining random kinetic energy can produce breeding and so, with $\boldsymbol{c}$ being the random speed for zero flow, the above, with re-arrangement, needs to be modified to:

$$\delta V/\delta t = C_R\left(1 - (v_r/c)^2\right)V \qquad [2.5.2]$$

Then at any arbitrary radius $\boldsymbol{r}$ between $\boldsymbol{R_0}$ and $\boldsymbol{r_c}$ the primaries will implode on the outer surface of an imaginary cylinder of radius $\boldsymbol{r}$ with implosion velocity $\boldsymbol{v_r}$. Then in moving a further distance $\boldsymbol{\delta r}$ this speed will have increased to $\boldsymbol{v_r + \delta v_r}$ due to $\boldsymbol{r}$ reducing to $\boldsymbol{r - \delta r}$ and to breeding in the volume $\boldsymbol{2\pi L r \delta r}$ where $\boldsymbol{L}$ is the length of the element. So by equating flows from outside to inside of the element:

$$2\pi L(r-\delta r)(v+\delta v)=(2\pi r L)v+C_R\left(1-(v/c)^2\right)2\pi L r\,\delta r$$

Multiplying the bracketed term and ignoring the second order term $\delta r \delta v$ after cancellation of the $2\pi vLr$ terms results in:

$$-(v_r/r)\delta r+\delta v_r=C_R\left(1-(v_r/c)^2\right)\delta r$$

Dividing throughout by c and re-arranging:

$$\frac{\delta v_r}{c}=\frac{C_R}{c}\left(1-\left(\frac{v_r}{c}\right)^2\right)\delta r+\frac{v_r}{c}\frac{\delta r}{r} \qquad [2.5.3]$$

A simple computer code assumes C_R = 4×10^{71} J/m^3 as evaluated later in §7.3 for the ratio V_B/V = 0.05 where a V_B is the total volume occupied by all primaries within a total volume V - it being argued in that section as about correct. The code then starts with $\delta r/R_0$ = 0.001, $r = R_0 + \delta r/2$ and $v_r = 0$.

A computational loop follows with $r = r - \delta r$, $v_r = v_r + \delta v_r$ to continue until $v_r = c$. This could give the limiting inside radius of the breeding zone since c cannot be exceeded.

This would be the case if the core were totally permeable to primaries. This is not the case since it has to be a composite structure of both kinds of energy. When an incoming primary hits one of its own kind at the surface of the core, it will bounce away. Only those meeting their opposites can enter to be annihilated by being squeezed out of existence. This will happen as all primaries move under their own inertia toward the centre point

(Continued after the following table)

TABLE 2.I Radial velocity profiles: V_B/V=.05: C_R= 4×10^{71} J/m^3

	R_0/σ= 10		R_0/σ= 20		R_0/σ= 30		R_0/σ= 40	
r/R_0	T/T_T	v_r/C_U	T/T_T	v_r/C_U	T/T_T	v_r/C_U	T/T_T	v_r/C_U
0.99	.442	.005	.461	.011	.463	.014	.495	.021
0.98	.542	.011	.566	.022	.588	.032	.608	.043
0.94	.703	.033	.733	.066	.762	.099	.788	.131
0.90	.777	.056	.810	.112	.842	.167	.871	.222
0.83	.851	.093	.888	.198	.924	.290	.957	.382
0.80	.873	.120	.912	.236	.949	.347	.983	.450
0.777	.891	.140	.927	.266	.969	.401	1.0	.5
0.718	.918	.178	.960	.348	1.0	.5		
0.7	.926	.192	.968	.374				
0.617	.956	.263	1.0	.5				
0.419	1.0	.5						
$T_T$sec	4.86E-18		4.76E-18		4.58E-18		4.43E-18	

Consequently $v_r/c = 0.5$ is the maximum within the converging flow itself. This determines the ratio of the diameter σ of primaries to cell radius R_0 and core radius R_C. To this extent theory of 1992 has been updated. Time T is seconds from R_0 : T_T from R_0 to R_C.

The previous table, TABLE 2.I, obtained in this way for the example selected above, shows how slowly a primary migrates from the outer rim of a cell with a very rapid increase of radial velocity as the core is approached. So to introduce the methodology in a simple way, the concepts of collision breeding, followed by annihilation in cellular arrays, has been presented.

It is the collapse into such an ordered state that provides the switch off mechanism the big bang theory lacks. In this way the problem of the cosmological constant is resolved. This is a first stage of organisation from the power of chaos.

As TABLE 2I shows the core radius r, where $T/T_T = 1.0$, increases disproportionately as cell radius R_0 increases and therefore sets a limit on the stable size of cells. The creation rate, however, ignored random bunching that is likely to reduce creation rate by a factor of about 10. Then for $R_0/\sigma = 40$ the core radius reduces from $.777 \times R_0$ to $.200 \times R_0$ with T_T increased to 5.07E-17 (i.e. 5.07×10^{-17}) seconds.

In the Big Breed theory only a minute fraction of the present energy of the universe was created in the inflation phase. This phase is the initial period of rapid exponential growth. This precedes the collapse into flow cells containing cores of annihilation. As the cells develop, first at the origin point, where density is highest, and then spread outward in spherical rings, growth is progressively slowed. By the time the rings of flow cells have reached the outer growing edge, growth will have been brought almost to a halt.

Growth will not have been completely stopped, however, a minute net creation will have remained that will go on everywhere forever. As will be shown later this minute net growth can create an ever-accelerating expansion without the need for postulating the existence of 'Dark Energy' with mysterious repulsive power.

Later chapters go into these matters in greater detail since this chapter mainly concerns the solution of 1992. In the next chapter the general case for collision breeding will be considered.

CHAPTER 3

COLLISION BREEDING GENERAL EQUATIONS

In the previous chapter a special case was considered in which no primaries had any rest energy and so travelled at their ultimate speed. All had numerically equal energies. But no attempt was made to justify these assumptions.

In this chapter the method is extended to the general case in which primaries have arbitrary rest energies and speeds. The hope is to find a way to establish the correct primary speeds.

Note that suffix $_P$ now means 'positive and $_n$ 'negative'

3.1 The general case of unequal masses will be addressed.

The ultimate speed of primaries is now to be represented by symbol C_U since as will be shown later, this must exceed the speed of light. In FIG.3.1 a velocity triangle is shown at F1 in the X,Y plane and in which a primary of positive energy and inertial mass m_p has velocity v_p that collides at an angle θ with a negative primary of inertial mass m_n travelling at velocity v_n. Both are shown moving in an absolute frame of reference. It is arranged so that their relative velocity v_R lies on the X axis. Only in this frame of reference can the collision appear as if head on or permit scattering to be modelled. (Although m_n is a negative quantity its sign is not included since the negative effect is taken into account by the reversed direction of momentum arrow.)

On this same X axis at F4 (far right) a primary, assumed of spherical shape, is shown in collision with the spherical negative primary at a collision angle α in the X,W plane where direction $\boldsymbol{W}$ is inclined at angle β to the Y direction as shown at F2..This is a view looking in the direction of v_R and so shows the Z,Y plane. This is because there is an additional Z component of scatter velocity and momentum perpendicular to the X,Y plane.

A momentum triangle is shown at F3, also in the X,W plane, in which the incident relative momentum p_R points along the X axis: the same direction as relative velocity v_R. The combined incident momentum as viewed in the relative frame at closing velocity v_R is p_R and is independent of α. This momentum is the first item needing to be considered.

If, for example, $m_n < m_p$ then the momentum $m_n v_R$, observed from the positive mass, will be less than that seen by an observer at the negative one. The only frame of reference that avoids this difference is one that makes both components equal.

If proportion x of v_R is apportioned to mass m_p giving $p_p = m_p x\, v_R$ then the momentum contribution from negative mass is

$p_n = m_n(1 - x))v_R$,

NOTE: the sign of m_n has been included i.e. read m_n as $\lfloor m_n \rfloor$

Then the sum of these is:

$$p_R = m_p v_R x + m_n v_R (1 - x) \qquad [3.1.1]$$

This is alternatively expressed as:

$$p_R = m_p v_R \left(x(1 - m_n/m_p) + m_n/m_p\right) \qquad [3.1.2]$$

(noting that the two momentum components p_n and p_p point in the same direction since the negative mass has its momentum arrow pointing opposite its own motion, but is moving in the opposite direction to the positive mass.

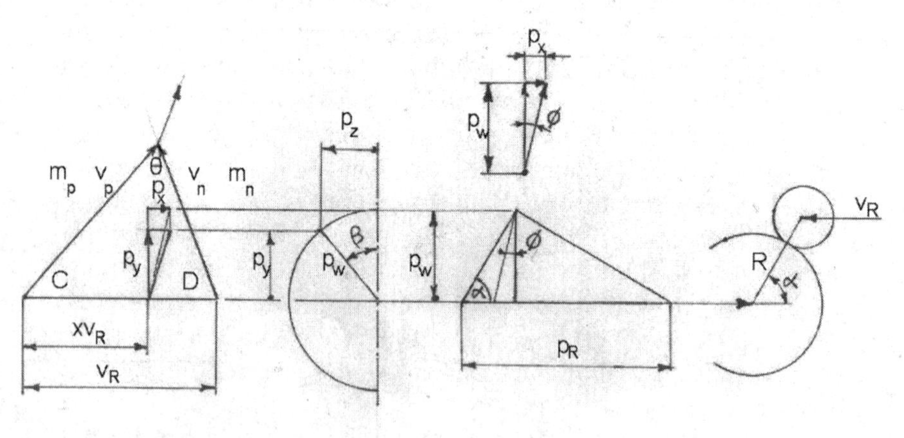

F1 velocity triangle F2 momentum disc viewed on v_R F3 scatter momentum triangle F4 primaries colliding

FIG.3.1 MECHANICS FOR THE GENERAL CASE OF COLLISION BREEDING

From [3.1.2] p_R varies linearly from $m_n v_R$ at $x = 0$ to $m_p v_R$ at $x = 1$. So in general p_R as viewed from m_p will be accorded a different value from that accorded by viewing from m_n. The correct value will have to accord equal amounts to both positive and negative components by the choice of position defined by x. This condition is found by equating the two components of [3.1.1] to yield:

$$x = \frac{m_n}{m_p + m_n} \qquad [3.1.3]$$

Substituting for x in [3.1.1] from [3.1.3] then gives the value of the combined relative momentum p_l as:

$$p_R = 2v_R \frac{m_p m_n}{(m_p + m_n)} \qquad [3.1.4]$$

This is shown as the base of the 'scatter momentum triangle' in FIG.3.1 at F3. This is designed for evaluating the momenta added by the collision.

It needs to be realised that in general, as shown in FIG.3.1 that depicts the X,Y plane, the resultant momentum does not lie in the X,Y plane. There will be an additional resultant relative momentum component in the Z direction (perpendicular to the page and given by $\boldsymbol{p_z} = \boldsymbol{p_W sin\beta}$). If the collision were head on then no scattering would occur and the same combined momentum would need to apply after collision. The only solution is that both primaries emerge without change of velocity. They must pass through one another to temporarily mutually annihilate followed by both reconstituting as they emerge. For collisions not head on the same argument applies for directions parallel to the direction $\boldsymbol{v_R}$ but now scattering must occur due to transverse impulses.

The transverse impulse is evaluated by the aid of right angle triangle F3 that is shown with one side parallel to the surfaces in contact as shown at F4. The third side is at angle α to the X direction: the direction of initial forces during contact of the two primaries. So the length of this momentum vector defining the impulse caused by contact is $p_R \cos(\alpha)$. However, this cannot represent net momentum change due to collision since its X component would prevent the X component of momentum from being conserved (as explained in Chapter 2 with reference to FIG.2.4F). This implies that the two primaries will partially interpenetrate and partially annihilate as they move into each other but will then mutually reconstitute as they move away. Then other forces come into play so that the X components of the initial contact forces are largely cancelled. Only in this way can momentum and energy be conserved overall.

The overall momentum added to either primary is limited to a vector in the W direction -perpendicular to the v_R vector in the X direction, as argued for symmetrical collisions in the previous chapter. However, to allow for this not being applicable for the general case the overall direction of impulse caused by collision is modified. For cases of numerically dissimilar masses the allowed vector will have an angle ϕ measured from the W direction as shown in the inset above F3. The value of ϕ is to be determined from the overall energy balance. As illustrated there will be a transverse momentum component $\boldsymbol{p_w}$ added where:

$$p_w = p_R \cos(\alpha)\sin(\alpha) \qquad [3.1.5]$$

-and a much smaller component p_x in the x direction equal to:

$$p_x = p_w \tan\phi = p_R \cos(\alpha)\sin(\alpha)\tan(\phi) \qquad [3.1.6]$$

Now the collision can occur with equal probability at any angle β with respect to the Y,Z plane of F2. To represent this it is imagined that a view is taken along the X axis so that the p_w momentum vector can lie anywhere on a 'momentum circle' of radius p_w. as illustrated at F2 pointing at an angle β measured from the plane of velocity triangle F1. Clearly there will be two momentum components of p_w. The one lying in the plane of the velocity triangle is p_y so that substituting from [3.1.6] its value is:

$$p_y = p_R \cos(\alpha)\sin(\alpha)\cos(\beta) \qquad [3.1.7]$$

And there will be a component in the Z direction perpendicular to the plane given by:

$$p_z = p_R \cos(\alpha)\sin(\alpha)\sin(\beta) \qquad [3.1.8]$$

But p_R depends on relative velocity v_R. The cosine law gives v_R as:

$$v_R^2 = v_p^2 + v_n^2 - 2v_p v_n \cos(\theta) \qquad [3.1.9]$$

Then using the cosine law again for angle C and substituting from [3.1.9]

$$\cos(C) = (v_p - v_n \cos(\theta))/v_R \qquad [3.1.10]$$

And $$\cos(D) = (v_n - v_p \cos(\theta))/v_R \qquad [3.1.11]$$

Sines are best obtained from the sine rule to avoid square roots with their troublesome + or – signs and become:

$$\sin(C) = \frac{v_n}{v_R}\sin(\theta) \quad \& \quad \sin(D) = \frac{v_p}{v_R}\sin(\theta) \qquad [3.1.12]$$

It is now possible to determine absolute values for the three momentum components by adding the appropriate velocities so that using [3.1.2] to [3.1.8] and noting that $\sin(\alpha)\cos(\alpha) = \frac{1}{2}\sin(2\alpha)$, the absolute momentum components for positive mass become:

$$p_x = m_p v_p \cos(C) + (p_R/2)\sin(2\alpha)\tan(\phi)$$

$$p_y = m_p v_p \sin(C) + (p_R/2)\sin(2\alpha)\cos(\beta)$$

$$p_z = (p_R/2)\sin(2\alpha)\sin(\beta) \qquad [3.1.13]$$

Which combine, noting that: $\sec^2\phi = 1 + \tan^2\phi$ to yield:

$$p_{P2}^2 = p_P^2 + (p_R/2)^2\sin^2(2\alpha)\sec^2\phi$$
$$+ p_P(p_R/2)\sin(2\alpha)(\sin C\cos\beta + \cos C\tan\phi)$$

[3.1.14]

For negative mass the absolute momentum components become:

$$p_{nx} = m_n v_n\cos(D) - (p_R/2)\sin(2\alpha)\tan(\phi)$$
$$p_{ny} = -m_n v_n\sin(D) - (p_R/2)\sin(2\alpha)\cos(\beta)$$
$$p_{nz} = -(p_R/2)\sin(2\alpha)\sin(\beta) \qquad [3.1.15]$$

Which combine to yield:

$$p_{n2}^2 = p_n^2 + (p_R/2)^2\sin^2(2\alpha)\sec^2\phi$$
$$+ p_n(p_R/2)\sin(2\alpha)(\sin D\cos\beta - \cos D\tan\phi)$$

[3.1.16]

In which $\boldsymbol{p_R}$ is given by [3.1.4]

The negative sign appearing in the equation for p_{n2} allows an energy balance to be achieved by the variation of ϕ.

3.2 Energy Gains

Equation [2.2.5] can now be applied to obtain both positive and negative energy changes, which can be an increase or a decrease. For either positive or negative changes we can write:

$\delta E = E_2 - E_1$ and using [2.2.5] becomes:

$$\delta E = \sqrt{E_0^2 + (p_x^2 + p_y^2 + p_z^2)C_U^2} - \sqrt{E_0^2 + (mvC_U)^2}$$

[3.2.1]

With set [3.1.13] and m_pv_p substituted for mv for positive gains: and with set [3.1.14] and m_nv_n substituted for mv for negative gains. The values of E_{0n} and E_{0p} do not need to be equal and are related to m_p and m_n by v_p and v_n according to [3.1.1] (noting that the masses given there are inertial masses m: where $m = m_0 + m_K$).

Then by trial the values of ϕ in sets [3.1.14] and [3.1.16] that make $\Delta E_n = \Delta E_p$ given by [3.1.17] can be found in order to satisfy the conservation of energy.

3.3 Non-dimensionalisation

Analysis is simplified by dividing [3.1.14] and [3.1.16] by p_P^2 (the absolute incident momentum of the positive primary) and dividing equation [3.2.1] by E_0 to give non-dimensional forms. Then after simplification m_p/m_0 and m_n/m_{0n} can be replaced by speed functions from [2.2.1], so that:

$$\frac{m_p}{m_0} = \frac{1}{\sqrt{1 - \left(v_p/C_U\right)^2}} \quad \& \quad \frac{m_n}{m_{0n}} = \frac{1}{\sqrt{1 - \left(v_n/C_U\right)^2}}$$

[3.3.1]

Hence the ratio m_p/m_n becomes:

$$\frac{m_p}{m_n} = \frac{m_0}{m_{0n}}\sqrt{\frac{1 - \left(v_n/C_U\right)^2}{1 - \left(v_p/C_U\right)^2}} = \frac{E_p}{E_n} \qquad [3.3.2]$$

Only m_0 and m_{0n} or E_0 and E_{0n} remain constants if speeds vary. The quantity E_{0n}/E_0 is a 'rest energy asymmetry', which was thought necessary in order to explain gravitation and to permit the i-ther to provide a source of power. So the negative sum energy ratio, needed for providing the same denominator as that used for positive primaries, then becomes:

$$\frac{E_n}{E_0} = \frac{E_n}{E_{on}}\frac{E_{0n}}{E_0} \qquad [3.3.3]$$

The non-dimensional equivalent of [3.2.1], noting that $E_0 = m_0\, C_U^2$ and also $\frac{m}{m_0}\frac{v_P}{C_U} = \frac{p_P\, C_U}{E_0}$. can be written as:

$$\frac{\delta E_P}{E_0} = \sqrt{1 + \left(\frac{p_{P2}}{p_P}\right)^2\left(\frac{p_P\, C_U}{E_0}\right)^2} - \sqrt{1 + \left(\frac{p_P\, C_U}{E_0}\right)^2}$$

[3.3.4]

and for negative primaries:

$$\frac{\delta E_N}{E_0} = \sqrt{\left(\frac{E_{0n}}{E_0}\right)^2 + \left(\frac{p_{n2}}{p_P}\right)^2\left(\frac{p_P\, C_U}{E_0}\right)^2} + \qquad [3.3.5]$$

$$-\sqrt{\left(\frac{E_{0n}}{E_0}\right)^2 + \left(\frac{p_{n2}}{p_P}\right)^2\left(\frac{p_P\, C_U}{E_0}\right)^2}$$

p_{P2} & p_{n2} are given by equations [3.1.14] and [3.1.16] respectively.

3.4 Integration

Finally the following equations need to be solved in order to obtain the average collision energy gains for the mixture. For the numbers colliding for relevant values of β, α, θ the equation [2.4.6] developed in 1992 is still valid and, with the differential co-efficient d replaced by δ for finite difference integration becomes:

$$\frac{\delta\xi}{\xi} = \frac{\delta\beta}{\pi}\sin(2\alpha)\delta\alpha\frac{3}{4}\sin\left(\frac{\theta}{2}\right)\sin\theta\,\delta\theta$$

Then:

$$\frac{\Delta E_p}{E_0} = \int_0^{\pi}\int_0^{\pi/2}\int_0^{\pi}\frac{\delta E}{E_0}\frac{\delta\xi}{\xi} \qquad [3.4.1]$$

The numerical value of energy gain has to be the same for both collision partners at each step of the integration, for energy conservation to be satisfied. So the two energy gains $\delta E_P/E_0$ and $\delta E_N/E_0$ from [3.3.4] and [3.3.5] respectively are equated at each calculation by varying the value of $\tan\phi$.

Finally it is the incident kinetic energy that is responsible for the energy gain. Incident energy is supplied by both partners. Of course they have opposite energies and so if simply added they would partially cancel. This is not what is required. A means for comparison at different values of v/C_U is required that concerns both partners. A suitable 'total kinetic energy parameter' E_{KT} is therefore used which ignores the negative sign of E_n to be given by:

$$\frac{E_{KT}}{E_0} = \frac{1}{\sqrt{1-(v_{P1}/C_U)^2}} - 1 + \frac{E_n}{E_0}\left[\frac{1}{\sqrt{1-(v_{n1}/C_U)^2}} - 1\right] \qquad [3.4.2]$$

This is then divided into the result of evaluation of [3.4.1] to give the result in form $\Delta E_P/E_{KT}$.

3.5 Matching the two phases

For the general case all primaries in each phase need to have equal and opposite average sum energies but includes the asymmetry $\boldsymbol{E_{0n}/E_0}$. It is necessary to find the average speed of the negative phase for a given value of average $\boldsymbol{v_P/C_U}$ for the positive phase. This is derived as follows:

Since energy gains must balance at each collision the average energies of both phases must always balance but rest energies differ for the two phases. Hence kinetic energies must also have a matching imbalance so that:

$$E_{0P} + E_{KP} = E_{0n} + E_{Kn}$$

Or:

$$\frac{E_{0P}^{\ 2}}{1-(v_P/C_U)^2} = \frac{E_{0n}^{\ 2}}{1-(v_n/C_U)^2}$$

And this can be re-arranged to yield:

$$\left(\frac{v_n}{C_U}\right)^2 = 1 - \left(\frac{E_{0n}}{E_{0P}}\right)^2\left(1 - \left(\frac{v_P}{C_U}\right)^2\right) \qquad [3.5.1]$$

If Maxwell's speed distribution law is applied then [3.5.1] still has to be used but in this case refers to the most probable speed, usually denoted as $\boldsymbol{v_0}$. All probabilities will be changed in the same ratio as given by Maxwell's distribution law, when based on $\boldsymbol{v/v_0}$. Maxwell's distribution law given by Jeans (1887) page 31 is given by:

$$\frac{dn_v}{n} = \frac{4}{\sqrt{\pi}}\left(\frac{v}{v_0}\right)^2 \exp\left(-\left(\frac{v}{v_0}\right)^2\right)\frac{dv}{v_0} \qquad [3.5.2]$$

Here $\boldsymbol{dn_v}$ is the number of primaries in speed range $\boldsymbol{v}$ to $\boldsymbol{v + dv}$ out of the total number $\boldsymbol{n}$. It needs to be stressed that this equation was derived for identical molecules in thermodynamic equilibrium and so may not be exactly applicable for primaries since the latter may never reach equilibrium when breeding and, as mentioned earlier, primaries must exist as a mixture of mass values in which the largest have double the mass of the smallest. However, at least some similar equation must apply but could only be determined by a very complex computer analysis. It will be demonstrated later that the exact form of this random speed distribution law is not critical for our purpose and so [3.5.2] can be regarded as sufficiently accurate.

Equation [3.5.2] shows that when $\boldsymbol{v/v_0}$ is either 0.1 or 2.5 the probability is only 10% of that for $\boldsymbol{v/v_0} = 1$. Above $\boldsymbol{v/v_0} = 3$ the probability is negligible. Hence a range $\boldsymbol{v/v_0} = 0$ to 3 needs to be considered in analysis.

This does not yet represent a complete analysis, however, since primaries in each phase will not have uniform rest energy. As each collision of opposites occurs a gain in kinetic energy arises. After each primary has reached a critical size further collisions will cause splitting. The end result is to increase the numbers of primaries. Clearly they will ideally exist in a two to one mass range. So in addition to speeds being randomised, the rest masses will exist with equal numbers in each hypothetical "bin" over a 2 to 1 mass range. This case has not yet been included in the analyses made so far but the way ϕ is affected by collisions between unequal masses is illustrated in the last of the tables that appear at the end of this chapter. In this way primaries can be considered to breed by repeated collision.

There are other complications so further explanation is needed. All the equations presented so far are exact based on the conservation of linear momentum only. However, primaries will also rotate and so exhibit rotational kinetic energy and will possess angular momentum. It will be shown in a later chapter that pair collisions result in additional spin energy gains. The primaries will grow collision by collision of opposites as they gain kinetic energy and so increase their inertial mass. This high-grade energy will also be continually decaying to the lower grade rest energy due to subsequent collisions with primaries of the same energy sign. For example, such collisions will tend to convert linear kinetic energy to the rotational kind until 'thermal equilibrium' of both kinds is reached. Rotational (spin) kinetic energy will add mass like added rest energy and indeed it is reasonable to posit that decay to rest energy occurs at each of these collisions, possibly enhanced by frictional effects.

At the macroscopic scale friction causes a gain in entropy and the high grade energy loss re-appears as low grade heat: random molecular motion. At the level of i-ther no such means is available for absorbing this energy release and so has to re-appear as a gain of rest energy in the primaries themselves. Rest energy is the lowest grade to which the energy at the i-ther level can appear.

There may be some other form of decay not yet considered that causes decay from kinetic to rest forms of energy as time proceeds.

The net effect will be continual growth of rest energy of the mixture as a whole due to continual increase in number of primaries with average linear and rotational speeds remaining unchanged: due to the repeated collisions of opposites.

The positive and negative energy gains also need to be numerically equal in order to satisfy the revised law of conservation of energy. This has to be achieved by iteration by change of **tan**ϕ.

With rotation ignored the variables can be listed as:

$$E_0 : \frac{E_{0n}}{E_0} : \frac{v_p}{C_U} : \frac{v_n}{C_U} : \alpha : \beta : \theta$$

The second item, E_{0n}/E_0, could have been expressed simply as E_{0n} but in total seven variables need to be considered and is far too large a number to permit the usual numerical multiple integration. First attempts therefore provided a code that specified the first four variables as constants and provided probabilities for the angles α,β **&** . Then the triple integral could be evaluated by finite difference methods.

3.6 Integration using random numbers

Then to produce a complete solution a code was developed based on a random choice approach that permitted all seven variables to be accommodated. For this purpose the random number generator provided with the GWBASIC language used was first adopted. It was found hopelessly inaccurate. Therefore a better method was adopted with all computations made in the double precision that uses 16 digits. The procedure adopted took one of the sines or cosines from the end result of one of the variables. These functions were chosen: since their numbers are limited to the range –1 to +1. Next care was taken to ensure the sign was positive using the 'abs' function. This real number, now limited to the range 0 to 1, was multiplied by 10^8 and the resulting real number saved. Then this number was truncated to yield an integer and this integer subtracted from the saved real number. The result was a random number between 0 and 1 to 8 places of decimals. A different variable was used to get each random number. This gave very good randomness and it was found that this way all variables could give consistent results using about 10,000 runs.

It was assumed that $\boldsymbol{v_p}$ and $\boldsymbol{v_n}$ each obeyed Maxwell's distribution law, as derived for the molecules of a gas, but since speeds close to light were being examined the distribution law was assumed to apply to a momentum distribution instead of a speed distribution. The end result for average $\boldsymbol{v/C_U} = 0.99$ gave an energy gain of 16% of the averaged incident energy values. For these early runs it had been assumed that $\boldsymbol{C_U} = \boldsymbol{c}$: an assumption later found to be wrong. However, the equations remain valid and have been used to explore lower $\boldsymbol{v/C_U}$ ratios, as will be detailed in the next chapter.

3.7 Computation of energy gains by collision breeding

Note that $\boldsymbol{E_K}$ is the kinetic energy based on [2.2.2]. However, since both positive and negative energies act to give the energy gains, the kinetic energy is defined by the equation:

$$\frac{E_K}{E_{0p}} = \left[\frac{1}{\sqrt{1 - \left(v_p/C_U\right)^2}} - 1\right] + \frac{\left|E_{0n}\right|}{E_{0p}}\left[\frac{1}{\sqrt{1 - \left(v_n/C_U\right)^2}} - 1\right]$$

[3.7.1]

The following table shows how $\boldsymbol{\phi}$ and $\boldsymbol{\Delta E/E_K}$ vary for some chosen parameters.

TABLE 3 I GIVING ϕ and $\Delta E/E_K$ with ϕ and θ measured in degrees.

$v_{p0}/C_U = 0.1$: $\gamma = v_P/v_{P0}$ and $\eta = v_N/v_{N0}$

For E_{N0}/E_{P0} = 1/1.01 & $\alpha = 45^0$ all cases

Case 1 $\theta = 15^0$ Values in the columns are ϕ and $\Delta E/E_K$

← ϕ degrees →← $\Delta E/E_K$ →

η↓	γ = 0.2	γ = 1.0	γ = 2.2	β	γ = 0.2	γ = 1.0	γ = 2.2
2.2	-0.955	-1.0622	-2.389	0	0.2275	0.1888	0.1549
2.2	-0.956	-1.0325	-1.3674	90	0.2150	0.1293	0.0390
2.2	-0.957	-1.0005	-0.3257	180	0.2025	0.0694	-0.069
1.0	-0.076	-0.3479	-3.118	0	0.0222	0.1529	0.1326
1.0	-0.075	-0.1449	-0.4077	90	0.1930	0.0404	0.0147
1.0	-0,075	0.0589	2.292	180	0.1636	-0.0724	-0.104
0.2	0.0997	0.0047	-0.2973	0	0.1524	0.1796	0.2068
0.2	0.1057	0.0596	-0.2189	90	0.0396	-0.1012	0.1693
0.2	0.138	0.1155	-0.1402	180	-0.073	0.0227	0.1317

Case 2 $\theta = 90^0$

← ϕ degrees →← $\Delta E/E_K$ →

c	γ = 0.2	γ = 1.0	γ = 2.2	β	γ = 0.2	γ = 1.0	γ = 2.2
2.2	-0.9516	-1.0732	-1.894	0	0.2865	0.4657	0.6435
2.2	-0.955	-1.0264	-1.317	90	0.2383	0.2397	0.2422
2.2	-0.9576	-0.9639	-0.699	180	0.1900	0.0095	-0.176
1.0	-0.076	-0.2697	-1.125	0	0.3610	0.6772	0.7133
1.0	-0.0751	-0.1442	-0.427	90	0.2477	0.2485	0.2447
1.0	-0.0737	-0.0187	0.2602	180	0.1343	-0.184	-0.233
0.2	0.1258	-0.0268	-0.437	0	0.6855	0.5518	0.3858
0.2	0.1244	0.062	-0.22	90	0.2499	0.2483	0.2405
0.2	0.1356	0.1499	-0.005	180	-0.186	-0.056	0.0947

Case 3 $\theta = 165^0$

← ϕ degrees →← $\Delta E/E_K$ →

η↓	γ = 0.2	γ = 1.0	γ = 2.2	β	γ = 0.2	γ = 1.0	γ = 2.2
2.2	-0.953	-1.0297	-1.3882	0	0.7741	0.4075	0.5400
2.2	-0.954	-1.0218	-1.3079	90	0.2616	0.3492	0.4368
2.2	-0.955	-1.0132	-1.2261	180	0.2492	0.2906	0.3326
1.0	-0.075	-0.1617	-0.5227	0	0.3318	0.5665	0.5939
1.0	-0.075	-0.1441	-0.4293	90	0.3025	0.4557	0.4727
1.0	-0.075	-0.1265	-0.3362	180	0.2731	0.3447	0.3510
0.2	0.1286	0.0468	-0.2647	0	0.5729	0.4738	0.3492
0.2	0.1340	0.0613	-0.2213	90	0.4603	0.3952	0.3115
0.2	0.1283	0.0764	-0.1778	180	0.3475	0.3166	0.2738

The values of ϕ are all very small, mostly between ±1 degree, and the values of $\Delta E/E_K$ rarely fall below zero. Logic seemed to suggest that energy changes would become negative when β was between 90^0 and 180^0 but the computation using the exact equations show few negative values. For this reason considerable checking was made by hand calculation but these confirmed that the results were correct.

The energy gains ΔE, which of course are numerically equal for both collision partners, appear to vary over a surprisingly small range. This, however, is because of the division by kinetic energy as defined by equation [3.1.30]
Of course this reference kinetic energy varies considerably with $\gamma = v_P/v_{P0}$ and $\eta = v_N/v_{N0}$. The variation is expressed in the following table so that corresponding values of $\Delta E/E_0$ can be obtained by multiplication with the values of $\Delta E/E_K$ given above.

TABLE 3.II Showing how E_K/E_0 varies with γ and η

	γ →	0.2	1.0	2.2
	v_P/C_U →	0.02	0.1	0.22
η v_n/C_U	2.2 0.3779	0.07950	0.08434	0.1044
η v_n/C_U	1.0 0.1718	0.01514	0.01998	0.04005
η v_n/C_U	0.2 0.03436	7.849×10^{-4}	5.623×10^{-3}	0.02570

TABLE 3.I also shows that the velocity change due to collision is almost perpendicular to the direction of the relative velocity vector v_R (by the small values obtained for ϕ) but in all cases the value of $E_{0P}/E_{0n} = 1.01$ which is close to unity.

As previously explained at least a 2 to 1 range of rest energy or rest mass will exist in the mixture due to growth to maximum size. It is therefore necessary to explore the effect of consequent rest mass variation. Some values of ϕ are given in the following TABLE 3. III showing also how energy gains per collision are affected.

TABLE 3.III EFFECT ON ϕ OF VARIATION IN E_{0n}/E_{0P}

For $\theta = 15^0$, $\alpha = 45^0$, $\beta = 0^0$, $v_P/C_U = 0.1$, $v_n/C_U = 0.3779$

E_{0P}/E_{0n}	ϕ *degrees*	$\Delta E/E_K$	E_K/E_0
1.2	1.391	0.1848	0.07178
0.8	-4.34	0.1895	0.10516
1.4	3.582	0.17879	0.062249
0.7	-6.197	0.18758	0.11946
2.0	8.57	0.15655	0.045086
0.5	-10.685	0.176355	0.16523
3.0	13.942	0.12062	0.031736
0.3333	-15.655	0.15385	0.24557

TABLE 3 IV as TABLE 3. III but with θ, ϕ & β also varied

θ *degrees*	β *degrees*	E_{0P}/E_{0n}	ϕ	$\Delta E/E_K$	E_K/E_0
90	0	3.0	13.56	0.51230	0.031736
"	"	0.3333	-15.197	0.29525	0.24557
180	"	3.0	13.855	0.3655	0.031736
90	90	3.0	13.933	0.20017	0.031736
"	"	0.3333	-15.708	0.18420	0.24556
90	180	3.0	14.045	-0.11587	0.31736
"	"	0.3333	-15.92	0.06986	0.24556
165	0	3.0	13.786	0.44025	0.031736
"	"	0.3333	-15.531	0.27196	0.24557
165	90	3.0	13.858	0.35995	0.031736
"	"	0.3333	-15.628	0.24318	0.24557

3.8 CONCLUSION TO CHAPTER 3

It can be concluded from TABLES 3.III AND 3. IV that the velocity change in the X direction due to collision of opposite primaries is still fairly small – as indicated by ϕ being limited to about ±15 degrees from the direction of the relative velocity vector. This is even with a ratio of rest energy range of 3 to 1/3, which is greater than the anticipated range of 2 to ½.

We are still no nearer to finding a way of establishing the most probable speed ratios: the speeds of primaries divided by the ultimate speed C_U. If this value is small then an analytical solution becomes possible as shown in the next chapter.

CHAPTER 4

ANALYTICAL SOLUTIONS FOR BREEDING COLLISIONS

4.1 Why an Analysis for Low Speeds is Worthwhile

For applying the exact equations derived in Chapter 3 to evaluation of average energy gains by the collision of opposite primaries a triple integral had to be evaluated by finite element computer code. However, it has not yet been established that v/C_U has a high enough value to justify an exact approach. Furthermore, it is useful to have an analytical solution as a check on the computed values. For low values of v/C_U mass increase due to speed can be neglected for the accuracy required. The approximation so permitted then allows integrable expressions to be derived as will now be shown.

The approximate equations [2.2.3] or [2.2.4] now become acceptable with the second and higher order terms ignored. Only this simplest case will therefore be considered further but now the probability of collisions will be explored.

A scattering collision between two primaries of opposite energy can now be considered in which both are travelling in opposite directions but on parallel paths before collision. The positive energy primary moves rightward with velocity $\boldsymbol{v_{p1}}$ with the negative one moving leftward with velocity $\boldsymbol{v_{n1}}$. Changes of momentum components are to be evaluated in the parallel and transverse directions. *X* and *Y* respectively.

Now impulse $\boldsymbol{i}$ is defined as force×time of action: $\boldsymbol{F\delta t = m\delta v}$:

If in the *X* direction the impulse is positive on the positive particle then the *X* component of velocity changes to $\boldsymbol{v_{p2X}}$:

$$v_{p2X} = v_{p1} + i_X/m \qquad [4.1.1]$$

But for the negative primary, suffix *n*, the impulse is $-\boldsymbol{i_X}$ and so points in the same direction as velocity $\boldsymbol{v_{n1}}$. Since responses are reversed the resulting *X* velocity component changes to $\boldsymbol{v_{n2X}}$:

$$v_{n2X} = v_{n1} - i_X/m_n \qquad [4.1.2]$$

What is important to note is that, with momentum defined by arrows, both *X* momentum components point in the same direction and so add up. For the case of zero scattering this means that, for energy to be conserved as well as momentum, no changes in either can arise. The primaries have to pass through each other, first mutually annihilating and then reconstituting.

Then with scattering in the *Y* direction alone:

$$v_{pY} = i_Y/m \quad \& \quad v_{nY} = -i_Y/m_n \qquad [4.1.3]$$

Since negative primaries move in opposite direction to impulse, both deflect in the same *Y* direction.

From [2.2.4] p.91 the positive kinetic energy gain:

$$E_{K2} - E_{K1} = \tfrac{1}{2}\, m_0 (\, v_{2X}^2 + v_{py}^2 - v_{p1}^2)$$

Energy gains can now be expressed and then equated in order to comply with the revised law of energy conservation. These become after simplification:

$$E_{K2} - E_{K1} = i_x v_1 + (i_x^2 + i_y^2)/(2m) \qquad [4.1.4]$$
$$E_{Kn2} - E_{Kn1} = -i_x v_{n1} + (i_x^2 + i_y^2)/(2m_n): \qquad [4.1.5]$$

The problem has, however, been oversimplified.

The Value of i_Y for the General Case

In general collisions will not occur from primaries moving toward one another along parallel paths but will have an angle θ between 0 and 180° between their paths. Hence it is necessary to work from a frame of reference that gives the impression of collision from parallel paths. This is given by the relative velocity vector v_R, as shown at F1 in FIG.3.1 p.106 and given by setting up a velocity triangle in which v_{p1} & v_{n1} intersect at angle θ with v_R opposite θ, v_{p1} is opposite angle *D* and v_{n1} opposite angle *C*. Then v_R is given by the cosine rule:

$$v_R^2 = v_{p1}^2 + v_{n1}^2 - 2\, v_{p1}\, v_{n1} \,Cos(\theta) \qquad [4.1.6]$$

It is appropriate to choose a reference point along vector v_R that will give a consistent value of i_Y. Putting the momentum of the positive primary as $x\, m\, v_R$, then that of the negative one becomes $(1\text{-}x)m_n v_R$ and so the total *p* becomes:

$$p = ((m - m_n)x + m_n)v_R$$

Hence *p* varies linearly with *x*. Hence, for consistency it is necessary to choose a value of x that makes the momentum contribution equal for both primaries.

So to make these equal:

$$x = m_n/(m + m_n) \qquad [4.1.7]$$

And the relative momentum P_R before impact becomes:

$$P_R = 2\frac{v_R m_n m}{m + m_n} \qquad [4.1.8]$$

It will be assumed that primaries are spherical since it is difficult to treat mathematically any other shape although this is not impossible. If the angle of contact is α measured from the relative vector v_R then the interaction momentum component will be $P_R \cos(\alpha)$ and i_Y will be sin(α) times this. Since 2 cos(α) sin(α) = Sin(2α) then using [4.1.8], i_Y becomes:

$$i_Y = \frac{m_n m}{m + m_n} v_R Sin(2\alpha) \qquad [4.1.9]$$

If the v_R vector is viewed end on, as shown at F2 in FIG.3.1, then the primaries will be seen to have a scattering probability that is uniform in any direction and can have angle β as measured from the plane of the velocity triangle (the plane) with $\beta = 0$ taken pointing toward θ. There will be three mutually perpendicular components of impulse to be considered: $i_Y Cos(\beta)$ in the plane pointing towards A, i_X parallel to v_R and $i_Y sin(B)$ perpendicular to the plane.

Then i_X the impulse component parallel to v_R has to be determined so that positive energy gain ΔE is exactly balanced by negative energy gain ΔE_n based on the absolute frame of reference. With kinetic energy $E_K = ½ mv^2$ as based on only the first term of equation [2.2.4] we can write:

$2E_K/m = v^2$. Hence to yield the kinetic energy E_{K2} after collision we can write:

$$\frac{2E_{K2}}{m} = \left(v_{p1}\sin(c) + \frac{i_Y}{m}\cos(\beta)\right)^2 + \left(v_{p1}\cos(c) + \frac{i_X}{m}\right)^2 + \left(\frac{i_Y}{m}\sin(\beta)\right)^2$$

Expanding and subtracting $v_{p1}{}^2$ then yields the positive energy gain ΔE given by.

$$\frac{2\Delta E}{m} = \left(\frac{i_X}{m}\right)^2 + \left(\frac{i_Y}{m}\right)^2 + 2v_{p1}\left(\frac{i_X}{m}\cos(C) + \frac{i_Y}{m}\sin(C)\cos(\beta)\right)$$

[4.1.10]

The impulse i_X has increased $v_{p1} \cos(C)$ in equation [4.1.10] and the opposite impulse must therefore produce a velocity change that reduces $v_{n1}\cos(D)$ so that $2\Delta E_n/m_n$ can be found in a similar manner. However this is multiplied by m_n/m throughout in order to permit $2\Delta E_n/m_n$ to be subsequently equated to $2\Delta E/m$ in [4.1.10]. The negative energy gain then becomes:

$$\frac{2\Delta E_n}{m} = \frac{m}{m_n}\left[\left(\frac{i_X}{m}\right)^2 + \left(\frac{i_Y}{m}\right)^2\right] + 2v_{n1}\left(-\frac{i_X}{m}\cos(D) + \frac{i_Y}{m}\sin(D)\cos(\beta)\right)$$

[4.1.11]

Conservation of energy dictates that [4.1.10] and [4.1.11] be equated. This enables i_X to be evaluated in terms of i_Y, the latter being data input as a guess, and yields a quadratic best expressed as:

$$\left(\frac{m}{m_n} - 1\right)\left(\frac{i_X}{m}\right)^2 - 2Q\frac{i_X}{m} + C_A = 0 \quad [4.1.12]$$

Where:

$$Q = v_{n1}\cos(D) + v_{p1}\cos(C) \qquad [4.1.13]$$

And:

$$C_A = 2\big(v_{n1}\sin(D) - v_{p1}\sin(C)\big)\cos(\beta)\frac{i_Y}{m} + \left(\frac{m}{m_n} - 1\right)\left(\frac{i_Y}{m}\right)^2$$

[4.1.14]

Whose solution is:

$$\frac{i_X}{m} = \frac{Q}{(m/m_n - 1)} - \sqrt{\left(\frac{Q}{m/m_n - 1}\right)^2 - \frac{C_A}{(m/m_n - 1)}}$$

[4.1.15]

The value of i_X/m obtained from equations [4.1.12] to [4.1.15] can then be used in both [4.1.10] and [4.1.11] to obtain the energy gains when m_n is not equal to m: both the latter should give the identical result of course. Resort to finite difference computer analysis is then required to integrate over the ranges of α, β and θ in order to find what net gains occur. Values of β from 90 to 180 degrees can clearly give negative gains so a complete integration is required.

The Case of Equal Magnitudes of Mass

Since $(m/m_n - 1) = 0$ simplifies [4.1.12] and [4.1.14] the solution then becomes:

$$\frac{i_X}{m} = \frac{\left(v_{n1}\sin(D) - v_{p1}\sin(C)\right)}{\left(v_{n1}\cos(D) + v_{p1}\cos(C)\right)}\cos(\beta)\frac{i_Y}{m} \qquad [4.1.16]$$

By the sine rule we can write:

$$\sin(C) = \frac{v_{n1}}{v_R}\sin(\theta): \quad and \quad \sin(D) = \frac{v_{p1}}{v_R}\sin(\theta)$$

[4.1.17]

Substituting [4.1.17] in [4.1.16] shows the numerator is zero. Hence i_X is zero when $m_n = m$ for all values of v_{n1}/v_{p1}. The impulse is perpendicular to the relative velocity vector for all values of θ. This provides the simplification that enables an analytical solution to be formulated.

Substituting from [4.1.9] and [4.1.17] to [4.1.10] for this case yields:

$$\frac{2\Delta E}{m} = \left(\frac{v_R\sin(2\alpha)}{2}\right)^2 + v_{p1}v_{n1}\sin(\theta)\cos(\beta)\sin(\alpha)$$

[4.1.18]

Now the probability of collision between angles α and $\alpha + d\alpha$ is equal to the projected area of the ring element between these angles divided by the entire projected collision area, a disc, and is readily shown to be: $sin(2\alpha)\,d\alpha$ with α ranging between 0 and $\pi/2$ radians.

The probability of collisions between angles β and $\beta + d\beta$ is $d\beta/\pi$ if integration is restricted to the range 0 to π on grounds of symmetry (no need to integrate 0 to 2π).

Finally for collisions between angles θ to $\theta + d\theta$ it is best to think of a sphere, as shown in FIG.2.5 p.99, having a radius proportional to v_{p1}, about which vector v_{n1} can be rotated at collision angle θ to form a cone of velocity vectors. The number available for collision will be proportional to an element of the surface of that sphere equal to $2\pi\, v_{p1}\, sin(\theta)\, v_{p1}\, d\theta$ divided by the total surface of that sphere, which is $4\pi\, v_{p1}^{\ 2}$ yielding $\sin(\theta)d\theta/2$. The probability of collision will also be proportional to v_R the relative velocity of approach. So for the final integration for collision within the range of angle θ to $\theta + d\theta$ will be proportional to:

$$v_R\, sin(\theta)\, d\theta / 2$$

We will add a constant of proportionality k so that when $\boldsymbol{d\beta/\pi}$ and $\boldsymbol{\sin(2\alpha)d\alpha}$ are included, this number element $\boldsymbol{dN_A}$ becomes:

$$dN_A = k\,(d\beta/\pi)\,\sin(2\alpha)d\alpha\,(v_R\,sin(\theta)\,d\theta/2 \qquad [4.1.A19]$$

The total number N colliding can then be written:

$$N = k\int_0^{\pi}\int_0^{\pi/2}\int_0^{\pi} \frac{d\beta}{\pi}\sin(2\alpha)d\alpha\frac{v_R\sin(\theta)d\theta}{2} \qquad [4.1.19]$$

Putting $z = v_R^2$ in equation [4.1.6] p.120 and differentiating yields:

$$dz = 2\,v_{p1}\,v_{n1}\,sin(\theta)d\theta$$

Substituting in [4.1.19] and performing the first two integrals we are left with:

$$N = \frac{k}{4v_{p1}v_{n1}}\int_0^{\pi}\sqrt{z}dz = \frac{k}{6v_{p1}v_{n1}}[z]^{3/2}$$

Then with v_R given by [4.1.6] the result is:

$$N = \frac{k}{6v_1v_{n1}}\left[(v_{p1}+v_{n1})^3 - (v_{p1}-v_{n1})^3\right] \qquad [4.1.20]$$

For the energy gain the value $N\Delta E$ is given by combining [4.1.A19] with [4.1.18] and noting that the term containing $cos(\beta)d\beta/\pi$ integraters to yield zero so that:

$$\frac{N\,2\Delta E}{m} = \int_0^{\pi}\int_0^{\pi/2}\left[\frac{v_R\sin(2\alpha)}{2}\right]^2\sin(2\alpha)d\alpha\frac{v_R\sin(\theta)d\theta}{2}$$

Which integrates to yield, noting that $v_R^3\sin(\theta)d\theta = z^{3/2}dz/(2v_{p1}v_{n1})$:

$$\frac{N2\Delta E}{m} = \frac{k}{60\,v_{p1}v_{n1}}\left[(v_{p1}+v_{n1})^5 - (v_{p1}-v_{n1})^5\right] \qquad [4.1.21]$$

Then dividing [4.1.21] by [4.1.20] the required average energy gain becomes:

$$\frac{2\Delta E}{m} = \frac{1}{10}\frac{(v_{p1}+v_{n1})^5 - |(v_{p1}-v_{n1})^5|}{(v_{p1}+v_{n1})^3 - |(v_{p1}-v_{n1})^3|} \qquad [4.1.22]$$

The energy of the negative primary is just as important as that of the

positive one and so the sum of incident kinetic energy will be defined as: $E_{K1} = ½ (m\, v_{p1}^{\,2} + | m_n v_{n1}^{\,2} |)$
(but $m = m_n$ for this case). So dividing into [4.1.22] yields:
:

$$\frac{\Delta E}{E_{K1}} = \frac{1}{10} \frac{\left(1 + v_{n1}/v_{p1}\right)^5 - \left|\left(1 - v_{n1}/v_{p1}\right)^5\right|}{\left\{\left(1 + v_{n1}/v_{p1}\right)^3 - \left|\left(1 - v_{n1}/v_{p1}\right)^3\right|\right\}\left(1 + v_{n1}^{\,2}/v_{p1}^{\,2}\right)}$$

[4.1.23]

For this case of equal mass, ($m_n = m_P$,) on which equation [4.1.23] is based, a relativity basis could have been adopted, in which momenta are evaluated from relative vector $\boldsymbol{v_R}$. A much simpler derivation was found to yield the identical result. This is because the impulse given by scattering collisions is perpendicular to the direction of the relative vector v_R . When masses are unequal, so that the impulse produced is no longer in this direction owing to a finite I_X, then the relative and absolute frames yield different gains. Then, for unequal masses, only an evaluation using the absolute frame can then be considered valid.

Equation [4.1.23] shows that a maximum energy gain of 20% appears when $v_{n1} = v_{p1}$. The gain falls to a 17.7% when $v_{n1} = 3\, v_{p1}$ or 0.33 v_{p1}.. So predicted energy gains remain high over a wide range of speeds. The variation with mass ratio m_n /m_p has been evaluated by computer and some results of these calculations are presented in the following table. Again it is clear that energy gains do not vary greatly over a wide range of variables. Speeds will have a random distribution and will probably be Maxwellian but for present purposes it is reasonable to assume that an average energy gain of about 18% will occur.

TABLE 4.I OF ENERGY GAINS

$\eta = \Delta E/E_{K1}$ in columns 2 to 4, rows 2 to 6
1st column are values of v_{n1}/v_{P1}
1st row are mass ratios m_n /m_P

v_{N1}/v_{P1}	1.0	1.4	2.0
.333	.177	.161	.134
.614	.192	.179	.157
1.0	.2	.189	.169
1.63	.192	.182	.165
3.0	.177	.171	.157

Clearly despite the wide ranges of m_n /m_P and v_{n1}/v_{P1} the energy gain ratios $\Delta E/E_{K1}$ vary by a surprisingly small amount.

CHAPTER 5

EVALUATING POSSIBLE PARTIAL ANNIHILATION

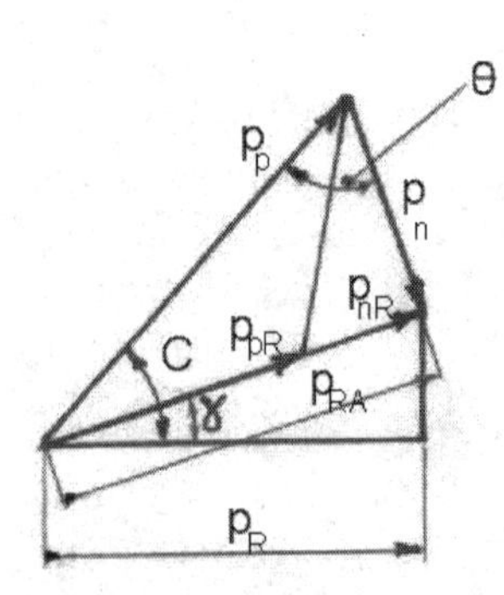

F5 Annihilation momentum triangle

In Chapters 3 and 4 breeding by collision appeared showing substantial average energy gains. The possibility exists, however, for a partial annihilation to occur that has not yet been considered. We also need to find the most probable speeds of primaries.

It is concluded that net creation always occurs on average when two primaries of opposite energy collide and that the most probable average speed is 75% of the ultimate value C_U.

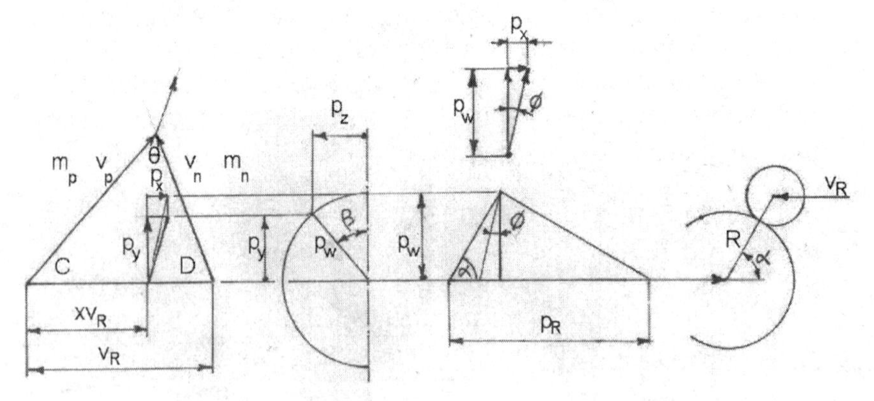

F1 velocity triangle F2 momentum disc viewed on v_R F3 scatter momentum triangle F4 primaries colliding

FIG.5.1 COLLISION BREEDING WITH PARTIAL ANNIHILATION

We recommend going straight to §5.5.2 p.138 on a first reading

5.1 Deriving a General Equation including Annihilation

The collision details were described with reference to FIG.3.1 and so only a brief reminder summary description will be given. In FIG.5.1 at F1 a velocity triangle is shown with the momentum gains due to collision superimposed shown as momentum vectors $p_Y = p_W cos(\beta)$ in the Y direction and $p_X = p_W tan(\phi)$ in the X direction. These are derived from the collision momentum triangle F3 based on the relative collision momentum vector p_R measured in the X direction and with primaries colliding with their centres at angle α to the X direction. The net collision vector is shown as in FIG.3.1 at angle ϕ to p_W. This vector then operates through the momentum disc in the Y-Z plane, shown at F2 again just as described with reference to FIG.3.1.

In addition, however, a partial annihilation momentum vector triangle is shown at F5. The absolute positive and negative momentum vectors p_P and p_n add to yield the resultant momentum vector p_{RA}. It has angle γ with respect to the X direction that is also the direction of the relative momentum vector p_R.

The p_{RA} vector is composed of components of p_P and p_n, labelled p_{PR} and p_{nR}, that are additive but clearly the components almost perpendicular to p_{RA} are opposed and could cancel. Then the collision gains need to be added to the reduced absolute vectors p_{pR} and p_{nR}.

It is first necessary to find the values of p_{PR} and p_{nR} such that energy conservation is satisfied.

5.2 Evaluating the Annihilation that could occur

For the general case the numerical values of the positive and negative inertial masses in collision will be unequal. F5 shows a momentum vector diagram for the case of partial annihilation in which the negative value is much smaller than the positive one so that p_n is smaller than p_p. Partial annihilation could occur such that, as shown at F5, the junction of the absolute momentum vectors is transferred from the apex, at angle θ, to the junction between p_{PR} and p_{nR}. The actual values of p_{PR} and p_{nR} are to be determined by the need for energies of the two collision partners to change by equal amounts in order to satisfy the conservation of energy.

The general case, as will be shown, is rather complex and can only be solved by iteration. However, for the special case of both primaries moving at the ultimate speed the solution is simple. For this reason this case will be considered first.

5.2.1 Case 1. Both primaries move at speed C_U

Since for this case $E = mC_U^2 = pC_U$ and C_U is a constant, it follows that energy change is directly proportional to change of momentum. With suffix $_L$ meaning 'loss' we can equate the loss of both positive momentum p_{PL} to that of negative momentum p_{nL} so that we can write:

$$p_{nL} = p_P - p_{nR} \text{ and } p_{nL} = p_n - p_{nR} \qquad [5.1.1]$$

It follows that: $p_{nR} = p_{nR} - p_P + p_n$ and also $P_{RA} = p_{PR} + p_{nR}$

Hence by substitution p_{nR} can be eliminated to yield:

$$p_{nR} = (p_{RA} + p_P - p_n)/2 \qquad [5.1.2]$$

So enabling the loss to be obtained by substitution in [5.1.1] since p_{RA} is given by application of the cosine rule.

Then by combining equations [5.1.1] with the probability of collision given by [2.4.6] the average loss can be calculated. Since the condition that primaries travel at the ultimate speed is a special case that will be shown later to be inapplicable, further analysis of this case will not be recorded except for quoting end results.

Even though collision gains are added a net loss is always returned. For example with $\Delta E_G/E_P$ the ratio of energy gain due to collision and $\Delta E_A/E_P$: the net loss after allowing for the annihilation, the following results appeared

m_n/m_P	$\Delta E_G/E_P$	$\Delta E_A/E_P$
0.5	0.1216	-0.03502
0.8	0.1434	-0.07757
0.99	0.1587	-0.09015

Clearly annihilation is dominant so that under these assumptions no collision breeding could occur. If the Big Breed theory is to be fully validated the condition where primaries travel at speeds other than the ultimate needs to be explored.

5.2.2 General Equation for Partial Annihilation

The positive primary has inertial mass m_P and speed v_P

The negative primary has inertial mass m_n and speed v_n

It will be assumed initially that rest energies E_{0P} and E_{0n} will not be altered by annihilation. The analysis will determine the truth of this assumption.

If a start is made by assuming a value for p_{PR}/p_P then:

$$\frac{p_{nR}}{p_P} = \frac{p_{RA}}{p_P} - \frac{p_{PR}}{p_P} \qquad [5.2.1]$$

If E_{P2} is the positive energy after annihilation, then from equation [2.2.5] we can write:

$$\left(\frac{E_{p2}}{E_{0p}}\right)^2 = \left(\frac{E_{02}}{E_{0p}}\right)^2 + \left(\frac{p_{pR}}{p_p}\frac{p_p C_U}{E_{0p}}\right)^2 \qquad [5.2.2]$$

Now:

$$\left(\frac{p_p C_U}{E_{0p}}\right)^2 = \left(\frac{m_p v_p C_U}{E_{0p}}\right)^2 = \left(\frac{E_p}{E_{0p}}\frac{v_p}{C_U}\right)^2 = \frac{(v_p/C_U)^2}{1-(v_p/C_U)^2}$$

[5.2.3]

The energy 'gain' by the positive primary then becomes

$$\frac{\Delta E_{p2}}{E_{0p}} = \sqrt{\left(\frac{E_{02}}{E_{0p}}\right)^2 + \left(\frac{p_{pR}}{p_p}\right)^2\left(\frac{p_p C_U}{E_{0p}}\right)^2} - \sqrt{\frac{1}{1-(v_p/C_U)^2}}:$$

[5.2.4]

The 'gain' will of course be negative. For the negative primary it is necessary to relate to the same positive rest energy E_{0p} in order to provide the same basis, since the negative gain and the positive gain need to be equated. For the negative gain the equation becomes:

$$\frac{\Delta E_{n2}}{E_{0p}} = \sqrt{\left(\frac{E_{0n2}}{E_{0p}}\right)^2 + \left(\frac{p_{nR}}{p_p}\right)^2\left(\frac{p_p C_U}{E_{0p}}\right)^2} - \sqrt{\left(\frac{E_{0n}}{E_{0p}}\right)^2 + \left(\frac{p_n}{p_p}\right)^2\left(\frac{p_p C_U}{E_{0p}}\right)^2} \qquad [5.2.5]$$

Then by iteration the values of p_{pR} and p_{nR} can be found such that the values of $\Delta E_{p2}/E_{0p}$ and $\Delta E_{n2}/E_{0p}$ are equal. The solution has to begin by assuming that $E_{02}/E_{0p} = 1$ and $E_{0n2}/E_{0p} = E_{0n}/E_{0p}$.

If a solution appears on this basis then no change in rest energies will occur. If none is possible then rest energies need to be reduced in balanced amounts until equations [5.2.4]. and [5.2.5] can be successfully equated. The solution to be presented from a computer study (code PrColC08) suggests that rest energies will not change.

5.3 Allowing for Collision Gains

The positive primary gain will add to p_{pR} and the negative gain to p_{nR}. The collision energy gains were provided by equations on p.106 to 109: [3.1.4], [3.1.13], [3.1.14]&[3.1.15] or the non-dimensional equivalents [3.3.1] to [3.5.2] p.110 to 112.

Also referring to FIG.5.1 and putting $p_3^2 = p_x^2 + p_y^2 + p_z^2$ The overall energy gains are then given by:

Positive Primary:

$$\frac{\Delta E_{p3}}{E_{0p}} = \sqrt{1 + \left(\frac{p_{p3}}{p_p}\right)^2 \left(\frac{p_p C_U}{E_{0p}}\right)^2} - \sqrt{1 + \left(\frac{p_p C_U}{E_{0p}}\right)^2} \qquad [5.3.3]$$

Negative Primary

$$\frac{\Delta E_{n2}}{E_{0p}} = \sqrt{\left(\frac{E_{0n}}{E_{0p}}\right)^2 + \left(\frac{p_{n3}}{p_p}\right)^2 \left(\frac{p_p C_U}{E_{0p}}\right)^2} - \sqrt{\left(\frac{E_{0n}}{E_{0p}}\right)^2 + \left(\frac{p_n}{p_p}\right)^2 \left(\frac{p_p C_U}{E_{0p}}\right)^2} \qquad [5.3.4]$$

These equations [5.3.3] and [5.3.4] assume no change in rest energies since evaluation of equations [5.2.4] and [5.2.5] showed that this assumption always holds true. Again an iterative solution is involved that makes $\Delta E_{p3} / E_{0p} = \Delta E_{n3} / E_{0p}$. This can be achieved by the correct choice of $\tan\phi$ in equations [5.3.1] and [5.3.2] where the same value for $\tan\phi$ needs to be used in both of these equations. It should be noted that $\sec^2\phi = 1 + \tan^2\phi$.

5.4 Results of Computation for collision energy gains with partial Annihilation (GWBASIC code "PrColC08")

First Case 1 is presented to give some idea of the way ϕ changes for annihilation only and ignoring collision gains. Details for computation of a particular case are presented. Energy 'gains' are presented as ratios of E_{0p} by equations [5.2.4] and [5.2.5]. However, this is not as meaningful as when presented as a ratio of total incident kinetic energy E_{KT} of the two colliding primaries. The equation for E_{KT} / E_{0p} is given by equation [3.7.1] p.115 and so the results given by [5.2.4] and [5.2.5] p.129, after iteration by choice of $\tan\phi$ to equate these and so satisfy the conservation of energy, are divided by E_{KT} / E_{0p}.

The second, Case 2, uses the same input data as for Case 1 but now presents the values of ϕ when collision gains from equations [5.3.3] and [5.3.4] are added to the annihilation values.

Then thirdly summaries of integrations covering all possible ranges of angles β, α and θ to give overall energy 'gains' are presented.

Case 1 Details for the case $v_p/C_U = 0.7$, $v_n/C_U = 0.7$ and $E_{0n}/E_{0p} = 0.5$ are presented $\Delta\theta = 20^o$

TABLE 5.4 I $E_{KT}/E_{0p} = 0.6004$ ANNIHILATION ONLY

θ deg.	γ deg.	ϕ deg.	p_R/p_{p1}	p_{pR}/p_{p1}	$\Delta E_3/E_{kT}$
10	75.29	-30.23	0.515	0.817	-0.1987
30	51.21	-21.41	0.6197	0.6840	-0.3269
50	36.55	-15.34	0.7793	0.6899	-0.3216
70	25.46	-11.34	0.9529	0.7426	-0.2690
90	18.43	-8.541	1.118	0/8164	-0.1990
110	13.14	-6.247	1.2618	0.8831	-0.1294
130	8.835	-4.113	1.3758	0.9375	-0.7030
150	5.104	-2.439	1.4547	0.9771	-0.0255
170	1.670	-0.684	1.495	0.9973	-0.0031

Case 2 As Case 1 but with collision gains added

TABLE 5.4 II ANNIHILATION + COLLISION GAINS

β	α	θ	ϕ	$\Delta E/E_K$	β	α	θ	ϕ	$\Delta E/E_K$
30	15	30	-49.2	-.3047	30	75	90	-14.57	-.1447
90	"	"	-6.84	-.3145	90	"	"	1.146	-.1472
150	"	"	44.42	-.3074	150	"	"	17.74	-.1432
30	45	"	-45.0	-.2669	30	15	150	-2.291	.0557
90	"	"	-0001	-.2940	90	"	"	2.290	.0553
150	"£	"	49.72	-.2562	150	"	"	6.843	.0559
30-	75	"	-49.2	-.3048	30	45	"	-1.146	.2712
90	"	"	-6.84	-.3145	90	"	"	5.71	.2734
150	"	"	44.42	-.3074	150	"	"	10.20	.2794
30	15	90	-14.57	-.1447	30	75	"	-2.291	.0557
90	"	"	1.146	-.1472	90	"	"	2.290	.0553
150	"	"	17.74	-.1432	150	"	"	6.843	.0559
30	45	"	-11.31	-.0066	15	7.5	15	-73.56	-.2419
90	"	"	5.71	-.0015	75	"	"	-59.97	-.2457
150	"	"	21.8	.0115	165	"	"	53.67	-.2463

In Case 2 given in TABLE 5.4 II (above) collision gains using equations [5.3.3] and [5.3.4] are given. The collision gains are added to the losses due to partial annihilation. It will be noticed, particularly at low values of θ, that ϕ varies from high negative to high positive values. In the bottom right hand corner of that table values for the ϕ variation are huge when θ is as low as 15 degrees. This contrasts sharply with the small variations of ϕ given in

Chapter 3 and recorded in TABLE 3.I p.115 for the case of collision energy gains without partial annihilation.

The assumption on which Case 2 rests is that no changes in rest energies occur. If E_0 reduced then less reduction in E_K would be needed to give a match between the positive and negative energy losses that TABLE 3.II p.116 shows will arise, at least at low values of θ. It is therefore necessary to deduce what would happen if a small reduction in rest energy occurred whilst momentum remained unchanged.

If a low v/C_U case is considered for simplicity then with suffix $_1$ representing the initial state and $_2$ the state after a change in m_0

$$p = m_0 v = \frac{E_{01}}{C_U^{\;2}} v_1 = \frac{E_{02}}{C_U^{\;2}} v_2$$

then: $v_2/v_1 = E_{01}/E_{02}$ [5.3.5]

It follows that the ratio of sum energies becomes:

$$\frac{E_{02}}{E_{01}} = \frac{E_{02}\left(1 + 0.5(v_2/C_U)^2\right)}{E_{01}\left(1 + 0.5(v_1/C_U)^2\right)}$$

Substituting from [5.3.5] results in

$$\frac{E_{02}}{E_{01}} = \frac{E_{02}/E_{01} + 0.5(E_{01}/E_{02})(v_1/C_U)^2}{\left(1 + 0.5(v_1/C_U)^2\right)} \qquad [5.3.6]$$

It is easiest to interpret this equation by putting $E_{02} = E_{01} + \Delta E$ and, using a binomial expansion with all high order terms ignored, yields:

$$\frac{E_{02}}{E_{01}} = 1 + \frac{\Delta E}{E_{01}} \frac{\left(1 - 0.5(v_1/C_U)^2\right)}{\left(1 + 0.5(v_1/C_U)^2\right)} \qquad [5.3.7]$$

It is clear from this approximate expression that if ΔE is negative, meaning that a partial reduction in rest energy has occurred, then a reduction in sum energy will also result. In consequence a balance of positive and negative energy gains can appear without requiring such huge values of tan ϕ. at low values of θ.

Case 3 Integration for overall energy gain

Now Case 3 needs to be considered to determine the overall energy gains by integration over all possible collision angles.

This requires equations [5.2.4] and [5.2.5] p.129 to be equated for any chosen value of θ to determine partial annihilation. Then, for every value of β, equations [5.3.1] and [5.3.2] need to be equated. The result is then multiplied by the collision rate equation [2.4.6] p.100 giving $d\zeta/\zeta$. so that integration can be carried out in order to deter-mine overall energy gains. Results are given in following TABLE 5.4.III

TABLE SET 5.4 III OVERALL COLLISION ENERGY GAINS WITH PARTIAL ANNIHILATION

See explanation following the tables on page 134

$\Delta\theta = 5^0$, $v_p/C_U = 0.1$, $v_n/C_U = 0.1$

E_{0n}/E_{0p}	E_{KT}/E_0	$\Delta E_{AV}/E_{KT}$		E_{0n}/E_{0p}	E_{KT}/E_0	$\Delta E_{AV}/E_{KT}$
0.4	0.007053	-0.00240		0.9		
0.6	0.008060	+0.001589		0.95	0.009824	-0.000098
0.8	0.009068	+0.000671		0.99	0.010025	-0.00015

$\Delta\theta = 5^0$, $v_p/C_U = 0.7$, $v_n/C_U = 0.7$

This case is given in more detail since, as will be shown in the next chapter, it is close to the region that seems the most probable.

E_{0n}/E_{0p}	E_{KT}/E_0	$\Delta E_{AV}/E_{KT}$		E_{0n}/E_{0p}	E_{KT}/E_0	$\Delta E_{AV}/E_{KT}$
0.2	0.4803	-0.01854		0.7	0.68048	-0.01041
0.3	0.52036	-0.01391		0.8	0.72050	-0.01142
0.4	0.56039	-0.01074		0.9	0.76053	-0.01212
0.5	0.60042			0.95	0.78055	-0.01231
0.6	0.64045	-0.009539		0.99	0.79656	-0.01237

$\Delta\theta = 5^0$, v_p/C_U & v_n/C_U as in columns 1 & 2

v_p/C_U	v_n/C_U	E_{0n}/E_{0p}	E_{KT}/E_0	$\Delta E_{AV}/E_{KT}$
0.65	0.75	0.4	0.52065	-0.01725
0.75	0.65	0.4	0.6382	-0.001177
0.65	0.75	0.6	0.6230	-0.01374
0.75	0.65	0.6	0.7014	+0.00296
0.65	0.75	0.8	0.7254	-0.00774
0.75	0.65	0.8	0.7446	+0.001129
0.65	0.75	0.95	0.8022	+0.004169
0.75	0.65	0.95	0.8120	-0.001899
0.65	0.75	0.99	0.8226	errors
0.75	0.65	0.99	0.8246	-0.002825

Errors occur when iteration fails

$\Delta\theta = 5^0$, $v_p/C_U = 0.9$, $v_n/C_U = 0.9$

E_{0n}/E_{0p}	E_{KT}/E_0	$\Delta E_{AV}/E_{KT}$		E_{0n}/E_{0p}	E_{KT}/E_0	$\Delta E_{AV}/E_{KT}$
0.4	1.812	-0.01557		0.9	2.459	-0.01847
0.6	2.071	-0.01579		0.95	2.524	-0.01863
0.8	2.329	-0.01781		0.99	2.575	-0.01870

$\Delta\theta = 5^0$, $v_p/C_U = 0.95$, $v_n/C_U = 0.95$

E_{0n}/E_{0p}	E_{KT}/E_0	$\Delta E_{AV}/E_{KT}$		E_{0n}/E_{0p}	E_{KT}/E_0	$\Delta E_{AV}/E_{KT}$
0.4	3.084	-0.01654		0.9	4.185	-0.01965
0.6	3.524	-0.01045		0.95	4.295	-0.01982
0.8	3.965	-0.01943		0.99	4.383	-0.01987

$\Delta\theta = 5^0$, $v_p/C_U = 0.99$, $v_n/C_U = 0.99$

E_{0n}/E_{0p}	E_{KT}/E_0	$\Delta E_{AV}/E_{KT}$		E_{0n}/E_{0p}	E_{KT}/E_0	$\Delta E_{AV}/E_{KT}$
0.4	8.524	-0.01628		0.9	11.57	-0.01952
0.6	9.742	-0.01722		0.95	11-87	-0.01966
0.8	10.96	-0.01897		0.99	12.12	-0.01971

Explanation of the foregoing tables

$\Delta E_{AV}/E_{KT}$ is the average of energy gain of the positive and negative primaries divided by the incident total kinetic energy of the two. Iteration should make these equal but the iteration routine is not quite perfect.

E_{KT}/E_0 can be multiplied by the above to obtain $\Delta E_{AV}/E_{-0}$

5.5 Partial Annihilation Revised

The previous analysis assumed total interpenetration of primaries during collision. This is only true, however, for collisions that are almost head-on as viewed from the relative frame. For most collisions interpenetration will be only partial. The ratio η will be defined as the volume suffering interpenetration divided by the total volume of the smaller primary concerned. The volume of interpenetration is the same for both partners, but in general the two will have different total volumes. So the two colliding primaries will have different values of η.

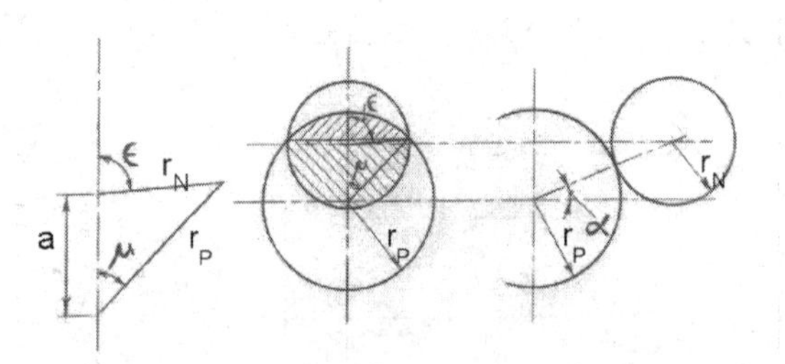

C detail of triangle shown in B

B end view hatching shows interpenetration

A side view primaries in contact

FIG. 5.5 INTERPENETRATION OF A SMALL NEGATIVE PRIMARY WITH A LARGER POSITIVE PRIMARY

It is also to be remembered from *QUANTUM GRAVITY*, where gravity as negative buoyancy force was considered, that all primaries have equal density. This means both energy density and mass density. Then primaries have a volume proportional to their inertial mass.

It will be assumed that the actual annihilation debited from any collision gain will be proportional to η multiplied by the full partial annihilation considered previously.

The cross hatched areas shown in FIG.5.5 represent the volumes of the two 'caps of spheres' that need adding to give the total interpenetration volume. Both primaries deflect in the same direction so that when the two are the same size no relative motion, in the vertical direction of FIG.5.5, will occur during the interaction. When of unequal size the smaller moves away faster than the larger since the transverse momenta induced are equal and opposite. This means $v_n = v_P m_P / m_n$. Consequently the amount of interpenetration will be reduced. This reduction is not taken into account in the following derivation.

5.5.1 Derivation of Partial Annihilation

From FIG.5.5 at **A** it is obvious that:

$$a = (r_P + r_n \sin\alpha) \qquad [5.5.1]$$

Then directly from **C** and using the cosine law:

$$\cos\mu = \frac{a^2 + r_P^2 - r_n^2}{2ar_P} \quad \text{and:}$$

$$\cos\varepsilon = \frac{a^2 + r_n^2 - r_P^2}{2ar_n} \qquad [5.5.2]$$

The volume ΔV of a cap of the negative sphere is given by:

$$\Delta V = \pi \int_0^{\varepsilon} r_n^3 \sin^3 \varepsilon \, d\varepsilon$$

Which integrates to yield, after dividing by volume V of the complete sphere, the ratio $\Delta V/V$, so that for the negative sphere:

$$\frac{\Delta V_n}{V_n} = \frac{3}{4}\left[\frac{2}{3} - \left(\cos\varepsilon - \frac{\cos^3\varepsilon}{3}\right)\right] \qquad [5.5.3]$$

Replacing $_n$ by $_P$ and ε by μ gives the cap for the positive sphere. To obtain the total volume of interpenetration both cap volumes need to be added and divided by that of the primary concerned. Then the value of the interpenetration ratio η_n for the negative primary becomes:

$$\eta_n = \Delta V_n / V_n + (\Delta V_P / V_P)(V_P / V_n) \qquad [5.5.4]$$

And for the positive primary the value η_P becomes:

$$\eta_P = \Delta V_P / V_P + (\Delta V_n / V_n)(V_n / V_P) \qquad [5.5.5]$$

Since equations [5.5.4] and [5.5.5] give different values for η if a wrong choice of ϕ has been made, it follows that final iteration will yield the correct value of ϕ needed to obtain an overall balance of energy gain.

In the following TABLE 5.5.I the integrated results for energy gain over the entire range β=0 to π, α= 0 to $\pi/2$, θ=0 to π are presented The gains by collision alone are more than cancelled by the total annihilation first derived. However, when only applied to the volume of primaries that actually interpenetrate a substantial net energy gain remains for all cases considered.

(Continued after the tables)

TABLE 5.5.I COLLISION ENERGY GAINS WITH ANNIHILATION - AS PERCENTAGES

Note that column 4 is the annihilation from §5.5 based on partial interpenetration and is the only valid assessment.

CASE 1. $\Delta\theta = 10^0$: $v_P/C_U = 0.7$: $v_n/C_U = 0.7$:

(1) E_{0n}/E_{0P} equals m_{0n}/m_{0P}	(2) E_K/E_{0P}	(3) $\Delta E_A/E_K$ Ann'n Tot	(4) $\Delta E_A/E_K$ Ann'n net	(5) $\Delta E_G/E_K$ Col'n.Gain	(6) $\Delta E_{G1}/E_K$ Cols (3)+(5)	(7) $\Delta E_{G2}/E_K$ Cols (4)+(5)
0.3	0.5204	-11.76	-3.660	10.47	-1.333	6.77
0.4	0.5604	-13.98	-4.047	12.14	-1.837	8.09
0.5	0.6004	-15.65	-4.212	13.33	-4.212	9.12
0.6	0.6404	-16.85	-4.240	14.14	-2.712	9.903
0.7	0.6805	-17.71	-4.180	14.67	-3.04	10.49
0.8	0.7205	-18.26	-4.063	14.99	-3.27	10.93
0.9	0.7605	-18.57	-3.909	15.15	-3.420	11.24
0.95	0.7805	-18.64	-3.823	15.19	-3.45	11.37
0.99	0.7966	-18.66	-3.75	15.20	-3.75	11.45

CASE 2. $\Delta\theta = 10^0$: $v_P/C_U = 0.9$: $v_n/C_U = 0.9$

0.5	1.941	-14.20	-3.825	10.40	-3.80	9.69
0.7	2.2	-16.01	-3.782	11.48	-4.537	9.49
0.9	2.459	-16.76	-3.529	11.86	-4.9823	8.895
0.99	2.575	-16.85	-3.385	11.90	-4.948	8.58

CASE 3. $\Delta\theta = 10^0$: $v_P/C_U = 0.3$: $v_n/C_U = 0.3$

0.5	0.07243	-16.57	-4.47	16.90	+0.333	12.44
0.7	0.08208	-18.81	-4.44	18.47	-0.339	14.03
0.9	0.09174	-19.77	-4.161	19.03	-0.74	14.87
0.99	0.09609	-19.86	-3.992	19.09	-0.776	15.10

CASE 4. $\Delta\theta = 10^0$: $v_P/C_U = 0.1$: $v_n/C_U = 0.1$

0.5	7.557E-3	-16.73	-4.50	17.70	+0.275	13.2
0.7	0.008564	-18.99	-4.48	19.30	+0.317	14.8
0.9	0.009572	-19.94	-4.20	15.67	-0.074	15.67
0.99	0.010025	-20.04	-4.03	19.92	-0.112	15.9

CASE 5. $\Delta\theta = 10^0$: $v_P/C_U = 0.8$: $v_n/C_U = 0.6$

0.5	0.7917	-8.80	-2.60	9.90	+1.096	7.30
0.7	0.8417	-11.22	-2.964	11.91	+0.69	8.94
0.9	0.8917	-13.09	-3.14	13.25	+0.159	10.11
0.99	0.9142	-13.82	-3.18	13.7	-0.121	10.52

CASE 6. $\Delta\theta = 10^0$: $v_P/C_U = 0.6$: $v_n/C_U = 0.8$:

E_{0n}/E_{0P} *equals* m_{0n}/m_{0P}	E_K/E_{0P}	$\Delta E_A/E_K$ *Ann'n Tot*	$\Delta E_A/E_K$ *Ann'n net*	$\Delta E_G/E_K$ *Col'n.Gain*	$\Delta E_{G1}/E_K$ *+Ann'n.Tot*	$\Delta E_{G2}/E_K$ *+Ann'n.Net*
0.5	0.5833	-17.80	-4.293	15.08	-2.72	9.078
0.7	0.7167	-16.34	-3.38	14.88	-1.46	11.5
0.9	0.85	-14.65	-2.666	14.16	-4.835	11.5
0.99	-.91	-13.95	-2.41	13.78	-0.166	11.37

CASE 7. $\Delta\theta = 10^0$: $v_P/C_U = 0.9$: $v_n/C_U = 0.1$

0.5	1.297	-0.156	-.0464	4.175	+4.02	4.13
0.99	1.299	-0.273	-0.0743	7.19	+6.92	7.12

CASE 8. $\Delta\theta = 10^0$: $v_P/C_U = 0.1$: $v_n/C_U = 0.9$:

0.5	0.6521	-0.518	-0.090	11.23	0.1071	11.14
0.99	1.286	-0.265	-0.029	7.3	7.03	7.27

(Continued from before the tables)

The table covers a wide range of mass ratios m_{0n}/m_{0P} and speed ratios v_P/C_U and v_n/C_U. The rest energy ratios E_{0n}/E_{0P} are equal to the rest mass ratios just mentioned. Energy gain ratios $\Delta E/E_{0P}$ are first calculated but are divided by E_K/E_{0P} to give ratios in the form $\Delta E/E_K$ in order to try and find gain parameters (pure numbers) that do not change very much over a wide range of conditions. E_K is of course the total incident kinetic energy of the collision partners as given by equation [3.7.1] p.115.

5.5.2 Discussion regarding this section and Cases 1 to 8

The most striking feature is provided by column 7 in each case 1 to 8 giving the most probable net energy gain ratios $\Delta E_{G2}/E_K$. Calculated as a percentage of the incident total kinetic energy of the colliding partners, the values generally range between almost 7% and nearly16%. This is quite a surprisingly small range considering the entire wide range of conditions specified. This column is the sum of columns 4 and 5, the net annihilation and collision gain.

Cases 7 and 8 cover the important speed range ratios of the statistical speed distribution given by Maxwell. This means that, for evaluating energy gains, the speed distribution is not very important. It may well depart from Maxwell's equation by a substantial amount since collision breeding is taking place. But whatever the true statistical equation the computations show that the speed distribution will have only a secondary effect on the final outcome.

Collision gains cannot annihilate because these are of the explosive type that cannot produce annihilation. This gain occurs first and then, during interpenetration some annihilation can occur. As first evaluated, annihilation is given in columns 3. In general this slightly exceeds collision gains, given in column 5 so that, as given in columns 6, a small net annihilation is returned given by $\Delta E_{G1}/E_K$.

However, as derived in section 5.5.1, total interpenetration of primaries during collision occurs only close to the condition when primaries meet head on. This factor drastically reduced the annihilation as shown in columns 4. Then a positive net energy gain is always returned as shown in columns 7.

In all cases no change in rest energy or rest mass has occurred.

5.6 Could partial annihilation of rest mass occur?

This matter has been given very careful consideration but was finally discounted. The reason is the need to conserve momentum and the matter is best understood by considering motion at low values of v/C_U.

Momentum $p = m_0 v$ but kinetic energy is $E_K = ½m_0v^2$. Substituting $v = p/m_0$ in the energy equation then results in:

$$E_K = \frac{1}{2}\frac{p^2}{m_0} \qquad [5.6.1]$$

Now the resultant momentum of the two primaries is p and remains constant since this momentum is conserved across the collision event. If rest mass reduced during any mutual annihilation process, then equation [5.6.1] shows the kinetic energy would increase. Since nature always tends to achieve the lowest possible energy state it follows that rest mass cannot reduce during collision. This argument validates the assumption on which TABLE 5.5 I p.136 is based: - that rest mass and therefore also rest energies are conserved during any collision of opposites.

5.7 THE MOST PROBABLE SPEED OF PRIMARIES

At last it has become possible to find a logical derivation for the most probable speed for these ultimate particles. A start is to be made by considering the natures of rest energy E_0 and kinetic energy E_K. A rotating dumbbell, consisting of two point objects each of zero rest mass and connected by a weightless link, would appear and behave as a single particle at rest and having finite rest mass. Each object would be moving at the ultimate speed and so would be made entirely from kinetic energy. If each of the two objects now had a rest mass $m_0/2$ and a kinetic energy $E_{SK}/2$ due to this spinning motion then the pair would have an equivalent rest energy given by $E_{E0} = E_0 + E_{SK}$.

In addition the dumbbell could have a linear kinetic energy E_K. From these considerations the equation for linear kinetic energy, derived from equation [2.2.1] p.90 can be re-written as:

$$E_K = \left(E_0 + E_{SK}\right)\left[\frac{1}{\sqrt{1 - \left(v/C_U\right)^2}} - 1\right] \qquad [5.7.1]$$

From thermodynamic considerations relating to diatomic gases it is known that linear and spinning kinetic energies equilibrate. The same can be expected of primaries and so $E_{SK}=E_K$ can be assumed. This enables [5.7.1] to be expressed in the ratio E_K/E_0 and re-arranged to give a cancellation of E_K as shown below:

$$E_K = E_K\left(\frac{E_0}{E_K} + 1\right)\left[\frac{1}{\sqrt{1 - \left(v/C_U\right)^2}} - 1\right]$$

This can be rearranged to solve for v/C_U and yields:

$$\frac{v}{C_U} = \sqrt{1 - \left(\frac{1 + E_0/E_K}{2 + E_0/E_K}\right)^2} \qquad [5.7.2]$$

From equation [5.7.2] the following table is calculated:

TABLE 5.7 I The most probable speed of primaries

E_0/E_K	0	1	2	3	4
v/C_U	0.866	0.745	0.661	0.6	0.5528

Clearly primaries can never travel at speed C_U since spin energy acts like a finite rest energy. It is now possible to narrow the search still further. With no rest energy the primary would need to exist as the dumbbell first imagined and is not a practical case since no weightless link is possible. A single object at this condition could only exist as a thin ring or cylinder of $v = C_U$ throughout and nothing else. So $v/C_U = 0.866$ is ruled out

Now the collision breeding process involves gain in kinetic energy until a critical size is reached. On splitting at the next collision a transmutation from kinetic to rest energy must occur in order to provide two primaries in the same state as their parent began with. The breeding cycle would not be completed otherwise.

So we now see that rest energy is likely to equilibrate with kinetic energies. A three way split into linear kinetic energy, spin kinetic energy and rest energy is the most probable outcome, in which each category has the same value. This makes $E_0/E_K = 1$ and so the most probable value *is* $v/C_U = 0.75$ in rounded figures. This is surprisingly close to the author's first guess of 0.7, made in 2005 and used in the previous TABLE 5.5 I case 1.

We use code PrColC09 with the new most probable speeds:

TABLE 5.7 II Net Collision Energy Gains with Partial Annihilation for a Range of Rest Mass ratios for:
$v_P/C_U = 0.75$, $v_n/C_U = 0.75$ (Date Jan 2009)

E_{0n}/E_{0P} & m_{0N}/m_{0P}	E_K/E_{0P}	$\Delta E_A/E_K$ *Ann'n.Tot* %	$\Delta E_A/E_K$ *Ann'n pet* % §5.5	$\Delta E_G/E_K$ *Col'n.Gain* %	$\Delta E_{G1}/E_K$ *+Ann'n.Tot* %	$\Delta E_{G2}/E_K$ *+Ann'n.Net* %
0.5	0.7178	-15.42	-3.68	12.706	-2.71	8.84
0.7	0.8703	-17.43	-3.84	13.99	-3.43	10.15
0.9	0.9725	-18.26	-3.88	14.45	-3.81	10.58
0.99	1.0186	-18.36	-3.67	14.50	-3.86	10.83

The last column again demonstrates that net collision energy gains have to occur, even allowing for partial annihilation, when only linear motion is considered.

5.8 CONCLUSION TO CHAPTER 5

General equations were derived for the collision breeding process with annihilation covering all possible ranges of incident conditions. The equations permitted all speeds up to the ultimate value C_U to be accommodated and with all possible ratios of negative to positive incident rest masses.

From TABLE 5.4 II p.131 mutual annihilation is shown to be very large especially at small angles of collision. In TABLE 5.4 III it is clear that, instead of the energy gains required that could make collision breeding possible, a net loss of about 2% occurs on average. This is based on the total incident kinetic energy of the collision partners and ignores the fact that, only in collisions that are purely head-on, can total interpenetration occur. On this analysis gains due to collision are more than offset by the annihilation that seems to be permitted.

However, as shown in section 5.5.1 p.135 the annihilation process can only apply to those parts of the primaries that interpenetrate during the collision interaction. When this is taken into account net overall gains appear as shown in column 7 of TABLE 5.5 I.

Results showed that speed distributions, and numerically unequal colliding masses, have only a small effect on the average collision energy gains.

Finally an attempt was made to determine the most probable speed of primaries. This assumed equilibration of rest, linear kinetic and spin energies. This was shown to yield the most probable value of $v_P/C_U = 0.75$. This gives the same value for v_n/C_U. However, in our companion volume, *QUANTUM GRAVITY via an Exact Classical Mechanics*, negative primaries need to have higher average speeds than positive ones in order to produce the gravitational force. Therefore the 0.75 value has to be regarded as an average of both kinds.

The analysis is, however, not yet complete. Energy gains due to only linear motions have been considered so far.

Could spin have negative effects to prevent energy gains from being realised? This question is considered in the next chapter.

CHAPTER 6

SPINNING MOTIONS OF PRIMARIES and the PHASE COUPLING FACTOR

The possibility of spinning motion being able to cancel the predicted growth needed to be assessed. First it is shown that when spin axes are inclined at an angle ε to the reference direction, then the angular momentum of the apparent elliptical path of every part of the rotating object is proportional to $\cos(\varepsilon)$ at every point in the orbit. The effective spin becomes $\cos(\varepsilon)$ of the true spin. Then it is shown that, taking probabilities of spin direction as well as collision probabilities into account, an additional energy gain of 8.33% of combined incident translational kinetic energy can be expected. This fully establishes the fundamental basis of the theory: that breeding by opposed energy collisions must arise

Another requirement is to establish the degree of coupling between the positive and negative phases of the mixture. The question to be answered is, "Will the inertia of the negative phase cancel that of the positive phase, so that acceleration of a large volume of the mixture can happen without any pressure gradients being required?"

The reader can bypass most of this chapter at a first reading, if found excessively difficult to comprehend, by going to page 172 giving the CONCLUSIONS for spin-gains and phase coupling.

6.1 Derivation of Angular Momentum Components for Evaluating Spin Effects

Primaries will have rotational as well as linear kinetic energies and the axes of such spin can have any direction. For analysis it is necessary to deduce the angular momentum exhibited at an angle ε measured from the true spin axis. This classical type of spin is not to be confused with the kind defined by quantum physicists that is limited to application at the quantum level of reality. At i-ther level spinning motion will be exactly the same as that at the macroscopic level with any rotational speed allowed.

FIG.6.1 looks complicated, being a composite of several views. However, we first look at view **A**, in the X,Y plane, which is a simple depiction of two spheres of opposite mass that we first

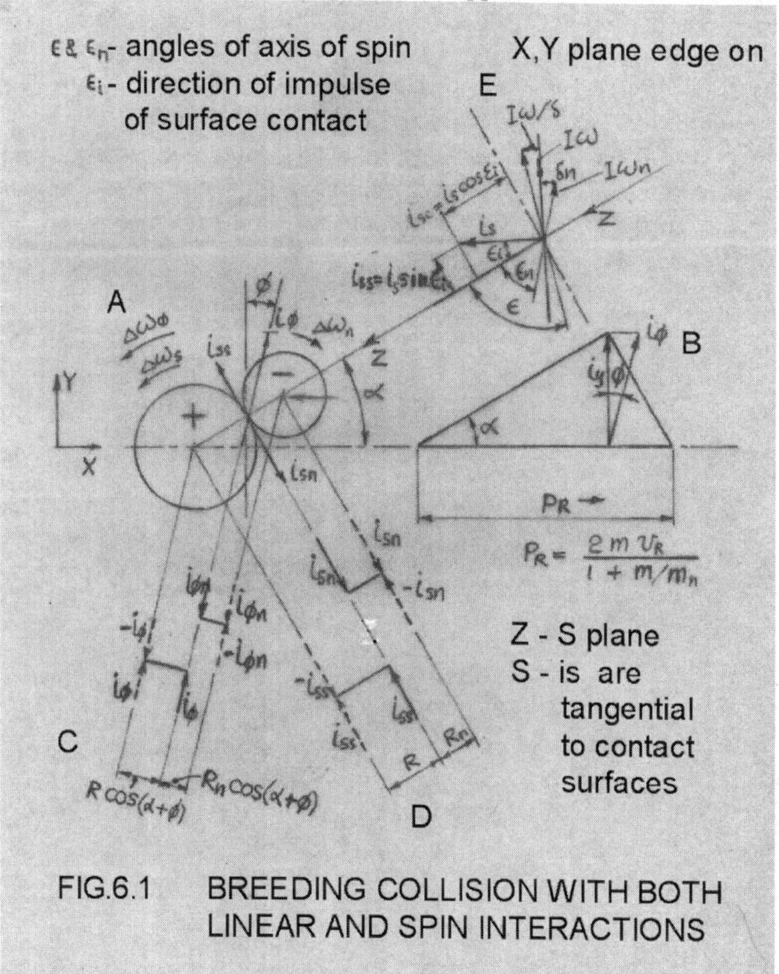

FIG.6.1 BREEDING COLLISION WITH BOTH LINEAR AND SPIN INTERACTIONS

imagine to be non-spinning. They collide at a scattering angle α, and then the line of action of the net impulse i_ϕ at angle ϕ to the Y direction will clearly not pass through the centres of those spheres. This is because, as found in previous chapters, the net impulse i_ϕ will be close to the normal of the relative momentum vector P_R as shown in the view **B** the angle ϕ not generally exceeding 10 degrees. Consequently, superimposed on linear changes of momentum, each will be subjected to an impulsive torque and will leave with a spinning motion. These torques are given by i_ϕ times the distance of the normal to the line of action of i_ϕ to the centres of each sphere as indicated at **C**. The problem is made more complex when the spheres have initial spinning motion in a random direction. Then the emerging angular momentum will be either increased or reduced. The problem of calculating the net change in the kinetic energy of spinning motion is made complicated by the incident spin axis being, in general, different from the axis of spin added by the collision. It is therefore necessary to resolve the incident moment of inertia into two components, one of which has its axis in the same direction as the spin increment produced by the net impulse. The other component of spin being normal to the one affected, will remain unchanged. It is therefore first necessary to find a way of

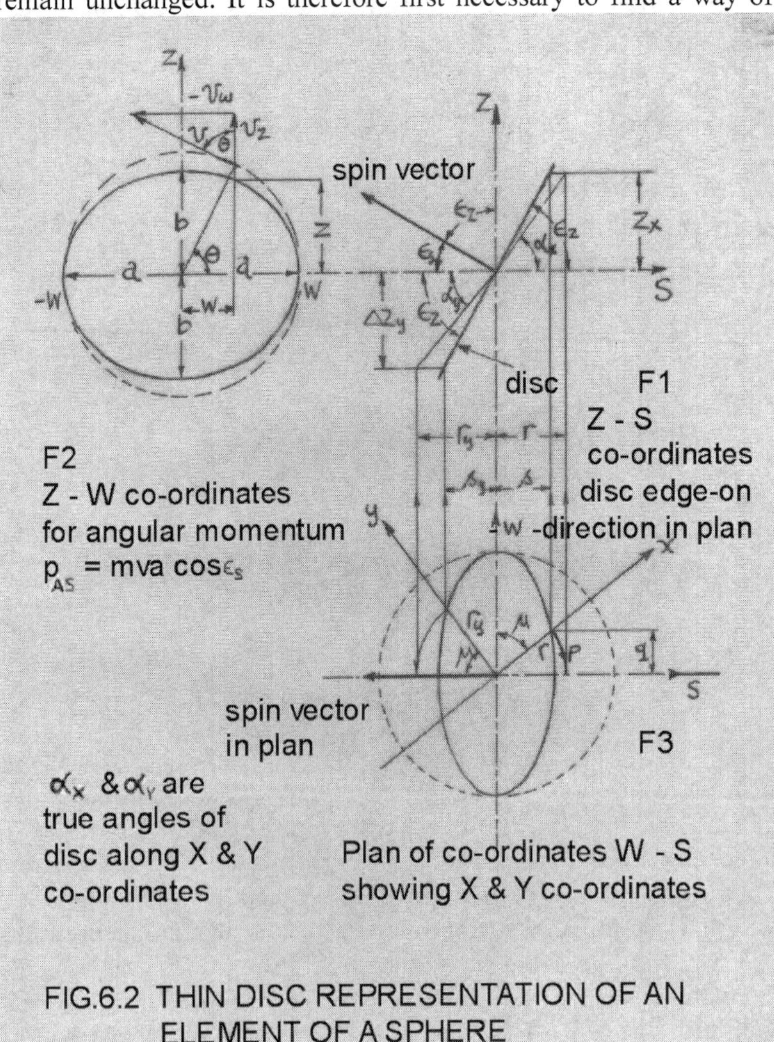

FIG.6.2 THIN DISC REPRESENTATION OF AN ELEMENT OF A SPHERE

resolving spinning motion into two mutually perpendicular spin components. This requires the evaluation of two mutually perpendicular components of moments of inertia. No gyroscopic motions will be induced since no gyroscopic torque is involved.

6.2 Resolving a Moment of Inertia into Two Components

The moment of inertia of a thin disc about an axis not normal to the plane of the disc is required first. Then by integration the moment of inertia of a sphere should be possible. In FIG.6.1 at E the Z direction is shown as being along the line joining the centres of the two spheres and direction S lies in the X,Y plane, being tangential to the contact surfaces of the spheres. In FIG.6.2 the same definitions apply and a disc is shown edge-on at F1 represented in the Z-S plane and a plan view along the Z direction is shown at F3. Direction W is perpendicular to the Z-S plane so F3 is the W-S plane and the X,Y co-ordinates are superimposed with the *X*-axis at angle μ to the W-axis. The reason for this will appear later when the direction of all three orthogonal spin vector components is to be derived. First the elevation at F2 will be considered which depicts the W-Z plane. This allows the angular momentum viewed along the S-axis to be investigated.

A mass m is to be considered spinning at a radius R about an axis at an angle ε_S to the viewing direction S so that the trajectory appears as an ellipse of semi major axis $a = R$ and minor semi axis b where $b = a\cos(\varepsilon_S)$. The semi major axis lies in the W direction (horizontal) with the minor axis in the Z direction. The tangential velocity of the mass is v at radius R representing true motion in a "base circle", shown in dashed lines and perpendicular to the spin axis. The dimensions W and Z are measured from the central point of the base circle or the ellipse.

When the mass has moved anticlockwise (initially in the Z direction from the S-axis) by angle θ it will have W and Z velocity components for the base circle of:

$$v_W = -v\sin(\theta) \quad and \quad v_Z = v\cos(\theta)$$

As viewed at angle ε_S, the W component will remain unchanged but the Z component is reduced so that:

$$v_W = -v\sin(\theta) \quad and \quad v_Z = v\cos(\theta)\cos(\varepsilon_S) \qquad [6.2.1]$$

The values of w and z, as viewed from angle ε_s for this position are:

$$w = a\cos(\theta) \quad and \quad z = a\sin(\theta)\cos(\varepsilon_s) \qquad [6.2.2]$$

And the perceived angular momentum P_ω for any value of θ, is the sum of two components written as:

$$P_\omega = m(wv_Z - zv_W)$$

Substituting from [6.2.1] and [6.2.2] gives the result:

$$P_\omega = m a\cos(\theta) v\cos(\theta)\cos(\varepsilon_S) - a\sin(\theta)\cos(\varepsilon_S)(-v\sin(\theta))$$

Which simplifies, with R replacing a, to:

$$P_\omega = m v R\cos(\varepsilon_S)\left(\cos^2(\theta) + \sin^2(\theta)\right) = m R v\cos(\varepsilon_S)$$

[6.2.3]

This shows that the angular momentum perceived at angle ε_S to the axis of rotation does not change as θ varies: where the radius is small the speed is high and vice-versa.

For this example the moment of inertia I_ω (not to be confused with impulse i) is given by mR^2 and the angular velocity ω is defined as v/R. Hence [6.2.3] can be written in the alternative form:

$$P_\omega = m R^2 \frac{v}{R}\cos(\varepsilon_s) = I_\omega \omega\cos(\varepsilon_s) \qquad [6.2.4]$$

In general for objects of arbitrary shape the moment of inertia is given by: $I_\omega = m k^2$ where k is the, "radius of gyration".

For solid spheres of radius R the value of k^2 is 3/5 so

$$I_\omega = (3/5)mR^2. \qquad [6.2.5]$$

Unfortunately this is only correct for low values of v/C_U where the mass increase due to speed is small enough to be ignored. Since the peripheral speed increases from zero at the centre to a maximum at the equator, the mass increase is a function of radius. Integration has to resort to a series solution and, putting $\gamma = v/C_U$, this is readily shown to be:

$$.I_\omega = \frac{3}{5} m_0 R^2 \left(\begin{array}{l} 1 + \frac{5}{14}\gamma^2 + \frac{15}{72}\gamma^4 + \frac{25}{176}\gamma^6 + \frac{175}{1664}\gamma^8 + \\ \frac{21}{256}\gamma^{10} + 0.0664\gamma^{12} + 0.0551\gamma^{14} + \ldots \end{array} \right)$$

[6.2.6]

The number of terms indicated is sufficient for v/C_U up to about 0.75. In this case for a linearly moving object $m/m_0 = 1.5119$ but a rotating sphere for the same equatorial speed has $m/m_0 = 1.3104$ by way of comparison (from the bracketed part of [6.2.6]).

6.3 Vector Components of Spinning Motion

This section will be required when a computer code is devised. This needs to evaluate the collisions occurring between about 300 primaries needed to demonstrate the spontaneous organisation of the i-ther. However, spin vectors need to be applied in the remainder of this chapter. They will be defined as follows.

The "right-hand screw" convention will be adopted in which positive rotation of an object is assumed as clockwise when viewed in the direction of linear motion. So for an object moving in the *Z* direction, which is upwards as viewed looking down on the *X,Y* plane, positive spin will be anticlockwise.

Again the convention is to define the angular momentum *mvr* of a spinning object by representing this quantity to some scale by the length of an arrow lying along the axis of spin. This is the "angular momentum vector" or "spin vector" for brevity. The rim of a wheel would then move at right angles to the spin vector. When linear motion is absent positive spin is motion in a clockwise direction as observed when looking in the direction the spin arrow is pointing.

For positive mass both the angular velocity and angular momentum will be positive when rotation is clockwise looking in the direction of the arrow. If linear motion is also in the direction of this arrow then this also points in the direction of the object's linear momentum.

For negative mass the spin arrow again defines positive rotation if motion is clockwise as viewed in the direction of the arrow but now the angular momentum has an anticlockwise direction. If the spin arrow also points in the direction of linear motion then the linear momentum of the object points opposite linear motion..

The rim velocity v_S of a spinning disc of radius R will be ωR looking along the spin vector. It is important to prove that spinning motion can have components just as does linear momentum.
The following derivation can be omitted on a first reading, so it may be advisable to pass direct to the next section.

On the basis defined, motion in both the ellipses shown in Fig.6.2 will be anticlockwise. Now F3 is to be inspected. The *x*-axis cuts the rim of the disc at *p* and the *S* and *W* co-ordinates of this point can be found from the equations for the ellipse and the straight line and these are:

$$\left(\frac{w}{a}\right)^2 + \left(\frac{s}{b}\right)^2 = 1 : \quad \frac{b}{a} = \cos(\varepsilon_z) \qquad [6.3.1]$$

and $$s = w\tan(\mu) \qquad [6.3.2]$$

Substituting for s from [6.3.2] in [6.3.1] yields;

$$\left(\frac{w}{a}\right)^2\left(1 + \frac{\tan^2(\mu)}{\cos^2(\varepsilon_z)}\right) = 1 \quad \textit{which can be re-arranged as:}$$

$$\left(\frac{w}{a}\right)^2 = \frac{\cos^2(\varepsilon_z)}{\cos^2(\varepsilon_z) + \tan^2(\mu)}$$

[6.3.3]

And then substituting from [6.3.2] in the above:

$$\left(\frac{s}{a}\right)^2 = \frac{\cos^2(\varepsilon_z)\tan^2(\mu)}{\cos^2(\varepsilon_z) + \tan^2(\mu)} \quad [6.3.4]$$

Since $r^2 = s^2 + w^2$ and $1 + \tan^2(\mu) = \sec^2(\mu)$ it follows that the above combine to yield:

$$\left(\frac{r}{a}\right)^2 = \frac{\cos^2(\varepsilon_z)\sec^2(\mu)}{\cos^2(\varepsilon_z) + \tan^2(\mu)} \quad [6.3.5]$$

In the figure at F3 projections of both *S* and *r* are made to F1 and *S* intersects the disc at *p*. This gives the value of Δz_x the height of the intersection. Then projecting in the *S* direction to meet the *r* projection yields the true slope of the surface of intersection of the *X*-axis with the disc making angle α_x with the *x* co-ordinate. Now:

$$\frac{\Delta z_x}{s} = \tan(\varepsilon_z) \quad \textit{and} \quad \frac{\Delta z_x}{r} = \tan(\alpha_x) \qquad \text{Consequently:}$$

$$\tan(\alpha_x) = \frac{s}{a}\frac{a}{r}\tan(\varepsilon_z)$$

Substituting for *s/a* and *r/a* from [6.3.4] and [6.3.5] reduces to:

$$\tan(\alpha_x) = \tan(\mu)\cos(\mu)\tan(\varepsilon_z)$$

But the spin vector is at 90 degrees to the plane of the disc so $\varepsilon_x = 90^o + \alpha_x$ and $\tan(\varepsilon_x - 90^o) = -\cot(\varepsilon_x)$.
Consequently the solution is:

$$\tan(\varepsilon_x) = -\frac{1}{\sin(\mu)\tan(\varepsilon_z)} \quad [6.3.6]$$

By similar reasoning it is readily shown that:

$$\tan(\varepsilon_y) = \frac{1}{\cos(\mu)\tan(\varepsilon_z)} \quad [6.3.7]$$

Since all four quadrants in the *X,Y* plane are available, two positive and two negative, these expressions each yield two solutions and so they need careful checks to ensure the correct solution has been selected. However, [6.3.6] and [6.3.7] permit the three orthogonal directions of angular momentum vectors (spin vectors) to be determined.

With this derivation completed attention can now be given to the effect on energy gains, caused by the collision of primaries of opposite energy signs, with spin effects taken into account.

6.4 The Spinning Motion of Primaries and Effect on Energy Gains

Collisions will induce a spinning motion and collisions of primaries of the same energy sign will also tend to induce equilibrium between linear and rotational kinetic energies. The question to be answered is this, "Could spinning motion induce energy annihilation that could cancel all the gains due to linear motion?" If this were to be predicted then the entire creation theory would be invalidated.

In the following analysis the simplification is made that the centres of all primaries move at the same linear speed v, with

$m_p = |m_n|$ and equilibrium between spinning and linear kinetic energies is achieved. It then follows that:

$$\tfrac{1}{2}I\omega^2 = \tfrac{1}{2}mv^2 \qquad [6.4.1]$$

Angular velocity is ω radians/s and R is the radius of the spherical primary. It follows that:

$$\omega = \sqrt{\frac{5}{3}\frac{v}{R}} \qquad [6.4.2]$$

Where v = the linear velocity measured at the centre of any primary: not its surface speed.

When hard spheres, both of positive mass, collide and have a spinning motion, a tangential force induced by friction will add to the normal impact force. The resultant of these has a different direction from the normal force. So spinning motion affects the resulting trajectories following impact. However, as far as linear motion is concerned, this effect is only the same as collision at a different scattering angle and so linear momentum is conserved without reference to spin. Angular momentum is also conserved even though the spinning motions of the two spheres are both altered by frictional forces.

It follows that a similar weak coupling will occur between spinning and linear motions when opposite primaries collide so that it is justified to ignore this coupling in the first analysis, followed later by any correction that is found necessary. A start will therefore be made assuming changes of angular momentum caused by collision, do not affect the energy gains previously derived in Chapter 5 for linear motion alone. To simplify the analysis it will be assumed that $m_n = m_p = m$ and $v_n = v_p = v$. In this case the momentum gain was found to be transverse to the relative velocity vector v_R –the X direction. Furthermore this relative velocity reduces to:

$$v_R = 2v\sin(\theta/2)$$

So that P_R now becomes::

$$P_R = m(2v\sin(\theta/2))\cos(\alpha)\sin(\alpha) \qquad [6.4.3]$$

This will act at an effective radius of $R\cos(\alpha)$ and so an impulsive couple C_I will act on the primary to produce a change in angular velocity $\Delta\omega$ so that by multiplying by the torque arm $R\cos(\alpha)$ the impulsive couple becomes:

$$C_I = I_\omega \Delta\omega = mvR\cos^2(\alpha)\sin(\alpha)2\sin(\theta/2)$$

With I_ω substituted from [6.2.5] the value of $\Delta\omega$ becomes:

$$\Delta\omega = \frac{10}{3}\frac{v}{R}\cos^2(\alpha)\sin(\alpha)\sin\left(\frac{\theta}{2}\right) \qquad [6.4.4]$$

This can predict quite large values. For example, with ω from [6.4.2], $\theta = \pi$ and $\alpha = 40^0$, the ratio $\Delta\omega/\omega = 0.9739$.

However, primaries will be spinning with equal probability in all directions prior to collision and so the increment of angular velocity imposed will be added to an existing angular velocity measured in the Z direction, i.e. perpendicular to the X, W plane in which the momentum gain vector P_R is acting. The effective angular momentum as viewed in the Z direction is given by [6.4.4].

Vectors pointing along the axis of rotation represent spins. To represent uniform probability distribution in all directions of the vector ω a hollow sphere can be imagined whose radius is proportional to ω with a uniform surface density to represent uniformity of probability. From [6.4.4] the angular momentum component along the Z-axis is $I_\omega\omega\cos(\varepsilon)$ where ε is the angle made by vector ω with the z direction.

The probability of spins lying between angles ε and $\varepsilon + \delta\varepsilon$ will then be the number δn out of total number n given by considering an element of the surface of area $2\pi\omega.\sin(\varepsilon)\omega\delta\varepsilon$ out of total area $4\pi\omega^2$ and becomes:

$$\frac{\delta n}{n} = \frac{\sin(\varepsilon)}{2}\delta\varepsilon \qquad [6.4.5]$$

With the increment $\Delta\omega$ added by collision to $\omega\cos(\varepsilon)$ the spin kinetic energy gain becomes;

$$\Delta E_\omega = (I_\omega/2)\left[(\omega\cos(\varepsilon)+\Delta\omega)^2 - (\omega\cos(\varepsilon))^2\right]$$

Then simplifying and multiplying by probability $\delta n/n$ the result is:

$$\Delta E_{\omega}\frac{\delta n}{n} = \frac{I_{\omega}}{2}\left[2\omega\cos(\varepsilon)\Delta\omega + \Delta\omega^{2}\right]\frac{\sin(\varepsilon)\delta\varepsilon}{2} \qquad [6.4.6]$$

Since ε can vary from 0 to π the incident angular velocity will be as often negative as positive. Clearly [6.4.6] shows that when ε exceeds 90^0 the initial spin could result in a loss instead of an energy gain and so the aim is to determine the net result. The overall probability of collision η is compounded of the probability of collisions of primaries between angles
α and $\alpha + \delta\alpha$, ε *to* $\varepsilon + \delta\varepsilon$ and θ to $\theta + \delta\theta$.. In the latter case collision probability is proportional to $v_R = 2\,v\sin(\theta/2)$ out of the ratio of the number available to the total possible that approach within angle $\delta\theta$, which is $\sin(\theta)\delta\theta/2$. So the overall flux η can be written:

$$\eta = v\int_0^{\pi}\int_0^{\pi}\int_0^{\pi/2}\sin(2\alpha)d\alpha\frac{\sin(\varepsilon)d\varepsilon}{2}\sin\left(\frac{\theta}{2}\right)\sin(\theta)d\theta \qquad [6.4.7]$$

Which integrates to yield (4/3) v. Then the spin energy gain flux, after substituting from [6.4.2] and [6.4.4] into [6.4.6] and including [6.4.7] becomes:

$$Put \quad fn(\varepsilon,\theta) = \left(\frac{\cos(\varepsilon)\sin(\varepsilon)d\varepsilon}{2}\right)\left(\sin^{2}\left(\frac{\theta}{2}\right)\sin(\theta)d\theta\right)$$

and:

$$put \quad fn2(\varepsilon,\theta) = \left(\frac{\sin(\varepsilon)d\varepsilon}{2}\right)\left(\sin^{3}\left(\frac{\theta}{2}\right)\sin(\theta)d\theta\right)$$

Then the solution is:

$$\Delta E_{\omega}\eta = \frac{I_{\omega}}{2}\left(\frac{v}{R}\right)^{2} v\int_0^{\pi}\int_0^{\pi}\int_0^{\pi/2}\frac{20}{3}\sqrt{\frac{5}{3}}\left(\cos^{2}(\alpha)\sin(\alpha)\sin(2\alpha)d\alpha\right)fn(\varepsilon,\theta)$$

$$+ \frac{I_{\omega}}{2}\left(\frac{v}{R}\right)^{2} v\int_0^{\pi}\int_0^{\pi}\int_0^{\pi/2}\frac{100}{9}\left(\cos^{4}(\alpha)\sin^{2}(\alpha)\sin(2\alpha)d\alpha\right)fn2(\varepsilon,\theta)$$

[6.4.8]

The first term contains $\cos(\varepsilon)\sin(\varepsilon)d\varepsilon$ and this integrates to zero. Only the second term yields a finite gain in which the bracketed sections integrate to yield respectively (1/12)(1/2)(4/5). Since [6.4.7] integrated to (4/3)v then dividing the latter into the solution of [6.4.8] yields the result:

$$\Delta E_{\omega} = \frac{1}{2} I_{\omega} \left(\frac{v}{R}\right)^2 v \frac{100}{9} \frac{1}{12} \frac{1}{2} \frac{4}{5} \frac{3}{4v} = \frac{5}{36} I_{\omega} \left(\frac{v}{R}\right)^2$$

But $I_{\omega} = m\frac{3}{5}R^2$ so $\Delta E_{\omega} = \frac{1}{12}mv^2$

And mv^2 corresponds with incident kinetic energy

So a net gain is predicted due to the increase of energy in spinning motion that adds to the 0.2 mv^2 from linear motion to yield a total value of 0.2833 times the incident kinetic energy defined as $\frac{1}{2}(m_p + |m_n|)v^2$ and $m_p = |m_n|$ was an initial assumption. The implication of the term containing the incident spin (the first term) integrating to zero suggests that the average spin gains are independent of average rates of spin. The average energy gains by spin are, however, predicted to be less than half the gains due to linear motion. Since there will be only one collision between primaries of like kinds for every breeding collision and since the former alone tend to produce equilibrium between translational and spin energies, it follows that equilibrium will never be attained. Energies locked up in spin will always lag behind those of translation. The average spin energy gain becomes:

$\Delta E_{\omega} = 0.0833mv^2$

The result obtained is only strictly correct for spinning energy gain if both collision partners have equal but opposite $\omega\cos(\varepsilon)$ values. This must be so since as shown by [6.4.6] the energy gain for any particular collision varies according to $\omega\cos(\varepsilon)$ and the two primaries will in general have different values of ε. Hence in general for individual breeding collisions the energy gains will have different numerical values for the two collision partners. Energy conservation can only be restored within each individual collision by an equal but opposite imbalance in the energy changes caused by linear motion.

It has already been pointed out that tangential frictional forces will act so that the resultant need not be in the direction assumed, in this case perpendicular to the relative velocity vector v_R. As shown with reference to FIG.6.1 if the net impulse is directed at angle ϕ to this perpendicular direction, so as to provide an extra impulse component in the X direction, then positive gains will be increased and negative gains reduced. It therefore follows that interaction between spin and linear motion must occur, in general, to provide overall conservation of energy within individual collisions.

If ϕ is positive (clockwise see Fig.16.3) then the net impulse will increase from I to $I\sec(\phi)$ with $I = P_R$ from [6.4.3] but the effective moment arm for creating the couple C_I is reduced to $R(\cos(\alpha+\phi)$. It follows that the couple becomes:

$$C_I = P_R R\sec(\phi)\cos(\alpha+\phi) = P_R R(\cos(\alpha) - \sin(\alpha)\tan(\phi)) \quad [6.4.9]$$

Clearly a positive ϕ value will reduce the couple and consequently the value of spin energy gain from that previously derived. This occurs together with the increase of positive energy gain from linear motion and reduction in corresponding negative gain. By this logic further refinement could be introduced but this will not be considered since it appears unnecessary.

It is necessary to check the possibility that spin energy might be lost during the interpenetration that must occur during collision, with consequent partial mutual annihilation. However, a little analysis shows this to be impossible. Only the X and W components of angular momentum could be involved but if their spin energy was lost, during interpenetration and not subsequently restored during separation, then angular momentum would be disconserved. It can therefore be concluded that interpenetration cannot result in permanent loss of spin energy. So the question initially posed has been answered: spinning motion must provide net addition and cannot cancel the energy gains derived from linear motion.

This section was inserted after the majority of this PART II had been written. It will be shown that provided net energy gains do occur of even only one or two percent, then overall conclusions remain unchanged owing to the corrective effect of the annihilation process to be considered in the next Chapter. The reader may by now have realised that what might be an important factor has been ignored: that although positive and negative energy gains balance overall they will not be likely to balance when incident spins are in the same sense. The next section assumes every individual collision of opposites is restricted to energy gains or losses in exact balance.

6.5 Translation and Spinning Motion Combined 7/7/06

So far linear and spin energy gains have been treated in isolation but it could be that their interaction produces effects different from those given by simple addition. Furthermore the positive and negative spin energy changes were considered separately to render the problem integrable. In this section individual collisions will be studied in detail and imbalances in spin energy changes countered by matching imbalances in linear energy changes. This implies the adjustment of ϕ to achieve an overall balance in energy changes form collision. Furthermore their could by extra spin and linear changes caused by transient tangential forces, such as frictional ones, during impact. The derivation will also make provision for tangential impulse. In the previous section only a special case was considered in which $m_n = m_p$ and $v_n = v_p$. In general these conditions will not apply and so the general case for motion where $v \ll C_U$ will be considered.

In this section the worst cases are to be evaluated. Section 6.3 derived expressions for the directions of the three orthogonal spin vector component directions. For present purposes these were subsequently found to be too general. The worst case occurs when the positive and negative spin vectors, prior to collision, lie in parallel planes with the vectors pointing in almost opposite directions. So the derivation is to be limited to this special case.
A return to FIG.6.1 illustrates the problem. As a first impression it appears very complex but after a more detailed study the reader should find it adequately comprehensible. The aim was to determine worst case effects and so spin vectors are assumed limited to planes parallel to the tangent of the contact points of the two primaries. These planes are at the angle $\pi/2 - \alpha$ to the X direction. Either spin vector, at angles ε or ε_n measured from the z direction, can vary over the range 0 to π.

At A two opposite energy primaries are shown at the instant of contact and at B the linear momentum vector diagram is shown. Since previous studies have shown that when $v \ll C_U$ any frame of reference is equally valid, the study will be confined to the relative frame in which v_R is the relative velocity between the collision partners. So both A and B are shown in the X,Y plane (since no W plane is required as had to be introduced when absolute frames of reference were adopted). Primaries collide at angle α measured from the X direction and energy gains from this relative frame of reference will be equal for all asimuthal positions β around the v_R vector. The net impulse i_ϕ imparted by collision with reference to the positive primary (the larger one) is at angle ϕ measured from the y direction in the sense of the x direction as illustrated at both A and B.
From the latter the value of i_ϕ becomes:

$$i_\phi = P_R \cos\alpha \sin\alpha \sec\phi \qquad [6.5.1]$$

Where:

$$P_R = \frac{2 m_p v_R}{1 + m_p / m_n} \qquad [6.5.2]$$

Transferred to A it is clear that this impulse will produce both linear and angular momentum changes for which projections at C help to clarify. Impulse i_ϕ acts at radius $R\cos(\alpha+\phi)$ to produce a torque as illustrated by adding two extra forces i_ϕ and $-i_\phi$ at the centre of gravity of the positive primary so that they cancel and so do not affect the problem. However, this shows that a couple is produced of magnitude $i_\phi R\cos(\alpha+\phi)$ together with a pure impulse i_ϕ acting through the centre of gravity. The latter produce velocity increments in the x and y directions. All velocities are best expressed non-dimensionally using v, the absolute velocity of the positive primary as a reference velocity. Then incorporating [6.5.1 & 2] these velocity components are given by:

$$\frac{\Delta v_x}{v} = 2\frac{v_R}{v}\frac{\cos\alpha\ \sin\alpha\tan\phi}{1 + m_p/m_n} \quad \& \quad \frac{\Delta v_y}{v} = 2\frac{v_R}{v}\frac{\cos\alpha\ \sin\alpha}{1 + m_p/m_n}$$

[6.5.3]

Corresponding velocity increments for the negative primary will be m_p/m_n times these acting in the same directions.

Now for the torque components

With $I_\omega = 0.6mR^2$ for the positive spherical primary the torque impulse T_i results in an angular velocity increment in the anticlockwise direction of $\Delta\omega_\phi$ given by:

$$T_i = i_\phi R\cos(\alpha+\phi) = 0.6m_p R^2\Delta\omega_\phi$$

Again re-arranging and expressing in non=dimensional form:

$$\frac{R\Delta\omega_\phi}{v} = \frac{i_\phi\cos(\alpha+\phi)}{0.6m_p v} = \frac{2}{1+m_p/m_n}\frac{v_R}{v}\frac{\cos\alpha\sin\alpha\cos(\alpha+\phi)}{0.6\cos\phi}$$

And this can be expanded to yield:

$$\frac{R\Delta\omega_\phi}{v} = \frac{10}{3}\frac{1}{1+m_p/m_n}\frac{v_R}{v}\left(\cos^2\alpha\ \sin\alpha - \cos\alpha\ \sin^2\alpha\ \tan\phi\right)$$

[6.5.4]

And $R_n\Delta\omega_n/v = (m_p/m_n)\times \textit{equation}\ [6.5.4]$ with this angular motion increment made in the clockwise direction so the two spheres tend to rotate together like a pair of rolls.

Additional linear and angular momentum additions by 'scuffing'

Another set of linear and angular motion increments arises from interpenetration of the surfaces in contact that could arise from some form of frictional interaction. This is to be analysed using the *z-s* plane shown at E in FIG.6.1. The *Z*-axis is normal to the X,Y plane and so points upwards out of the paper. It also passes through the contact point between the two spheres, whose centres are in the *X,Y* plane. The *S*-axis also passes through the contact point and is tangential to the surfaces in contact. It follows that the *X,Y* plane also passes through this point to appear edge on as a single line. The spin vectors are projected onto the Z,S plane since they do not actually lie in that plane.

The axis of spin of the positive primary is at angle ε measured from the Z axis and the length of the spin vector is to be assumed proportional to the angular momentum $I\omega$. The negative primary is at angle ε_n and is similarly represented. If the two surfaces in contact could accommodate so that no relative motion existed between them, then the resulting total angular momentum of the pair needs to equal the sum if the initial angular momenta. Since surface motion is perpendicular to the direction of the angular momentum vector, the 'scuffing line', generating scuffing impulse i_S can then be found.

For spherical objects of surface speed v_S at their equators and noting that $v_S = \omega R$ we can write:

$$I\,\omega = 0.6mR^2\omega = 0.6m\,R^2(v_s/R) = 0.6\;m\,R\,v_s$$

The angular momentum can be expressed as two mutually perpendicular components in the Z and W directions. In the Z direction the combined angular momentum at angle ε_{si} from the Z axis will be:

$$0.6(m_pR_p + m_nR_n)v_{si}\sin\varepsilon_{si} = 0.6(m_pR_p\,v_{sp}\sin\varepsilon_p + m_nR_nv_{sn}\sin\varepsilon_n)$$

It follows that:

$$v_{si}\sin\varepsilon_{si} = \frac{(m_p\,R_p\,v_{sp}\sin\varepsilon_p + m_n\,R_n\,v_{sn}\sin\varepsilon_n)}{(m_p\,R_p + m_n\,R_n)} \qquad [6.5.5]$$

Similarly for the component in the w direction:

$$v_{si}\cos\varepsilon_{si} = \frac{(m_p\,R_p\,v_{sp}\cos\varepsilon_p + m_n\,R_n\,v_{sn}\cos\varepsilon_n)}{(m_p\,R_p + m_n\,R_n)} \qquad [6.5.6]$$

Then since v_{si} is the same in each case the value of ε_{si} is given by:

$$\varepsilon_{si} = \tan^{-1}(\sin\varepsilon_{si}/\cos\varepsilon_{si}) \qquad [6.5.7]$$

The scrubbing line will be perpendicular to this "accommodation" vector and so its angle ε_i, measured from the z direction will be:

$$\varepsilon_i = \varepsilon_{si} + \pi/2 \qquad [6.5.8]$$

With both angular momenta rotating clockwise angular velocities are opposed and this makes the surface speeds after accommodation have the same direction.

The mass density of all primaries will be assumed a universal constant in which case:

$$\frac{R_n}{R_p} = \left(\frac{m_p}{m_n}\right)^{\frac{1}{3}} \qquad [6.5.9]]$$

Then since primaries have to exist in a 2: 1 mass range to allow for the gain during collision breeding, followed by fission to create new primaries, primaries must exist in a 1.26 : 1 radius ratio according to [6.5.9].

Angular velocity change due to scuffing

Now the changes in angular velocity from the scuffing vector need to be considered. Here the projections in diagram D need to be used that add equal and opposing i_s and i_{sn} impulses to yield the impulsive torques T_{is} and T_{isn} . From the previous method yielding [6.5.4] the scuffing values can be written:

$$\frac{R\Delta\omega_s}{v} = \frac{5}{3}\frac{i_s}{m_p v}\sin\varepsilon_i :$$

$$\& \quad \frac{R_n\,\Delta\omega_{sn}}{v} = \frac{5}{3}\frac{i_{sn}}{m_p v}\sin\varepsilon_i\,\frac{m_p}{m_n} \qquad [6.5.10]$$

Of course $i_s = -i_{sn}$ but operating on negative mass yields the same direction of surface rotational speed increment, meaning that the opposite sense of rotational speed change is caused.

It is now necessary to add the increments of both linear and rotational components. The linear component adds [6.5.3] to [6.5.9] to yield for the positive mass:

$$\frac{\Delta v_x}{v} = \frac{2v_R}{v}\frac{\cos\alpha\,\sin\alpha\,\tan\phi}{1 + m_p/m_n} - \frac{i_s}{mv}\sin\varepsilon_i\,\sin\alpha$$

$$\frac{\Delta v_y}{v} = \frac{2v_R}{v}\frac{\cos\alpha\,\sin\alpha}{1 + m_p/m_n} + \frac{i_s}{mv}\sin\varepsilon_i\,\cos\alpha$$

$$\frac{\Delta v_z}{v} = \frac{i_s}{mv}\cos\varepsilon_i \qquad [6.5.11]$$

For the negative mass values are m_p/m_n times greater but have the same direction.

Net linear energy gains

Since the analysis has been based on the relative velocity vector v_R it is first necessary to apportion the correct value to each component. In the absolute analysis it was found necessary to make the two (additive) components of relative momentum equal so: $m_p v_{Rp} = m_n v_{Rn}$ and from this it follows that v_{Rp} for the positive primary and v_{Rn} for the negative one become:

$$v_{Rp} = v_R/(1 + m_p/m_n) \quad \& \quad v_{Rn} = v_{Rp}(m_p/m_n) \qquad [6.5.12]$$

If now the positive energy gain ΔE_ϕ is expressed as a ratio of the incident absolute positive kinetic energy $\frac{1}{2}mv^2$ then since only in the x direction is there an initial velocity v_{Rp} this ratio becomes:

$$\frac{2\Delta E_\phi}{mv^2} = \left(\frac{v_{Rp}}{v} + \frac{\Delta v_x}{v}\right)^2 + \left(\frac{\Delta v_y}{v}\right)^2 + \left(\frac{\Delta v_z}{v}\right)^2 - \left(\frac{v_{Rp}}{v}\right)^2$$

Then simplifying and noting that $E_0 = m_0 C_U^2$ where m is indistinguishable from m_0 for $v<<C_U$ the result becomes:

$$\frac{\Delta E_\phi}{E_0} = \frac{1}{2}\left(\frac{v}{C_U}\right)^2 \left[\frac{2}{(1+m_p/m_n)}\frac{v_R}{v}\frac{\Delta v_x}{v} + \left(\frac{\Delta v_x}{v}\right)^2 + \left(\frac{\Delta v_y}{v}\right)^2 + \left(\frac{\Delta v_z}{v}\right)^2\right] \qquad [6.5.13]$$

For the negative mass all velocities are multiplied in ratio m_p/m_n but to use the same denominator the result needs dividing by m_p/m_n . Furthermore for the negative mass the incident relative velocity has the reverse sense so the end result becomes:

$$\frac{\Delta E_{n\phi}}{E_0} = \frac{1}{2}\left(\frac{v}{C_U}\right)^2 \frac{m_p}{m_n}\left[\frac{-2}{(1+m_p/m_n)}\frac{v_R}{v}\frac{\Delta v_x}{v} + \left(\frac{\Delta v_x}{v}\right)^2 + \left(\frac{\Delta v_y}{v}\right)^2 + \left(\frac{\Delta v_z}{v}\right)^2\right]$$

[6.5.14]

For both [6.5.13] & [6.5.14], v_R is given by

$v_R^2 = v^2 + v_n^2 - 2vv_n\cos(\theta)$ (See FIG.3.1 for angle θ)

And Δv's are given by [6.5.11]..

Net rotational energy gains

There will be a net angular velocity increment $\Delta\omega$, being the sum of $\Delta\omega_\phi$ and $\Delta\omega_s$, that adds to the incident value of ω_1.which is equal to $\omega_p\, cos(\varepsilon_p)$ for the positive primary. The result will be a change in rotational energy given by:

$$\Delta E_\omega = 0.5\, I_\omega\left((\omega_1 + \Delta\omega)^2 - \varpi_1^2\right)$$

And is more conveniently expressed in the form of a ratio by dividing by $m_0 C_U^2$: ($m = m_0$ nearly) and re-arranging to give;

$$\frac{\Delta E_\omega}{E_0} = \frac{3}{10}\frac{R^2\,\omega_p^{\;2}}{C_U^{\;2}}\left(\frac{2\,\Delta\omega}{\omega_p}\cos(\varepsilon_p) + \left(\frac{\Delta\omega}{\omega_p}\right)^2\right) \quad [6.5.15]$$

In order to relate rotational to linear energies it is convenient, for present analysis, to express the rotational kind in terms of a ratio η_ω of the linear sort. Then it follows that

$$0.5\,I_\omega\omega^2 = 0.5\times 0.6\,m\,R^2\omega^2 = \eta_\omega 0.5\,m\,v^2$$

Hence;

$$(\omega_p R)^2 = \eta_\omega \frac{5}{3} v^2 \quad [6.5.16]$$

Then substituting in [6.5.15] results in:

$$\frac{\Delta E_\omega}{E_0} = \frac{\eta_\omega}{2}\frac{v^2}{C_U^{\;2}}\left(\frac{2\,\Delta\omega}{\omega_p}\cos(\varepsilon_p) + \left(\frac{\Delta\omega}{\omega_p}\right)^2\right) \quad [6.5.17]$$

Or so that [6.5.4] plus [6.5.10] can be substituted, a parameter that becomes:

$$\frac{R\,\Delta\omega}{v} = \frac{10/3}{1 + m_p/m_n}\frac{v_R}{v}\left(\cos^2\alpha\,\sin\alpha - \cos\alpha\,\sin^2\alpha\,\tan\phi\right) + \frac{5}{3}\frac{i_s}{m\,v}\sin\varepsilon_i$$

Then [6.5.17] can be arranged as: :

$$\frac{\Delta E_\omega}{E_0} = \frac{1}{2}\frac{v^2}{C_U^{\;2}}\left(2\sqrt{\frac{3\eta_\omega}{5}}\frac{R\,\Delta\omega}{v}\cos(\varepsilon_p) + \frac{3}{5}\left(\frac{R\,\Delta\omega}{v}\right)^2\right) \quad [6.5.18]$$

For the negative mass the parameter $R\,\Delta\omega/v$ for positive mass is multiplied by m_p/m_n and η_ω can have a different value from the positive one. It is also necessary to note that the scuffing angle εi was based on clockwise motion being positive and for the negative mass the angular momentum vector is reversed.

Then the total energy gain for positive or negative mass is given by:

$$\frac{\Delta E}{E_0} = \frac{\Delta E_\phi}{E_0} + \frac{\Delta E_\omega}{E_0} \quad [6.5.19]$$

Of course the value of ε_n will differ from ε_p since the object of the whole exercise is to determine the effect of such difference.

All the required equations are now available for analysis. Unfortunately there is no way to obtain the value of the parameter

i_s/mv. Indeed inspection suggests this should be zero since any value ascribed to it causes a rolling motion. This seems to preclude any frictional effect during the collision of opposite primaries. So a zero value is the obvious one to choose. Unfortunately when the two incident angular momenta are opposed, one primary gains rotational energy whilst the other loses that of its own sign. The only way an overall balance can be achieved is by arranging an equal and opposite imbalance for linear energy gains. This can only be achieved by the variation of ϕ. However, as will be seen in the following table, the result can be that energy gains approach infinity. So then it is useful to see if a value of i_s/mv can be inserted to give a minimum energy gain.

Hence the directions of spinning motion ε and ε_n are specified with $i_s/mv = 0$ and a value for ϕ obtained by iteration that yields equal combined linear and spin energy gains for positive and negative masses (the negative sign for energy has been ignored in the equations). Then other starts are made with other values of i_s/mv to determine the effect and find what the minimum possible gain could be.

The main issue is to see if ever an overall zero or negative gain could appear that would destroy the whole basis of the theory. . Evaluation is provided by GWBASIC code "ECM18906" from which the following tables were prepared using the equations in this section.

TABLE 6.I data input. Note v_0 is the most probable speed of the primaries.
Data $m/m_n = 1.01$: $v_0/C_U = 0.01$: $v_{0n}/C_U = 0.015$: $v/v_0 = 1$: $v_n/v_{0n} = 1$: $A = 90°$: $\varepsilon_p = 15°$: $\varepsilon_n = 7.5°$ increasing in 15° steps: $\eta_\omega = 1$: $\eta_{\omega n} = 1$: (assumes equilibrium between linear and angular incident kinetic energies). The upper value of ε_n in column (1) is the direction of the angular momentum vector and the lower one that of the angular velocity vector for this negative mass. Both are measured from the z-axis.

Collision angle α is assumed as 45° with $i/mv = 0$ until positive and negative energy gains fail to balance. For the case of TABLE 6.I it is then α that is varied until a balance is achieved with $i/mv = 0$ always. In this case the value of velocity in the Z direction is zero: i.e. $\Delta v_z = 0$

Upper figures for ΔE and Δv in each double row, columns (3),(4),(7) & (8) give values for positive mass and lower ones for negative in each row.

TABLE 6.I

$v_R/C_U = 0.018028$ and $E_K/E_0 = 1.6139\times10^{-4}$.

Where –3 appears after a number it means $\times10^{-3}$

(1) ε_n $180\text{-}\varepsilon_n$	(2) α	(3) $\frac{\Delta E_p}{E_K}$	(4) $\frac{\Delta E_n}{E_K}$	(5) $\frac{\Delta E_T}{E_K}$	(6) ϕ & ε_i	(7) $\frac{\Delta v_x}{v}$	(8) $\frac{\Delta v_y}{v}$
7.5 172.5	45	.2604. .2407	.6778 .6973	.9382	1.27 79.48	.0191 .0201	.8969 .9059
22.5 157.5	45	.2432 .2579	.7008 .6941	.9520	-0.7 70.52	-.011 -.011	.8969 .9059
37.5 142.5	45	.2084 .2967	.7768 .6886	.9852	-4.92 61.54	-.0768 -.0776	.8969 .9059
52.5 127.5	45	.1558 .3682	.8965 .6841	1.052	-12.18 52.49	-.1879 -.1897	.8969 .9059
67.5 112.5	45	.0881 .4972	1.099 .6899	1.187	-24.59 43.3	-.3636 -.3672	.8969 .9059
82.5 97.5	45	.2035 .7397	1.450 .7311	1.471	-49.7 33.87	-.6406 -.647	.8969 .9059
97.5 82.5	70	-.1145 .5316	.8529 .2068	.7385	-89.24 24.11	-.5765 -.5823	.5765 .5823
112.5 67.5	89	-8.8-3 9.69-3	.0134 -5.1-3	4.55-3	-33.5 13.8	-.0166 -.0167	.0313 .0316
127.5 52.5	85	.0781 -.0548	-.0755 .0574	2.61-3	54.97 2.59	.1191 .1203	.1557 .1573
142.5 37.5	80	.2025 -.0997	-.1415 .1607	.0610	69.78 -10.19	.2710 .3068	.2737 .3098
157.5 22.5	80	.1609 -.0755	-.1077 .1288	.0532	47.42 -25.95	.2121 .2142	.3068 .3098
157.5 22.5	78	.2431 -.1022	-.1488 .1964	.0942	65.09 -25.95	.3097 .3685	.3648 .3685
172.5 7.5	78	.2216 -.0912	-.1331 .1797	.0885	55.48 -82.55	.2806 .2834	.3648 .3685
172.5 7.5	75	.3742 -.1260	-.1933 .3069	.1809	89.4 -82.55	.4488 .4533	.4485 .4529

On this basis, when a solution is not found for the actual collision angle α occurring, it is then assumed that the collision needs to act as if the primaries strike at an increased angle, although the probabilities for the numbers of primaries colliding remains unchanged. The angle of $\alpha = 45^\circ$ is assumed where possible since it is the case maximising energy gains and therefore maximises the difficulties of finding a solution. There is always a solution when α

is increased toward 90° but not always when reduced toward zero. This is because as α is increased the torque producing change of angular momentum is reduced.

For $\varepsilon_n > 90°$ only a small range of α is available that allows a solution but this results in a large range of possibilities for ϕ. This is indicated in the last four double rows where two different values of α are used as input data for each value of ε_n. For example, when $\varepsilon_n = 157.5$, $\alpha = 80°$ yields $\phi = 47.42°$ rising to 65.09° as α is reduced to 78°. The total energy gain changes very dramatically. This is given by column (5) showing it is nearly doubled

It is therefore impossible to give an exact solution: only a range where a solution is possible. Fortunately for these cases the predicted gains are small as compared with overall average values so the percentage error seems tolerable. There are, however, always energy gains.

This table only deals with one possibility: changing α to obtain a solution with i/mv = 0. Another alternative keeps α unchanged but varies i/mv instead. This solution, based on the same assumptions, is provided in TABLE 6.II. Now $\Delta v_z = 0$ no longer applies

TABLE 6.II α = 45°: *i/mv* varied

(1)	(2)	(3)	(4)	(5)	(6)	(7)	(8)	(9)
ε_n $180\text{-}\varepsilon_n$	$\frac{i}{mv}$	$\frac{\Delta E_p}{E_K}$	$\frac{\Delta E_n}{E_K}$	$\frac{\Delta E_T}{E_K}$	ϕ	$\frac{\Delta v_x}{v}$	$\frac{\Delta v_y}{v}$	$\frac{\Delta v_z}{v}$
97.5 82.5	-2.7	1.829 1.969	.1360 -.004	1.97	-87.3	-.1080 -.1091	.1169 .1181	-2.464 -2.489
112.5 67.5	-4.7	6.434 6.547	.0859 -.027	6.52	-77.1	-.0243 -.0432	.1042 .1052	-4.564 -4.610
112.5 67.5	-4.6	6.141 6.300	.1248 -.034	6.266	-82.2	-.0869 -.0878	.1210 .1223	-4.467 -4.512
127.5 52.5	- 10.9	36.62 37.58	.8912 -.072	37.51	-85.6	-.5313 -.5366	.5487 .5542	-10.89 -11.00
142.5 37.5	6.4	12.44 12.31	-.067 .0635	12.37	-42.6	.2268 .2291	.0966 .0975	6.299 6.362
157.5 22.5	3.0	4.098 2.228	-.171 1.698	3.927	66.9	1.702 1.719	-.0314 -.0317	2.698 2.724
157.5 22.5	2.9	2.498 1.916	-.220 .362	2.278	-24.1	.5408 .5461	-4.3-4 -4.3-4	2.608 2.634
172.5 7.5	1.7	1.069 .1552	-.274 .640	.7949	-4.43	.8238 .8320	3.89-3 3.93-3	1.138 1.149

This alternative solution yields absurdly high-energy gains as shown in column (5). There are usually two small ranges of *i/mv* that give widely different solutions and the one giving the lower energy gain is given in the table. For the case of ε_n = 157.5° two such solutions are illustrated Here a small change in *i/mv* switches ϕ from a negative to a positive value. Mathematically both sets of solutions are equally valid but the generally excessive values of $\Delta v_z/v$, shown in column (9), suggests the solutions given in TABLE 6.II are unrealistic.

Both tables consider cases where the opposed primaries have almost equal and opposite masses and both move at the most probable speeds. It is now necessary to consider less favourable conditions and in TABLE 6.III one such case is considered.

TABLE 6.III $v_R/C_U = 0.0$ and $E_K/E_0 = 2.375\times10^{-4}$.(see next page)

ε_n / $180\text{-}\varepsilon_n$	$\frac{\Delta E_p}{E_K}$	$\frac{\Delta E_n}{E_K}$	$\frac{\Delta E_T}{E_K}$	ϕ	$\frac{\Delta v_x}{v}$	$\frac{\Delta v_y}{v}$
.7.5	.3920	.2002	.5922	20.65	.7014	2.028
	.1851	.4070		77.8	1.403	4.055
22.5	.3856	.2049	.5905	19.94	.6791	2.028
	.1914	.3990		72.2	1.358	4.055
37.5	.3728	.2146	.5874	18.52	.6339	2.028
	.2045	.3829		66.34	1.268	4.055
52.5						2.028
						4.055
67.5						2.028
						4.055
82.5						2.028
						4.055
97.5						2.028
						4.055
112.5						2.028
						4.055
127.5						2.028
						4.055
142.5						2.028
						4.055
157.5						2.028
						4.055
172.5						2.028
						4.055

TABLE 6.III data input

Data $m/m_n = 2.0$: $v_0/C_U = 0.01$: $v_{0n}/C_U = 0.015$: $v/v_0 = 0.5$: $v_n/v_{0n} = 2$: $\theta = 90^\circ$: $\varepsilon_p = 15^\circ$: $\varepsilon_n = 7.5^\circ$ increasing in 15° steps: $\eta_\omega = 1$: $\eta_{\omega n} = 1$: (assumes equilibrium between linear and angular incident kinetic energies). The upper value of ε_n in column (1) is the direction of the angular momentum vector and the lower one that of the angular velocity vector for this negative mass. Both are measured from the *z*-axis.

Collision angle α is assumed as 45° with $i/mv = 0$ found applicable all cases.

6.6 The "Phase Coupling Factors" C_P & C_{PF}

When primaries of like energies collide they deflect through a larger angle than when those of unlike energies collide. Therefore it is to be expected that bulk momentum does not completely transfer from one phase to the other. For example, it was shown in Chapter 3 that if all primaries have equal inertial mass then regardless of their speeds, the momentum increment due to collision, when unlike kinds collide, is perpendicular to the relative velocity vector v_R. In this case no bulk momentum will be transferred from one phase to the other. (This extreme example has ignored the effects of spinning motion.) However, this example will serve to illustrate the problem and in this case collisions of two opposite energy primaries would, en mass, produce breeding and nothing else. The two phases would move independently of each other, when acted on by pressure differentials, and indeed could flow unimpeded through each other. Each phase could have a Maxwellian speed distribution yet would not affect this conclusion. For this case $C_P = 0$ would apply.

The term C_{PF} is defined as: $C_{PF} = 1 - C_P$

So in this case $C_{PF} = 1$ meaning that the pressure gradients required to produce acceleration would be exactly the same as if each phase existed without the other.

However, when inertial masses are not numerically equal the two opposite primaries will be deflected from the perpendicular by angle ϕ as has already been shown in Chapter 3 and illustrated in TABLES 3 III and IV. Hence a degree of coupling will arise.

For complete coupling then $C_P = 1.0$ would apply. Each phase would eliminate the inertia of the other to produce a combined fluid requiring zero pressure differentials to cause flow.

Neither of these extreme situations will ever arise since C_P will lie between the two extremes 0 to 1. A complete study would include the effects of spinning motions but this has not so far been included. It is under study but this is not yet finalised. The assumption is made, for the cases considered here, that primaries do not spin.

For this analysis, forces due to collision need to be resolved in a single direction, defined as the J direction. Then C_P will be defined as:

C_P = the average force produced by unlike impacts divided by that produced by like impacts.

To set up the problem a primary of positive mass m and velocity v is considered to move at an angle μ from the reference direction J as illustrated in FIG.6.3 showing the velocity triangle at F. This trajectory, in which angle μ can vary from 0 to π radians, always lies in the reference plane and this primary is regarded as a target for a second set of primaries to impinge upon. One of these has mass m_A that can be either positive or negative. As illustrated separately, to avoid clutter, in the velocity triangle at FA this primary moves at velocity v_A in a direction μ_A again measured from reference direction J, but this trajectory lies in a plane at angle γ with respect to the reference plane. Angle γ can vary from 0 to π radians, either clockwise or anticlockwise, but since symmetry exists, only one of these is considered. If the effects of μ_A were integrated from 0 to π radians then the net force would be zero when integrated for all cases, so yielding an indeterminate result. For this reason integration is limited to the range 0 to π/2 radians for this impinging set of primaries.

For the general case when one phase flows through the other, and each is in motion relative to the undisturbed fluid, as when waves propagate, then extra velocities are added.. This effect is indicated by a general flow w shown at F, for the target primary. Also as shown at FA w_n is the general flow for fluid A. This only applies when the latter is of negative mass. The dashed lines indicate the modified velocity vectors. These refinements are shown for completeness only. They have been incorporated in an advanced formulation of the theory but will not be considered further in this book.

It is the relative velocity v_R that needs to be used for all collision events. This is the vector difference between velocity vectors v_A and v: indeed it is vector v_A – vector v and is:

$$v_R^{\,2} = (v_A \sin\mu_A \sin\gamma)^2 + (v_A \cos\mu_A - v\cos\mu)^2 + (v_A \sin\mu_A \cos\gamma - v\sin\mu)^2$$

Expanding each term yields:

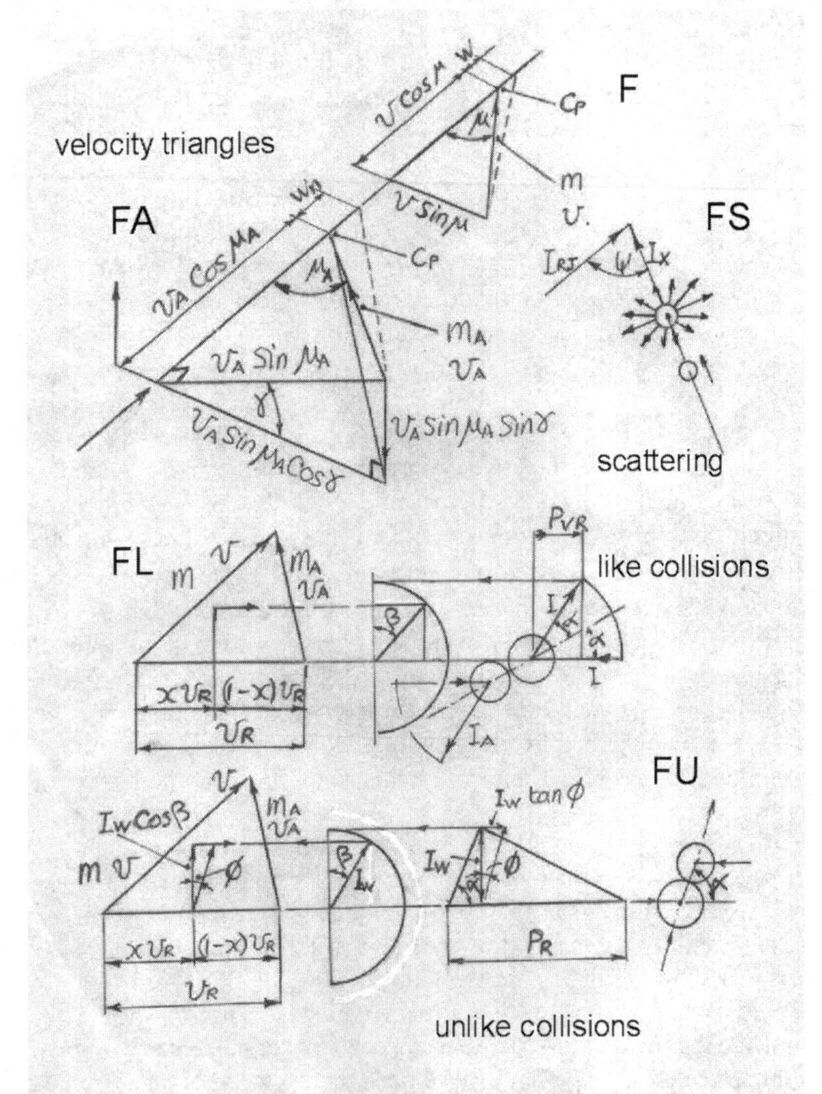

FIG.6.3 PHASE COUPLING FACTOR

diagrams for visualising the mathematics used

$$v_R^{\ 2} = v_A^{\ 2} \sin^2 \mu_A$$
$$+ v_A^{\ 2} \cos^2 \mu_A - 2 v_A v \cos \mu_A \cos \mu$$
$$+ v_A^{\ 2} \sin^2 \mu_A \cos^2 \gamma + v^2 \sin^2 \mu - 2 v_A v \sin \mu_A \cos \gamma \sin \mu$$

And this simplifies to:

$$v_R^{\,2} = v_A^{\,2} + v^2 - 2v_A v(\cos\mu_A \cos\mu - \sin\mu_A \cos\gamma \sin\mu)$$
[6.6.1]

We now need to consider the impulse produced on the target primary in direction J. The problem is illustrated at FS showing the primary A moving along relative vector v_R that is at an angle ψ to the direction J. After striking the target primary a component of impulse I_x remains in direction v_R and a component I_w remains that is perpendicular to v_R

Now as illustrated at FS this perpendicular component scatters with uniform distribution about an axis co-incident with v_R so that, averaged over a large number of collisions, no net impulse from this component is transferred to the J direction owing to mutual cancellation. Only the component I_X remains: the impulse in direction v_R.

When an infinite number of collisions are considered the remaining impulse I_x can be imagined distributed around a conical surface having the J direction as its axis so that the cone has half angle ψ. All components of impulse cancel in a direction perpendicular to J and so the only remaining impulse is $I_x \cos\psi$

The value of $\cos\psi$ is:

$$\cos\psi = v_{RJ}/v_R \qquad [6.6.2]$$

where v_{RJ} is the component of the relative velocity measured along the J axis and v_R is given by equation [6.6.1]. And v_{RJ} is simply given from the difference between triangles FA and F measured in the J direction and given by:

$$v_{RJ} = v_A \cos(\mu_A) - v\cos(\mu) \qquad [6.6.3]$$

What now remains is the assessment of I_x for both like and unlike collisions and the probabilities of these collisions.

First the impinging set of primaries is considered to have the same energy sign as the target to obtain the average force for like collisions and then the impinging set are considered to have the opposite energy sign. In FIG.6.3 the collision point is marked as CP on both sets of triangles. Of course CP is the same point as shown when both sets of velocity triangles F and FB are superimposed to coincide at point CP..

6.6.1 Like collisions

The diagram FL illustrates like collisions of either two positive energy primaries or two negative energy primaries. It is assumed that the reference frame is chosen at the point on the relative velocity vector that makes the incident momentum equal for both collision partners. Each then has an equal though opposite momentum P_R as given by:

$$P_R = I = \frac{m}{1 + m/m_A} v_R \qquad [6.6.4]$$

The centreline, defined as the line joining the centres of primaries, is at the scattering angle α with respect to the v_R vector, giving the v_R direction. A frame of reference has been chosen for evaluation of [6.6.4] that makes the incident relative momentum equal for both collision partners. Then each primary rebounds without change of speed relative to a frame of reference that does not move across the collision event and they rebound with the same angle α with respect to the line joining the centres of the two spheres (the centreline) as shown by the incident and rebound relative velocity arrows marked on the two spheres.. Hence mass m now leaves with a momentum component P_{VR} in the v_R direction of $I_X = I\cos(2\alpha)$. Since this needs to be added to the incident impulse I to obtain the change in impulse in direction x (i.e. v_R), the impulse exerted on the target primary is $I(1 + \cos(2\alpha))$ and I is given by [6.6.4]. . This is in the direction of v_R but what is required is the component in direction J.

Since I_J is given by the ratio v_{RJ} / v_R of the value of P_R, given by [6.6.4] it follows that with I_{JL} as the impulse for like collisions in the J direction:

$$I_{JL} = \frac{m}{1 + m/m_A} v_R \left(1 + \cos(2\alpha)\right) \frac{v_{RJ}}{v_R}$$

v_R cancels (but is needed later) and v_{RJ} is given by [6.6.3] leaving the result:

$$I_{JL} = \frac{m}{1 + m/m_A} \left(1 + \cos(2\alpha)\right)\left(v_A \cos(\mu_A) - v\cos(\mu)\right) \qquad [6.6.5]$$

At FB a "momentum circle" is shown whose axis is a projection of v_R. This represents momentum perpendicular to the direction of v_R. Clearly integration of impulse for all angles β around this circle will sum to zero and so the component of this in the J direction will also be zero. Hence all components perpendicular to the v_R direction can be ignored. This supports the conclusion made in the previous section.

Hence [6.6.5] provides the total impulse for like collisions.

6.6.2 Unlike Collisions

The impulse perpendicular to the v_R direction is I_W and that in the v_R direction is given by multiplying this value by $\tan(\phi)$ to become $I_W \tan(\phi)$. A study of TABLE 3 I cases 1 to 3 suggests that the variation of ϕ with β is very small. At this stage it is therefore sensible to use an approximation for tan(ϕ) in order to avoid excessive complication. The values in this table suggest the following simple expression as being adequate:

$$\tan(\phi) = Const \times \left(\frac{m}{m_A} - 1 \right) \qquad [6.6.6]$$

With *Const* evaluated by putting $\phi = 13^0$, when $m/m_A = 3$
Then the impulse in the J direction given as I_{JU} for unlike collisions can be written:

$$I_{JU} = \frac{m}{1 + m/m_A} \sin(2\alpha)\tan(\phi)(v_A \cos(\mu_A) - v\cos(\mu))$$

[6.6.7]

6.6.3 Collision Probability

The value of v_R is still required since it affects collision probability and is given by equation [6.6.1]

The total collision rate or flux ξ is the product of five components. First as shown in [6.6.8] (below) is the probability of collision between angles α and $\alpha + \delta\alpha$ and this is $\sin(2\alpha)\delta\alpha$.

Next γ can vary from 0 to π so this probability is $d\gamma/\pi$. Next the number of collisions targets moving within angle range μ to $\mu + \delta\mu$ is evaluated by considering an element $2\sin\alpha\delta\alpha$ as the proportion of a sphere of unit radius and becomes $\sin\alpha\delta\alpha$. The proportion of impinging primaries in range μ_A to $\mu_A + \delta\mu_A$ is $\sin\mu_A \delta\mu_A$. Then the probability of these two groups colliding will be proportional to v_R. Hence the value of ξ becomes

$$\zeta = \int_0^{\pi/2}\int_0^{\pi}\int_0^{\pi}\int_0^{\pi}\int_0^{\pi/2} \sin\alpha \, d\alpha \frac{d\beta}{\pi}\frac{d\gamma}{\pi} v_R \sin\mu \, d\mu \sin\mu_A \, d\mu_A$$

[6.6.8]

The variable v_R is included if v_A and v are made constant for a given integration since v_R is then a function of only γ, μ and μ_A. The first two integrations can be performed analytically so leaving a triple integral.

For like collisions the product of impulse and flux $I_{JL} \times \xi$ becomes the product of [6.6.2] and $d\xi$. Again the term in β becomes 1. The term in α now becomes:

$$\int_0^{\pi/2} (1 + \cos(2\alpha))\sin(2\alpha)d\alpha = \frac{\cos(2\alpha)}{-2} + \left[\frac{\cos(4\alpha)}{8}\right] = 1$$

$$\therefore \quad I_{JL}\zeta = \int_0^{\pi}\int_0^{\pi}\int_0^{\pi/2} \frac{m}{1+m/m_A} v_R^{\frac{3}{2}} (v_A \cos\mu_A - v\cos\mu)\sin\mu_A \, d\mu_A$$

$$\times \sin\mu \, d\mu \frac{d\gamma}{\pi} \qquad [6.6.9]$$

For unlike collisions for the product $I_{JU} \times \xi$ is the product of [6.6.5] and dξ.. Again the term in β becomes 1 and the term in α now integrates to:

$$\int_0^{\pi/2} \sin^2(2\alpha) d\alpha = \int_0^{\pi/2} \left(\frac{1-\cos(4\alpha)}{2}\right) d\alpha = \frac{\pi}{4}$$

[[6.6.10]

So

$$I_{JU}\zeta = \int_0^{\pi}\int_0^{\pi}\int_0^{\pi/2} \frac{m}{1+m/m_A} \frac{\pi}{4} v_R^{\frac{3}{2}} \tan\phi (v_A \cos\mu_A - v\cos\mu)$$

$$\times \sin\mu_A d\mu_A \sin\mu \, d\mu \frac{d\gamma}{\pi}$$

[6.6.11]

Then after integration of [6.6.8], [6.6.9] and [6.6.11] the average momentum imparted to the target primaries are given by:

$P_L = I_{JL}\zeta/\zeta \quad and \quad P_U = I_{JU}\zeta/\zeta$

The phase coupling factor then becomes: $P_C = P_U/P_L$ [6.6.12]

6.6.4 The Effect of Mass Range

So far only the effect of a single mass ratio m/m_A has been considered and this is inadequate. Primaries will have masses starting from about half their maximum values and increasing as they grow until, upon reaching the maximum value, split to start growing again. For every element of mass range for each phase collision with the entire range of mass of the other phase has to be considered. This increases the problem from a triple integral to a quintuple integral. Since adding Maxwellian speed distributions to both phases together with spin effects will add further integrations the present study is to be regarded as only a preliminary step. A complete solution will need to use the random number method described in Chapter 16.

It seems reasonable to consider only average speeds since ϕ has been shown to be dependent only on mass ratio and independent of the ratio of speeds. However, thermal equilibrium is to be assumed for all masses since this seems the best assumption to make. Then mv^2 will be equal for all primaries of either phase.

To limit the problem to a single extra integration without

too much loss of accuracy it will first be assumed that a limiting mass range m_{max}/m_{min} is specified. Then the maximum mass divided by average mass is taken as the square root of m_{max}/m_{min} and the minimum value as the inverse of this figure. With the number of elements N_M specified, the step change in mass δm_r is then given by:

$$\delta m_r = \left(\sqrt{m_{\max}/m_{\min}} - 1\right)/N_M \qquad [6.6.13]$$

At each step, starting from the centre point of the range, where masses are equal, one phase is given a mass increase with the other a decrease in the same ratio. Then with average speeds v_0 and v_{A0} specified at equal mass ratio the average speeds at other steps are taken as:

For this pair a value for impulse difference using equations [6.6.8] to [6.6.11] is then found. This is added to previous values until the full mass range has been covered. Only then is P_C evaluated from [6.6.12].

TABLE 6.IV giving C_P noting that C_{PF} the phase coupling factor is: $C_{PF} = 1 - C_P$

m_{max}/m_{min}	ϕ_D	v/v_0	v_A/v_{A0}	$\tan\phi$	C_P
2	13	0.8409	1.1892	0.1154	0.04558
3	"	0.7598	1.3161	0.23087	0.08504
4	"	0.7071	1.414	0.30463	0.1216
2	20	0.8409	1.1892	0.1820	0.07185
"	30	"	"	0.2887	0.1140
"	45	"	"	0.5	0.197
"	60	"	"	0.866	0.342

Results from code "ECMcrn06" giving C_P for specified mass ranges and ϕ_D assumption

The ϕ_D values assumed are for $m_{max}/m_{min} = 3$ and $\tan\phi$ values given are for the specified m_{max}/m_{min}. For the quadruple integral the number of steps was 18 per integral for the first three and 9 for the last covering mass range. Doubling these numbers made negligible difference but took excessive computing time.

The values of ϕ_D assumed are far greater than are predicted from linear momentum without spin but these high values might represent the effects of spinning motion. What the results seem to indicate is that the values of C_P will always be nearer zero than unity. Consequently the phase coupling factor C_{PF} will be close to unity since $C_{PF} = 1 - C_P$. This preliminary study shows the need for further study to determine accurate values for the phase coupling factor with the effects of spinning motion included. An accurate value needs to be sought since, as will be shown later, the P_{CF} factor affects the prediction of the radius of the universe from the Hubble constant.

However, in 2008 it was found that , as shown in PLATE 2 of PART I of this book bunching of primaries due to random motion never seems to occur in the same places for opposite primaries. This upsets the probability on which the previous assessment was made – indeed making the value even closer to unity.

The guess originally made of C_{PF} = 0.9 has been used in further computations in case the effects of spinning motion reduce the coupling. The analysis has to be regarded as incomplete owing to neglect of spinning motion but provides a first step in the analysis that may encourage others to complete.

6.7 CONCLUSION TO CHAPTER 6

The collision of opposite primaries in two's produces net energy gains due to spinning motion comparable to those of linear motion so that any possible net annihilation of rest energies is more than cancelled out. Consequently the analysis made in this chapter fully confirms that the breeding process will occur.

The positive and negative phases of the mixture have been shown to couple only weakly to each other as far as bodily motion is concerned. This means that the phase coupling factor C_{PF} must be close to unity, meaning that each is accelerated by pressure gradients of its own phase as if hardly affected by its companion phase. Only coupling due to linear motion has so far been considered and in case spinning motion produces a reducing effect a provisional value of C_{PF} =0.9 has been chosen for further analysis.

The hope is that sufficient interest will have been aroused so that others will be inspired to complete the analysis.

Sufficient detail of opposed energy dynamics for two opposed primary collisions has been presented for its effects on cosmology to be analysed. We feel this has justified the next step to be attempted. This is to consider the effects of multiple collisions arising when primaries of both kinds implode from all possible directions. A start on this analysis is made in the next chapter.

CHAPTER 7

ANNIHILATION CORES

Pure creation generates instability causing collapse with fragmentation into a myriad of cells. Each cell forms with a central annihilation core that cancels most of the creation that still continues unabated.

Creation drives an accelerating expansion requiring a pressure build up. The latter increases annihilation until balance is reached. The creation switch-off mechanism has appeared that the big bang theory lacks. The greatest difficulty is found to be the derivation of equations giving accurate prediction of i-theric growth due to the resulting minute net creation. This applies to both size and pressure development. This takes us on a 'drunken man's walk' occupying several chapters before a satisfactory resolution is achieved.

7.1 Conditions for Annihilation

The energy gain ratios $\eta = \Delta E/E$ predicted yield exponential growth of the universe at a far greater rate than astronomical observations of remote galaxies could possibly allow. Hence the conditions that allow annihilation centres to develop will now be explored. So far only collisions between two opposite energy primaries at a time has been considered and then energy breeding is dictated, from the need to conserve both energy and momentum, when collision forces are equal and opposite for the two concerned.

However, when multiple collisions from large numbers of primaries impinging from every possible direction are considered, the net momentum before collision is zero. Since zero net momentum has to apply when nothing exists, this is the condition favouring mutual annihilation. A uniform field of flow would therefore exist in a state of unstable equilibrium since a lower energy state can arise by annihilation. Chaos theory shows that under such conditions a self-organisation of the flow field into a pattern of cells can be expected in which primaries have a radial flow toward a central point or line in every individual cell. At such centres a core of annihilation will form in which primaries arrive at their surfaces as fast as annihilation proceeds internally.

It is first necessary to consider collision breeding and to provide numerical estimates of its rate. As a first necessary step the value of the ultimate speed C_U of primaries will first be considered.

7.2 Estimating Mean and Ultimate speeds of primaries

According to Clifford Will (1988), in 1974 Russell Hulse and Joe Taylor (1975) discovered the first 'binary pulsar' using the Arecibo radio telescope in Puerto Rico. They found that two neutron stars, each of 1.4 solar masses, were in close, though highly elliptical, orbit about one another. It was discovered, from progressive changes in orbital periods, that energy was being lost. This loss accurately fitted Einstein's prediction of the energy carried away by the gravitational waves that such close orbits should produce.

The present ECM theory explains such waves as longitudinal ones like sound waves in air, though now propagated through the fluid like component of the i-ther. Since ECM theory yields mostly the same end equations as general relativity, owing to the mathematical similarity of curved space-time and non-uniform i-theric density, the same prediction will be made by ECM theory.

Hence the equation for the propagation speed of gravity waves as longitudinal pressure waves moving at the speed of light is applicable to the present analysis.

From the kinetic theory of gases at temperature $\boldsymbol{T}$, gas constant $\boldsymbol{R}$ and average molecular speed $\boldsymbol{v}$ the perfect gas equations show $P = \rho RT = \rho v^2/3$ and the speed of sound $\boldsymbol{a}$ with γ as the ratio of specific heats at constant pressure and volume respectively is given by the well known equation:

$$a = \sqrt{\gamma RT} \quad \text{Hence} \quad a = v\sqrt{\gamma/3}$$

With $\boldsymbol{v}$ now representing the average speed of primaries $\boldsymbol{v_P}$ and $\boldsymbol{c}$ the speed of light (not C_U) substituted for $\boldsymbol{a}$ it follows that:

$$v_P = c\sqrt{3/\gamma} \qquad [7.2.1]$$

Primaries will behave like spinning molecules and for these the value of γ is 1.4. Hence the mean speed of primaries, according to [7.2.1], will be $\boldsymbol{v_p}$ = 1.464 times the speed of light or **4.392*10⁸ m/s.**

Now in Chapter 5 §5.7 p.139 the most probable value of v_P/C_U = 0.75 was deduced. Consequently $C_U = v_P/0.75$ giving:

$$C_U = \mathbf{1.952{\times}c} \text{ or } 1.952{\times}3{\times}10^8 = \mathbf{5.856{\times}10^8\ m/s.}$$

According to present scientific opinion nothing can travel faster than light. However, this only applies to matter and is a condition imposed by the special theory of relativity which forbids even information to travel faster than light. In the present theory relativity has been replaced by an exact classical mechanics that imposes no restriction on the propagation speed of information. Furthermore, primaries are not matter. They are the fundamental building blocks of matter and are not subject to the same restrictions.

7.3 THE RATE OF COLLISION BREEDING

To determine the conditions under which annihilation can occur it is necessary to introduce the rate of collision and find an expression for collision breeding. The collision rate is connected with the mean free paths of the primaries. Sir James Jeans (1887) gives the following equation for the rate of collision N_C for molecules in a gas and this can be equally applied to a mix of primaries. For James' analysis:

N_C is the number of collisions per unit time per unit volume occurring in the gas,

n is the number density: in molecules per unit volume.

σ is the diameter of each molecule assumed of spherical shape.

v is the average speed of the molecules.

Then he shows that:

$$N_C = \frac{\pi}{\sqrt{2}} n^2 \sigma^2 v \qquad [7.3.1]$$

Jeans also derives the following equation for the mean free path λ of such spherical molecules:

$$\lambda = \frac{1}{\sqrt{2}\,\pi n \sigma^2} \qquad [7.3.2]$$

The value of $n\sigma^2$ obtained from [7.3.2] can be substituted in [7.3.1] to yield an expression to be used at a later stage. It is:

$$N_C = \frac{nv}{2\lambda} \qquad [7.3.3]$$

Now η is the average ratio of energy gain ΔE per collision to the combined kinetic energy E_K and is adopted since, as shown in Chapter 5, this ratio varies little over a wide range of m_N/m_P and v/C_U. It is expressed as:

$$\eta = \frac{\Delta E}{E_K} \qquad [7.3.4]$$

Then since positive and negative energies balance and with negative mass numerically equal to positive mass, on average, the value of E_K in relation to the rest energy E_0 of the positive primary, from the logic given in pages 124 to 125 can be written:

$$\frac{E_K}{E_0} = \frac{1}{\sqrt{1-(v_P/C_U)^2}} - 1 + \frac{1}{\sqrt{1-(v_n/C_U)^2}} - 1$$

[7.3.5]

It will be assumed for initial analysis that $v_n = v_P$ but this can be made more general later. Furthermorc the random speed v of molecules has been replaced with an effective average random speed v_p of primaries for both negative and positive phases of the mixture.

The i-ther will be considered to have reached the state of 'i-theric liquidus' when the volume V_B of the combined volume of all its positive plus negative primaries is equal to one quarter of the space occupied of volume V. This is justified by analogy with liquids and Pryde (1966) gives this value.

At this state creation rate C_{RT} and annihilation rate C_A are equal so that net creation is zero. It follows that V_B/V is an important parameter, with the volume of a primary = $\sigma^3\pi/6$, it is given by:

$$\frac{V_B}{V} = n\frac{\pi}{6}\sigma^3 \quad So\ n^2 = \left(\frac{V_B}{V}\right)^2\left(\frac{6}{\pi}\right)^2\frac{1}{\sigma^6} \qquad [7.3.6]$$

Substituting for n^2 in equation [7.3.1] yields:

$$N_C = \frac{36}{\pi\sqrt{2}}\frac{v_P}{C_U}\frac{C_U}{\sigma^4}\left(\frac{V_B}{V}\right)^2 \qquad [7.3.7]$$

Only at the liquidus state is V_B/V = ¼ but [7.3.7] applies for any state but with V_B/V having a lower value. This rate of collision will be used later to obtain the breeding rate. This will cause an increase in both pressure P and volume V of an element of i-ther considered as a modification of the dynamical theory of gases that is presented by James Jeans (1887). The modification allows for the volume V_B occupied by primaries but will be entered as V_C since, as shown in *QUANTUM GRAVITY* § 3.5.8 and TABLE 3I, annihilation appears to cancel the effect of V_B so V_C is nearly zero. Then a perfect gas is closely simulated. The gas contains N primaries each of sum energy E_P in volume V and since $m = E_p/c^2$ it follows that:

$$P(V - V_C) = \frac{1}{3}Nmv_P^2 = \frac{1}{3}NE_P\left(\frac{v_P}{C_U}\right)^2\left(\frac{C_U}{c}\right)^2 \qquad [7.3.8]$$

Now overall, since primaries keep splitting, E_P remains constant on average but N, the total number in volume V, increases. Furthermore the average speed v_P will remain constant. Differentiating with respect to time yields:

$$(V - V_C)\frac{dP}{dt} + P\frac{dV}{dt} = \frac{1}{3}\frac{dN}{dt}E_P\left(\frac{v_P}{C_U}\right)^2\left(\frac{C_U}{c}\right)^2$$

Dividing the above by V & P and multiplying by dt results in:

$$\left(1 - \frac{V_C}{V}\right)\frac{dP}{P} + \frac{dV}{V} = \frac{1}{3}\frac{dN}{Vdt}\frac{E_P}{P}\left(\frac{v_P}{C_U}\right)^2\left(\frac{C_U}{c}\right)^2 dt \qquad [7.3.9]$$

Now *(dN/dt)/V* is the rate increase in number density of primaries caused by collision breeding. When multiplied by the average energy of each primary E_P the rate of energy density gain appears.

This is equal to the rate of collision of opposite primaries and to the average energy gain per collision. If of all collisions, proportion z are of opposites to produce breeding, the rate of energy gain per unit volume $\frac{dN}{Vdt}E_P$ of [7.3.9] can be replaced by the rate of energy gain per unit volume, with the rate of collision per unit volume N_C given by [7.3.7] which becomes:

$$\frac{dN}{V\,dt}E_P = z\,N_C\frac{\Delta E}{E_K}E_K = z\frac{\Delta E}{E_K}E_K\frac{36}{\pi\sqrt{2}}\frac{v_P}{C_U}\frac{C_U}{\sigma^4}\left(\frac{V_B}{V}\right)^2$$

This is more conveniently presented by introducing the rest energy density of a primary ε_{0P} since this will remain constant and is given by: $E_0 = \frac{\pi}{6}\sigma^3\varepsilon_{0P}$. This is to be substituted in the above for:

$E_K = \frac{E_K}{E_0}E_0$ so that: $E_K = \frac{E_K}{E_0}\frac{\pi}{6}\sigma^3\varepsilon_{0P}$. The result is:

The True Rate of Energy Creation per Unit Volume

$$\frac{dN}{V\,dt}E_P = \frac{6}{\sqrt{2}}z\frac{\Delta E}{E_K}\frac{E_K}{E_0}\frac{v_P}{C_U}\frac{C_U\varepsilon_{oP}}{\sigma}\left(\frac{V_B}{V}\right)^2 = C_{RT}\left(\frac{V_B}{V}\right)^2$$

Which can be expressed in units of J/m^3/s [7.3.10]

All constant terms have been condensed to C_{RT} suggesting that the rate of energy density gain is directly proportional to the square of $(V_B/V)^2$]. This means also approximately rising as the square of i-theric pressure. However, substituting this in equation [7.3.9] yields:

$$\left(1 - \frac{V_C}{V}\right)\frac{dP}{P} + \frac{dV}{V} = \frac{2z}{\sqrt{2}}\frac{\Delta E}{E_K}\frac{E_K}{E_0}\left(\frac{v_P}{C_U}\right)^3\left(\frac{C_U}{c}\right)^2\frac{C_U}{\sigma}\frac{\varepsilon_{0P}}{P}\left(\frac{V_B}{V}\right)^2 dt$$

[7.3.11]

The variable P has appeared so it is useful to introduce the constant P_L the pressure of the i-theric liqidus. Then this extra variable can be arranged to appear in the non-dimensional form P/P_L which is a function of V_B/V. Then further simplification should appear. First we turn attention to the connection between P/P_L and V_B/V.

Relation between P/P_L and V_B/V

The 'gas' of primaries will have a constant 'temperature' regardless of pressure change since v_P ,the average speed of primaries, does not change and so its law with C as a constant becomes:

$$(V - V_B)P = C \quad So \quad P = \frac{C}{V - V_B} \qquad [7.3.12]$$

For the liquidus where $P = P_L$: $V/V_B = 4$, as previously defined. So substituting in [7.3.12] gives $P_L = C/(3V_B)$. Dividing this into [7.3.12} yields:

$$\frac{P}{P_L} = \frac{3V_B/V}{1 - V_B/V} \qquad [7.3.13]$$

Then [7.3.13] can be re-arranged in the alternative form:

$$\frac{V_B}{V} = \frac{P/P_L}{(3 + P/P_L)} \qquad [7.3.14]$$

Substituting in [7.3.13] or [7.3.14] in [7.3.11] permits alternative forms to be provided and it is useful to first define the creation constant C_{BT} from [7.3.11] that is common to both as:

$$C_{BT} = z\sqrt{2}\frac{\Delta E}{E_K}\frac{E_K}{E_0}\left(\frac{v_P}{C_U}\right)^3\left(\frac{C_U}{c}\right)^2\frac{C_U}{\sigma}\frac{\varepsilon_{0P}}{P_L} \qquad [7.3.15]$$

The units are s^{-1}. Then the alternative equations become:

$$\left(1 - \frac{V_C}{V}\right)\frac{dP}{P} + \frac{dV}{V} = C_{BT}\frac{P/P_L}{(3 + P/P_L)}dt = C_T\,dt \qquad [7.3.16]$$

And

$$\left(1 - \frac{V_C}{V}\right)\frac{dP}{P} + \frac{dV}{V} = \frac{C_{BT}}{3}\left(1 - \frac{V_B}{V}\right)\frac{V_B}{V}dt = C_T\,dt \qquad [7.3.17]$$

Both alternatives are used in this book but [7.3.16] shows that when $P << P_L$ then the creation rate of an element of i-ther will be proportional to the pressure it has at that time.

These equations refer to the breeding rate prior to the onset of annihilation and so for general application to the history of the universe both need multiplying by the ratio of net creation to breeding rate given as C_N/C_R. This will be considered soon but first we need to show why annihilation has to exist for any creation scenario to be able to produce sensible predictions.

The energy density of i-ther

Before equations [7.3.15] to [7.3.17] can be used the ratio ε_{0P}/P_L has to be evaluated and we start with the semi-perfect gas law from which a start was made i.e..

$$3(V - V_B)P = N E_P (v_P/c)^2 \quad \& \quad \varepsilon = N E_P / V$$

It follows that the i-theric energy density ε is given by:

$$\frac{\varepsilon}{P} = 3\left(1 - \frac{V_B}{V}\right)\left(\frac{C_U}{v_P}\right)^2\left(\frac{c}{C_U}\right)^2 \qquad [7.3.18]$$

Now P_L occurs when $V_B/V = ¼$. Then substitution in [7.3.18] yields

$$\frac{\varepsilon_L}{P_L} = 3\left(1 - \frac{1}{4}\right)\left(\frac{1}{0.75}\right)^2\left(\frac{1}{1.952}\right)^2 \quad but \quad \frac{\varepsilon_{0P}}{\varepsilon_L} = 4$$

The last statement follows from $V_B/V = ¼$.at the liquidus. Then the ratio of rest energy of a primary to P_L becomes: $\boldsymbol{\varepsilon_{0P}/P_L = 4.199}$

An estimate of the Breeding Rate

It is useful to make an estimate of the value of C_{BR} as will now be attempted. The value of z was originally thought to be 0.5 but inadvertently was based on the assumption of a uniform distribution of primaries. The video described in Chapter 1, however, had primaries moving randomly and, as shown best on the front cover, the bunching of opposites never coincide. A value of z = 0.05 seems more appropriate.

In Chapter 5 §5.7 p.139 the most probable value of speed ratio $v_P/C_U = 0.75$ was deduced, in which case $E_K/E_0 = 1.0237$ according to equation [7.3.5] p.176. Then from Chapter 5 a reasonable value for $\Delta E/E = 0.08$ is taken. The value of $C_U = 1.952 \times 3 \times 10^8$ m/s for making gravity waves move at the speed of light as shown previously in §7.2 p.174.

The value of ε_{0P} can be estimated from the energy density of space given by Starobinskiiand Zel'dovitch (1988) as 10^{45} J/m^3. This was used in *QUANTUM GRAVITY* for estimating the size of quarks based on gravity being a negative buoyancy force, yielding for the quark diameter the value of 4×10^{-19} m. Since this seemed very reasonable it suggests this huge value for density cannot be far wrong.

Now Earth cannot be close to the liquidus state otherwise

no density increase could occur at the surface of a neutron star that is shown in *QUANTUM GRAVITY* §3.9 to be 1.476 times that at Earth. The black hole of ECM theory could be about 5 times that at Earth and could correspond with the liquidus state in which V_B/V =0.25. This makes the Earth value of V_B/V = 0.05. Now filaments, being composed of opposed energies crushing each other out of existence, will have V_B/V = 1 this makes ε_{0P} = $10^{45}/0.05$ J/m^3 = 2×10^{46} J/m^3 .

The value of σ remains and only a maximum value can be estimated. This cannot be greater than about 1/10th the diameter of the quark quoted above and so $\sigma = 4\times10^{-20}$ m will be assumed.

With these figures equation [7.3.10] yields:

$\mathbf{C_{RT} = 3.8\times10^{71}}$J/m^3/s.& $C_{RT}\times(V_B/V)^2 = \mathbf{C_{BR} = 9.5\times10^{68}}$ W/m^3

This is absolutely huge! Not only does it dwarf the heat output of the sun: it exceeds the output of the entire galaxy! It shows how vitally important is the introduction of a near-cancelling annihilation.

In [7.3.15] to [7.3.17], however, the creation rate C_{BR} is applicable to equations that enable pressure and volume development to be predicted. For the same V_B/V =0.05 and other data as used previously the value of:

$\mathbf{C_{BT} = 5.72\times10^{26}}$ s^{-1} results.

Then using [7.3.17] with V_B/V = 0.05 yields the breeding rate C_T of value:

$\mathbf{C_T = 9.06\times10^{24}}$. s^{-1}.

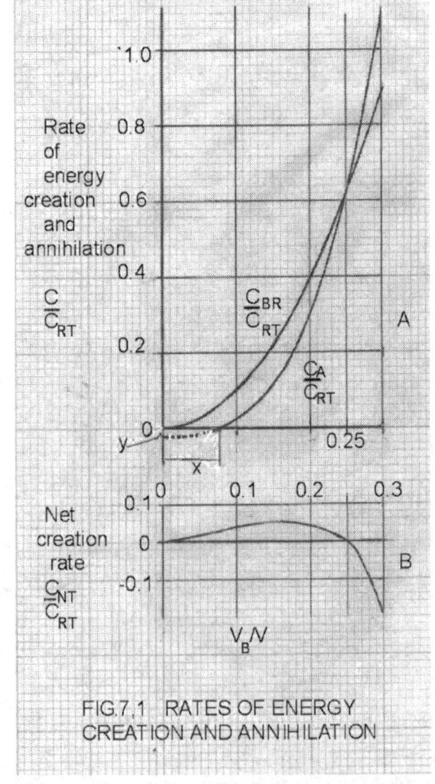

FIG.7.1 RATES OF ENERGY CREATION AND ANNIHILATION

7.4 ANNIHILATION

To be shown soon is $C_R = H_0/4$ ($H_0/3$ later) where H_0 is the Hubble constant and is about 2.302×10^{-18} s^{-1} so the value deduced above is 1.6×10^{43} times too large!

A massive kind of annihilation has to exist if the Big Breed theory is to survive!

Taking Annihilation into account began with the following

simple logic – DEFINING the I-THERIC LIQUIDUS

If no free space exists then $V_B/V = 1$. If primaries are spherical, however, then to fill all the space available, positive and negative kinds must overlap and cancel each other out in the overlapping regions. Clearly the ultimate density has been reached and is, in fact, the condition existing within annihilation cores.

These spherical primaries are now imagined arranged on a cubic lattice just touching one another at six points. Then since a sphere occupies about a half the volume of a containing cube a value of $V_B/V = 1/2$ is induced

The point is that at some critical value of V_B/V no net creation will obtain and at higher values annihilation will be dominant. The critical value of $V_B/V = 1/4$ has been chosen to define the **i-theric liquidus state** since, according to Pryde (1966), the molecules of a liquid have this value. Consequently at all lower densities a net creation will obtain.

This logic is retained for all analyses but all except one had to be superseded and so, before reading further, please note the following:

NOTE – bypass suggested to p.183 §7.5 - see why below

A first attempt to derive a working hypothesis for taking annihilation into account is illustrated in FIG.7.1. This was superseded later but it is assumed that the reader may wish to see how the final solution emerged. It is difficult to see the correct route in the initial stages of any pioneering effort. It should be useful; therefore, to see how wrong initial concepts can lead to much time wasted in going up blind alleys. I think it may also be useful to see how, what seems obvious from hindsight, can be missed.

Since the reader will probably not wish to waste time following maths that had to be abandoned later, advice on bypassing certain sections will be given on or near section headings.

.

The next section also assumed another point would exist where no net creation would occur: where pressure is zero. This led to FIG.7.2, based on [7.4.1] to [7.4.4] - maths that was later abandoned.

Equation [7.3.10] p.177 shows the breeding rate in W/m^3 proportional to $(V_B/V)^2$. Now up to a first critical value of (V_B/V), to be denoted x, no annihilation will occur since no cores will form. After that the annihilation rate will be assumed to increase at the rate proportional to $(V_B/V)^{(2+n)}$ where n is a small positive number. In this way annihilation rises more rapidly than creation and it is arranged that the two curves cross at $(V_B/V)=1/4$ - the i-theric liquidus state.

The proposal is illustrated in FIG.7.1 (Chapter 10 for final choice) [7.3.10] can be reduced to: $C_{BR} = C_{RT}(V/V_B)^2$ [7.4.1]
And with C_A as the annihilation rate, with C_{RA} introduced as the annihilation rate constant, the annihilation rate can be assumed as:

$$C_A = y + C_{RA}(V_B/V)^{(2+n)} \qquad [7.4.2]$$

These two curves are to cross at the liquidus state where $V_B/V=1/4$
Equation [7.4.1] then gives $C_{BR} = C_{RT}/16$ so here $C_A = C_{RT}/16$.
We assume values for x and n so that y and C_{RA} need to be evaluated. Starting with a false zero 0' by putting $y=0$ then from [7.4.2] the value of C_A obtained yields y as: $y = C_{RA}x^{(2+n)}$
The false zero 0' is below 0 by amount y and so y becomes negative, i.e. $y = -C_{RA}x^{(2+n)}$ [7.4.3]
Then substituting for C_A and y in [7.4.2] results in:

$$C_{RT}/16 = -C_{RA}x^{(2+n)} + C_{RA}(1/4)^{(2+n)}$$

Which can be rearranged to yield:

$$C_{RA} = \frac{C_{RT}/16}{(1/4)^{(2+n)} - x^{(2+n)}} \qquad [7.4.4]$$

Then y is given by equation [7.4.3] and together with C_{RA} from [7.4.4] gives a complete expression for C_A in [7.4.2]

FIG.7.1 is plotted with $n = 1.0$ and $x = 0.075$. This greatly exaggerates, by many orders of magnitude, the difference between the creation and annihilation curves C_{BR} and C_A respectively. The curves yield the net creation rate C_{NT} for the range of V_B/V of 0 to 0.3 and cross at 0.25. The net rate of creation is the difference between the two curves shown at **A**. At **B** this difference is plotted, noting that the dashed part of curve C_A is ignored since in this region of $V_B/V < x$. there is no annihilation.

Annihilation cores only arise at $V_B/V = x$.

In FIG.7.2 an enlargement of the lower part of FIG.7.1 is shown. The upper curve A is C_{NT} whilst the lower one B is multiplied by $(1-V_B/V)V_B/V$ (for a reason to appear later).

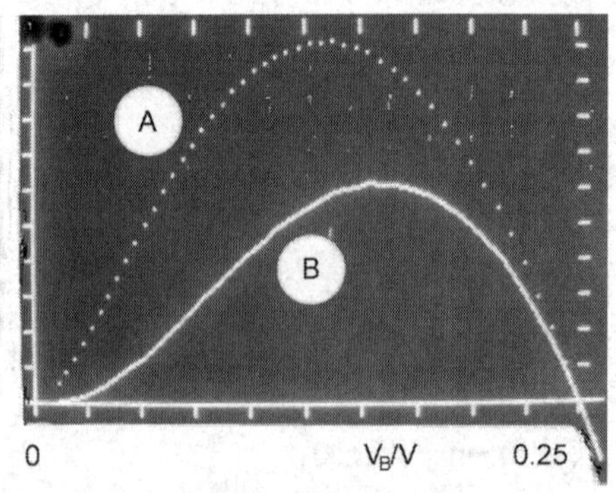

FIG.7.2 CREATION RATE CURVES

What has been ignored is the initial 'inflation'-like creation before annihilation cores have formed. This would appear as an instantaneous jump in pressure.

It needs to be understood that these curves are not rigorous derivations and can only serve as an initial guide. They do, however, seem to support the assum-ption that, in the initial stages, the effective creation rate will increase roughly linearly with pressure increase after that sudden jump. The next section suggests that in the present era the creation rate will be constant, since that is the assumption to be made, and yields a result supported by experiment.

A Numerical Approach is Attempted

At this stage it was thought that only a step by step computer simulation would be able to solve the formidable mathematical problem involved in creation. This was programmed but is left to APPENDIX V following PART II for presentation. The reason is that no satisfactory result was achieved at that time.

Starting with an initial assumed pressure/radius profile about six time steps would show no change in position of the edge of the universe, the initial profile would remain of unchanged shape except for pressures increasing at the centre point but not at the edge. They just pivoted about the edge and then the computation would go wildly unstable with plotted points scattering all over the screen. Consequently tedious approximate methods had to evolve before the numerical approach could be expected to achieve.

To obtain a solution to equation [7.3.17] in which the net creation rate $C_{NT} = C_{BR} - C_A$ is substituted for C_{BR} when $V_B/V > x$, a pressure gradient is needed to accelerate an element of volume. This will also be considered in the following derivations.

7.5 The Universe Growing due to Net Creation

This section is presented as it was conceived about 1995 before delving into the details of collision breeding given in the previous section. The simplest possible assumption was made of a net creation rate per unit mass that was constant.

As will be seen a startling result emerged!

The derivation is well worth studying since it makes an interesting connection with Hubble's law.

The effect of the space taken up by primaries in the i-ther is assumed cancelled by the way annihilation acts so that the perfect gas law can be assumed to apply given by:

$$PV = \tfrac{1}{3} n m_p v_p^{\,2} \quad \& \quad d(PV) = \tfrac{1}{3} m_p v_p^{\,2}\, dn = \tfrac{1}{3} v_p^{\,2}\, dm$$

Here v_p is the average speed of primaries that remains constant everywhere, m_p is the average mass of a primary and n is the number of primaries. Then $m_p dn = dm$ the change in mass. It then follows that with a constant net creation rate C_{NA} assumed:

$$\frac{dm}{m} = \frac{d(PV)}{PV} = C_{NA}\, dt$$

After differentiating PV by parts and dividing by PV we have:

$$\frac{dP}{P} + \frac{dV}{V} = C_{NA}\, dt \qquad [7.5.1]$$

Now the only possible shape for any medium, such as i-ther, driven to expand by a net creation everywhere, is a sphere. This is because, as will soon be proved by the following derivation, the expansion must accelerate: like an explosion in very slow motion. In order to provide the pressure gradients needed to produce the acceleration, pressures will always maximise at the central origin point, and fall off toward the growing edge. Hence for a spherical shape the volume of the sphere can be differentiated with respect to radius so that:

$$V = \tfrac{4}{3}\pi r^3 \quad \text{So } dV = 4\pi r^2\, dr$$
$$\therefore\ dV/V = 3\,dr/r \qquad [7.5.2]$$

It follows that by substitution in [7.5.1]:

$$\int_{P1} \frac{dP}{P} + 3\int_{r1} \frac{dr}{r} = \int_{0} C_{NA}\, dt$$

Which integrates to yield:

$$\frac{1}{3}\ln\left(\frac{P}{P_1}\right) + \ln\left(\frac{r}{r_1}\right) = \frac{C_{NA}}{3}t$$

Which can be re-arranged to yield:

$$r = r_1\left(\frac{P}{P_1}\right)^{-\frac{1}{3}} \exp\left(\frac{C_{NA}}{3}t\right) \qquad [7.5.3]$$

7.5.1 Deriving the Hubble Law

Unfortunately equation [7.5.3] cannot be solved unless a relation between r and P can be found. In the next section a first attempt to derive pressure/radius profiles are illustrated at the top of FIG. 7.7.p.189. It had seemed reasonable to conclude from [7.5.3] that profile shapes remain similar as time moves on and the radius of the universe increases. In consequence FIG.7.7 shows straight path lines radiating from the origin. These path lines are the trajectories of objects moving with the i-ther. As will be seen this was a false deduction that caused many years of wasted effort. Fortunately the main conclusion reached at this stage in equation [7.5.11] is still valid.

It follows, as shown by FIG.7.3 that for a short time step δt in which the radius increases by δr and the pressure increases by δP that in the limit as δt becomes dt:

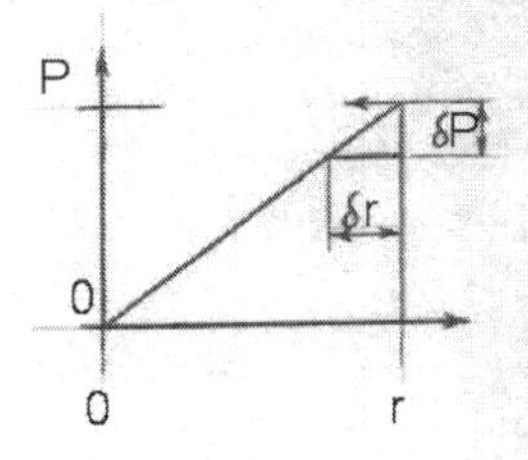

FIG.7.3 PATH LINE ON P/r PLANE

$$\frac{dP}{dr} = \frac{P}{r} \quad Consequently \quad \frac{dr}{r} = \frac{dP}{P} \qquad [7.5.5]$$

Substituting for dP/P from [7.5.5] and for dV/V from [7.5.2] in [7.5.1] results in:

$$\frac{dr}{r} + 3\frac{dr}{r} = C_{NA}\, dt : \; i.e. \; \int_{r1} \frac{dr}{r} = \frac{C_{NA}}{4} \int_0 dt \qquad [7.5.6]$$

And integrates to:

$$r = r_1 \exp\left(\frac{C_{NA}}{4} t\right) \qquad [7.5.7]$$

Which yields velocity $v = dr/dt$ on differentiation to yield:

$$v = r_1 \frac{C_{NA}}{4} \exp\left(\frac{C_{NA}}{4} t\right) \qquad [7.5.8]$$

Substituting from equation [7.5.7] in [7.5.8] yields the result:

$$v = \frac{C_{NA}}{4} r \qquad [7.5.9]$$

This is Hubble's Law $v = H_0 r$ and so the 'Hubble constant' H_0 is related to the net creation rate by:

$$C_{NA} = 4H_0 \quad \text{(revised later see below)} \qquad [7.5.10]$$

Differentiating [7.5.8] to obtain the acceleration f, after substitution from [7.5.9], yields:

$$f = H_0^2 r \qquad [7.5.11]$$

Hence it appears that when continuous creation is involved Hubble's law is predicted and also leads to the prediction of a universe in a state of ever-accelerating expansion.

Chapter 10 shows $C_{NA} = 3H_0$. A final solution that minimises errors of allowing for pressure increase is provided in Chapter 9.

7.5.2 Estimating the rate of net creation

In *QUANTUM GRAVITY* FIG.7.2 was plotted from equation [7.5.7] where it was shown that, from the same red-shift data, but using ECM theory instead of general relativity, the Hubble constant reduced from 71 km/s/Mpc to about 29 km/s/Mpc. These figures convert respectively to 2.302×10^{-18} s^{-1} & 9.4×10^{-19} s^{-1}

From [7.5.10] $C_{NA} = 2.302\times10^{-18}/4 = \mathbf{9.21\times10^{-18}}$ s^{-1}

Now this value of C_{NA} can be compared with the breeding rate having the same units that was evaluated using equation [7.3.12] yielding the value of $\mathbf{C_T = 9.06\times10^{24}}$ s^{-1}

This enables the operating annihilation ratio C_{NA}/C_T to be estimated. The result becomes: $\mathbf{C_{NA}/C_T = 10^{-42}}$

This means that only an incredibly small fraction of the breed creation rate remains after annihilation in the present era.

It also means that no computer can select a value of n in equations [7.4.2] to [7.4.4] small enough to match this result.

Consequently the smallest possible values of both x an n have been used in those equations to yield the smallest possible computed ratio. has been applied to reduce to a match Then a scale factor with observation. Fortunately the profile of the C_{AN} against V_B/V curve has been found to vary very little by the choice of x. As already described the net creation rates are shown in FIG.7.2

7.6 PRESSURES NEEDED FOR ACCELERATION

To accelerate each spherical shell element of volume $4\pi r^2 dr$ a pressure difference from inside to outside is required. Hence P will fall as shells having larger values of r are considered. It is therefore now necessary to derive the pressure profiles required.

A net accelerating creation will exist throughout the i-ther leading to its continual accelerating growth. Pressure will build up, being a maximum at the origin point and reducing toward the outer growing edge. The induced pressure gradients will cause acceleration of each element of volume. Since each such element is growing, its mass is also increasing and so the pressure gradients will need to take into account this rate of mass increase as well as just accelerating the existing mass. The hope is to obtain an equation that, together with equations [7.3.17] p.178 and [7.4.1] to [7.4.4] p.182 will enable a complete solution to be achieved.

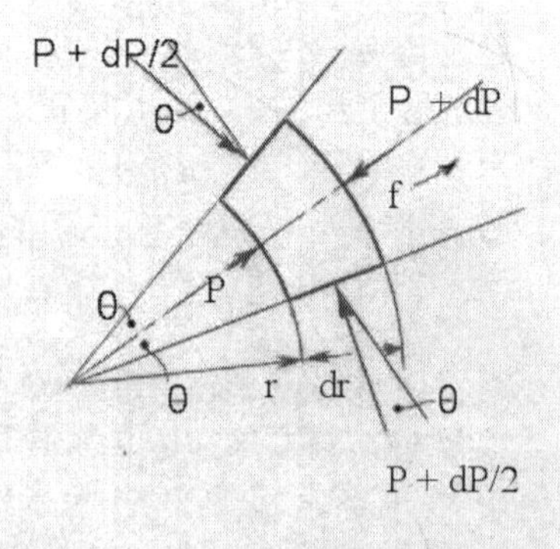

FIG 7.6 ACCELERATING AN ELEMENT OF I-THER BY A PRESSURE GRADIENT

An element is to be considered as part of the surface of a sphere of radius r whose extent is limited by the intersection of a conical surface of very small half-angle θ whose apex touches the sphere's centre, as shown in FIG.7.6. The pressure P acting outward on the surface area $\pi(r\theta)^2$ is opposed by a pressure $(P + dP)$ acting on the outer surface area $\pi(r + dr)^2\theta^2$. However, the rim of the element at radius $r\theta$ is part of the surface of a cone adding an extra outward force of $(P + dP/2)\ \pi\theta^2\ (r + dr/2)\ dr$. Summing these forces and eliminating terms greater than first order yields the result:

$$dF = -\pi(r\theta)^2 dP \qquad [7.6.1]$$

The next section shows the logic first used to derive the initial equations for accelerating expansion. This leads to a new maths theorem. It is therefore worthy of study even though superseded.

The force dF given by [7.6.1] produces a change in momentum mv in time dt and owing to creation both m and v increase. However, a 'phase coupling factor' C_{PF}, a number between 0 and 1 is introduced but the reason for this was explained in §6.6 p.164.

Hence applying differentiation by parts:

$$dF = C_{PF}\frac{d(mv)}{dt} = C_{PF}\, m\frac{dv}{dt} + C_{PF}\, v\frac{dm}{dt} \qquad [7.6.2]$$

Now m is the element of mass given by:

$$m = \rho\pi(r\theta)^2 dr \qquad [7.6.3]$$

The kinetic theory of gases is applicable as given in equation [7.3.8] p.176 but now using m as total mass content instead of Nm. From this the mass density ρ can be expressed as:

$$\rho = 3\frac{P}{v_P^{\ 2}}\left(1-\frac{V_C}{V}\right) \quad \text{since} \quad \rho = \frac{m}{V}$$

:Substituting this for ρ in [7.6.3] yields:

$$m = \frac{3\pi\theta^2}{v_p^{\ 2}}\left(P\left(1-\frac{V_C}{V}\right)r^2\, dr\right) \qquad [7.6.4]$$

The effective value of V_C/V can be almost zero since the effect of annihilation can almost cancel the volume occupied by primaries. With this simplification differentiation to obtain dm/dt yields:

$$\frac{dm}{dt} = \frac{3\pi\theta^2}{v_p^{\ 2}}\left(\frac{dP}{dt}r^2 dr + P\frac{d(r^2 dr)}{dt}\right) \qquad [7.6.5]$$

Since $v = dr/dt$ this can be substituted in the above. Then substituting the revised [7.6.5] and [7.6.4] (with $V_C/V = 0$) in [7.6.2] with dF replaced from [7.6.1] yields:

$$-\pi(r\theta)^2\,dP = \frac{3\pi\theta^2 C_{PF}}{v_P^{\ 2}}\left[P r^2 dr\frac{dv}{dt} + v\left(\frac{dP}{dt}r^2 dr + P\frac{d(r^2 dr)}{dt}\right)\right]$$

Which after dividing by $\pi P(r\theta)^2$ and noting that $dr/dt = v$, yields:

$$-\frac{dP}{P} = \frac{3C_{PF}}{v_p^{\ 2}}\left[v\,dv + v^2\frac{dP}{P} + v\frac{d(r^2 dr)}{r^2\,dr}\frac{dr}{dt}\right] \qquad [7.6.6]$$

re-arranging yields:

$$-\frac{dP}{P}\left(1 + 3C_{PF}\left(\frac{v}{v_P}\right)^2\right) = \frac{3C_{PF}}{v_P^{\ 2}}\left[v\,dv + v^2\frac{d(r^2\,dr)}{r^2\,dr}\right] \qquad [7.6.7]$$

The last term in the above presented a difficulty that needed to be addressed: An attempt was made in 2005 to treat this as the differentiation of a product but this did not succeed. Resort was made to an approximate resolution of the difficulty.

However, in 2010 I found a solution that might be considered a new maths theorem. This resolved the difficulty and fortunately showed the original attempt had given the correct solution. The revision is given in §7.9 p.193 from which:

$$v^2\frac{d(r^2\,dr)}{r^2\,dr} = v^2\frac{d(r^2)}{r^2} = v^2\frac{2\,dr}{r} \qquad [7.6.8]$$

Another difficulty is that integration cannot be carried out until v can be expressed in terms of r. To make a start the expedient of using Hubble's law will be introduced. This gives a linear relationship for the speed of recession v of remote galaxies and is given by:

$v = H_0\,r$ where H_0 is the 'Hubble constant', which as shown in [7.5.7] and [7.5.8] p.185, is consistent with the solution. Then the expression in [7.6.8] becomes:

$$v^2\frac{2\,dr}{r} = H_0^{\ 2}\,2r\,dr$$

Also differentiating $v = H_0 r$ yields $dv = H_0\,dr$ and so:

$$v\,dv = H_0^{\ 2}\,r\,dr$$

Substitution in [7.6.7] (with C_{PF} included as explained below) then yields;

$$-\int_{P0}\frac{dP}{P} = 3C_{PF}\left(\frac{H_0}{v_P}\right)^2\int_0\frac{3r\,dr}{1 + 3C_{PF}(H_0/v_p)^2 r^2}$$

This integrates, to yield:

$$\frac{P_0}{P} = \left(1 + 3C_{PF}\left(\frac{H_0}{v_p}\right)^2 r^2\right)^{\frac{3}{2}} \qquad [7.6.9]$$

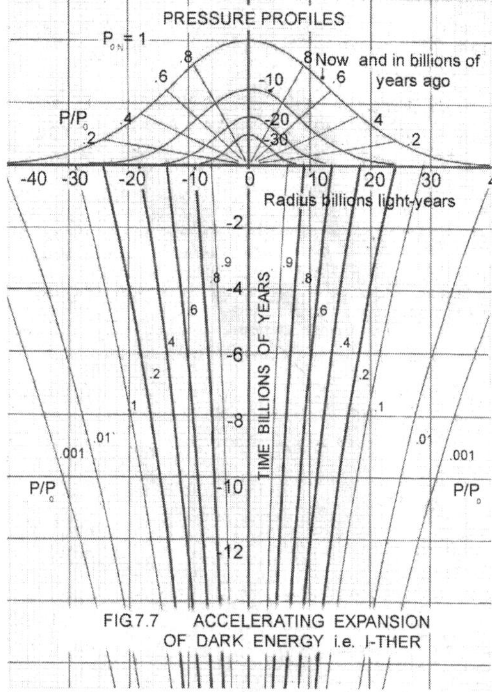

FIG.7.7 ACCELERATING EXPANSION OF DARK ENERGY i.e. I-THER

Where, if desired, from [7.5.10] p.185 H_0 can be replaced with $C_{NA}/4$.

Equation [7.6.9] gives the pressure profile where P_0 is the pressure at the origin point of the universe where r=0 and P is the pressure at any other radius r. The term C_{PF} is a "phase coupling factor" that was included in equation [7.6.2] p.187 as a multiplier for dF without being given an explanation. This will now be explained.

Primaries are not considered to be matter, though they form the ultimate real particles from which matter derives according to the theory presented in *CREATION SOLVED?*. They breed by the collision of opposites as described in Chapter 1 with positive and negative primaries having oppositely directed momentum arrows. If total momentum for mass motion were to be transmitted from one to the other then the entire mass could be accelerated without the application of any force or pressure differential.

Such a condition would define $C_{PF} = 0$

However, if the collision of opposites is confined to breeding and no coupling between the two phases exists for motions on a larger scale, then this defines $C_{PF} = 1$. Then each phase is accelerated by its own pressure differentials and is not affected by the other phase.

In 2005 the best estimates at the time indicated that a coupling factor $C_{PF} = 0.34$ would be a reasonable value.

FIG.7.7 is a plot showing expansion of the i-ther using this value. The figure is also based on a Hubble constant reduced from the value of 71 km/s/M_{PC}, taken from data by Wright (2005) using the derivation given in *QUANTUM GRAVITY* regarding its FIG.7.2.

FIG.7.7 is a plot of time in billions of years against a base of radius of the growing sphere of i-ther. The plot shows several path lines for particles fanning out and illustrating an accelerating expansion. At the top of the figure pressure profiles from equation [7.6.9] are shown for several instants 0, 10, 20 & 30 billion years ago. On these are also plotted the same path lines fanning out from the origin point. What this indicates is that pressures must continually rise everywhere in order to maintain the pressure gradients needed to produce accelerating expansion.

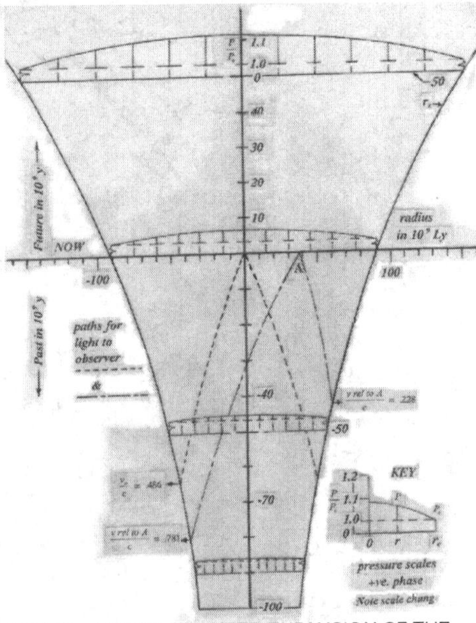

FIG.7.8 SHOCK-FRONTED EXPANSION OF THE UNIVERSE: CONSTANT CREATION RATE

The solution assumes a constant creation rate that is supported by astronomical data for the rate of recession of galaxies that are not too far distant. A worrying feature is that no definite edge for the universe is predicted.

One solution, produced about 2004, is presented in FIG.7.8. This assumed that the initial creation had no annihilation cores and produced a pressure jump. This meant the edge of the i-ther would be like a shock front raising the pressure suddenly to a value from which a gradual pressure increase would then occur. The net creation rate was assumed constant everywhere in space and time.

Either solution describes the accelerating expansion discovered by astronomers. They found this was consistent with the idea of remaining quiescent for many billions of years and then suddenly developing a long-range repulsive force. The nature of exponential growth always gives this false impression.

7.7 Is a solution with pressure constant possible?

Since the forgoing analyses can only be regarded as approximations it seems worthwhile to check that a simple and exact solution could obtain. Two cases will be considered.

CASE 1. Pressure at the origin point is assumed to remain constant and all elements of i-ther also remain at constant pressure as expansion continues. The shapes of the pressure/radius profiles will then all be similar at all instants but with all radii constantly increasing. It follows that the radial pressure gradients across any element of volume would keep reducing. Hence the acceleration of any element of i-ther would constantly reduce. Indeed for any spherical shell element of thickness b at radius r from the origin point and with A as a constant, the result of this logic yields:

$$r\frac{d^2r}{dt^2} = A \qquad [7.7.1]$$

This was the basis of the plot given in *QUANTUM GRAVITY* and presented as FIG.7.3. However, the derivation will not be included here since it has recently been realised that the derivation shows A cannot remain constant.

CASE 2 Shows why Fred Hoyle went wrong

The spherical shell element of thickness b at radius r will now be assumed to expand at a constant rate: no acceleration. In this way a mathematically exact solution can be guaranteed.

This allows $dr/dt = v = v_e\, r/r_e$ where v_e and r_e represent the speed of recession and radius at the outer edge of the growing sphere. Indeed for any element b both the thickness b and the speed v have to be directly proportional to radius for the whole sphere to grow at a uniform rate. No acceleration takes place in any element and so no pressure gradient applies. The outer edge moves into the void like a shock wave behind which a constant and uniform pressure prevails everywhere.

The problem is to determine the creation rate at any point that is required to permit such continuous expansion.

Since b/r and v remain constant with respect to time for any such element and $r = v \times t$, the volume V_b of any representative elemental spherical shell will be given by:

$$V_b = \frac{4}{3}\pi(vt)^3\left(\left(1+\frac{b}{r}\right)^3 - 1\right) \qquad [7.7.2]$$

Differentiating to find the rate of change of volume of element b we have:

$$\frac{dV_b}{dt} = 4\pi v^3 t^2\left(\left(1 + \frac{b}{r}\right)^3 - 1\right) \qquad [7.7.3]$$

But the rate of growth is given by equation [7.5.1] p.184 with $dP = 0$. It follows that:

$$\frac{dV_b}{V_b} = C_{NA}\, dt \qquad [7.7.4]$$

So substituting V_b from [7.7.2] into a re-arranged [7.7.4] yields:

$$\frac{dV_b}{dt} = \frac{4}{3}\pi v^3 t^3 \left(\left(1 + \frac{b}{r} \right)^3 - 1 \right) C_{NA} \qquad [7.7.5]$$

Equating [7.7.3] to [7.7.5] yields the simple result:

$$C_{NA} = \frac{3}{t} \qquad [7.7.6]$$

This shows the creation rate to reduce, everywhere in the growing sphere, inversely as time proceeds.

This is an impossible condition since if P remains constant so does V_B/V and therefore C_{NA} will not change with time Hoyle's theory had the universe growing due to the continuous creation of atoms: not i-ther, but the objection remains. He also assumed the universe had no beginning or end and was infinite in extent. Consequently this concept would have required an ever- reducing rate of creation and no acceleration anywhere.

Consequently the true solution has to be somewhat similar to the solution given in FIG.7.7 or FIG.7.8.

7.8 There are other difficulties

Our derivation contains two logical errors (not in the maths). These are corrected in the next chapter. Congratulations to any astute reader who has spotted these already! I did not appreciate their existence until about 2007. However, established physics must also be applying wrong analyses creating far greater error.

Galaxies must move with the expanding i-ther alias 'local space' and light from them must travel at a speed c relative to their local space. Viewed from Earth photons are travelling against the recession velocity v and so appear to move towards us at velocity $c - v$ or $c - H_O r$. It follows that light speeds up as it approaches and so the energy of photons needs to reduce so that momentum is conserved at any point. A huge red shift is involved. There is still a Doppler shift as well since light from the tail of a supernova is emitted at a greater distance from that of its head.

This means that cosmologists, basing their analyses on special relativity as described by Wright (2005), could have overestimated the Hubble constant. The value given by Wright is 71 km/s/M_{PC}. The analysis given in *QUANTUM GRAVITY* § 7.2 has shown that this value is reduced to about 29 km/s/M_{PC}. It is part of the solution for quantum gravity.

7.9 New Maths theorem appeared in May 2010

Equation [7.6.7] p.187 contains the term: $\int \frac{d(r^2\,dr)}{r^2\,dr}$ [7.9.1]

This presents a mathematical difficulty. To solve this substitute:

$$z = r^2\,dr \quad Then \quad dz = d(r^2\,dr) \quad Hence:$$

$$\int_{r1}^{r} \frac{d(r^2\,dr)}{r^2\,dr} = \int_{1}^{z} \frac{dz}{z} = \ln\left(\frac{z}{z_1}\right) = \ln\left(\frac{r^2\,dr}{r_1^{\,2}\,dr}\right) = \ln\left(\frac{r^2}{r_1^{\,2}}\right) = 2\ln\left(\frac{r}{r_1}\right).$$

$$\therefore \quad \frac{d(r^2\,dr)}{r^2\,dr} = 2\frac{dr}{r} \qquad [7.9.2]$$

This is the solution to [7.9.1] which assumes *dr* the same for all elements - which does not seem unreasonable.

7.10 CONCLUSION TO CHAPTER 7

The rate of breeding by the collision of primaries was estimated on page 180 as the huge value of 3.8×10^{71} J/m^3/s. This is equivalent to the entire rate of energy radiation from about 10^{27} galaxies like our own each consisting of about 10^{11} stars. And this in every cubic metre of space! However, after annihilation, although the rate is still huge, it reduces to the equivalent of only 2×10^{-6} Suns i.e two millionths of the total radiation of the sun.

Considerable effort was made to provide a convincing solution for the growth of the i-ther and therefore of the universe. Pressures needed to be greater at the origin point than at the growing edge as a consequence of a net creation existing everywhere. So far the solutions obtained cannot be regarded as sufficiently accurate.

However, a solution provided a derivation of Hubble's Law and showed that a universe growing by net energy creation is committed to a state of ever-accelerating growth. This was fundamental to the theory published in 1992 before the discovery of acceleration by astronomers as reported by Schwarzchild (1998). Pearson (1994) published this in Russia and in the peer-reviewed scientific journal, Frontier Perspectives, in 1997 – also before discovery.

A mathematically exact solution for a universe expanding, as described by Sir Fred Hoyle, showed the rate of creation needed to fall inversely with increase of radius and time. This is impossible.

In the next chapter we look for other reasons for the difficulties encountered, find a simplification due to a misconception, and then attempt to find a more satisfactory solution.

CHAPTER 8

REFINING ANALYTICAL THEORY FOR GROWTH OF THE I-THER

In Chapter 7 attempts were made to provide derivations for the expansion of the universe due to a net creation occurring everywhere. None were entirely satisfactory. This chapter begins by discovering a conceptual error that actually provides simplification and increases the estimated size of the universe.

Then an equation is found showing how the pressure P_0 at the origin point of the i-ther increases with time as its accelerating growth develops. This was not inserted until 2010.

Pressure/radius profiles are then refined by concepts that had been explored before the previous addition was made. These are included for completeness and to discourage others from following the false trail that is now apparent.

8.1 No Force is associated with mass increase

An important error arose in §7.6 p.187 when assuming $F = d(mv)/dt$ i.e. that force equals rate of change of momentum with mass increasing as well as speed. It is true that mass has to increase due to continuous creation but the reasoning failed to take account of the fact that the i-ther is a composite of both positive and negative mass.

As shown in Chapter 1, positive and negative primaries both constantly gain mass in balanced amounts as they repeatedly collide. Also they arise with cancelling momentum gains regardless of any collective accelerating motion. This means that no pressure differential is required associated with mass increase. Only the existing mass within any given element of volume requires a pressure differential to provide acceleration. This is because the two phases of the mixture, in bulk, do not couple strongly. If they did no pressure differential would be needed at all. The problem is actually simplified since each phase of the mixture is accelerated by its own pressure differential that is reduced in the proportion C_{PF}, the phase coupling factor. So now equation [7.6.2] p.187 can be revised for the force dF, as illustrated in FIG.7.6 p.186 and acting on an element of i-ther of thickness dr, gives rise to:

$$dF = -\pi(r\theta)^2 dP = C_{PF} dm\, dv/dt \qquad [8.1.1]$$

And $dm = \rho\pi(r\theta)^2 dr$ where $\rho = 3P/v_P^{\,2}$

Showing that for the i-ther, treated as a perfect gas, the density ρ is related to pressure P and the average speed v_P of the primaries. Combining with [8.1.1] yields:

$$-\int_{P0} \frac{dP}{P} = \frac{3C_{PF}}{v_P^{\,2}} \int_0 dr \frac{dv}{dt} \quad \& \quad \frac{dr}{dt} = v \qquad [8.1.2]$$

$$\text{So } -\int_{P0} \frac{dP}{P} = \frac{3C_{PF}}{v_p^{\,2}} \int_0 v\, dv \qquad [8.1.3]$$

It will be noted that this is far simpler than equation [7.6.6] p.188 that wrongly assumed dm/dt would involve an extra force.
Since the Hubble law is $v = H_0 r$, integration of [8.1.3] yields:

$$\ln\left(\frac{P_0}{P}\right) = \frac{3C_{PF}}{v_P^{\,2}} \frac{v^2}{2} = \frac{3}{2} \frac{C_{PF}}{v_P^{\,2}} (H_0\, r)^2 \qquad [8.1.4]$$

Now we come to see how preconceived ideas can lead the researcher onto a false track. This is of importance, not just for this book, but because I am convinced that a parallel diversion has happened in theoretical physics in the classical domain: the domain of my own expertise.

By this time (about 2000) I was convinced that all pressure/radius profiles would be of similar shape – as illustrated in FIG.7.7 p.189. Consequently an important yet very simple analysis, which should have been made about the year 2000 was not carried out until April 2010.

For readers who are not interested in the historical record, a bypassing of most of this chapter is therefore advised. Then in Chapter 9 an important new way of interpreting red shift data is considered that is needed for taking into account the existence of any background medium through which light must travel.

The remaining sections of this chapter are included only for the record and to highlight the dangers of being thrown onto false tracks that can lead the pioneer astray. It is my opinion that this illustrates how a parallel misdirection of effort has led physics into the difficulties its members now admit it faces. It may be of interest to see the magnitude of errors involved in the present analysis by comparison with the finally derived theory given in Chapter 10.

Therefore please only scan through the rest of this chapter having a look at the computed pressure profiles and then go straight to: 8.4 CONCLUSION TO CHAPTER 8 page 206.

THE NEXT ATTEMPT AT REFINEMENT circa 2005

Adding the Gamma factor

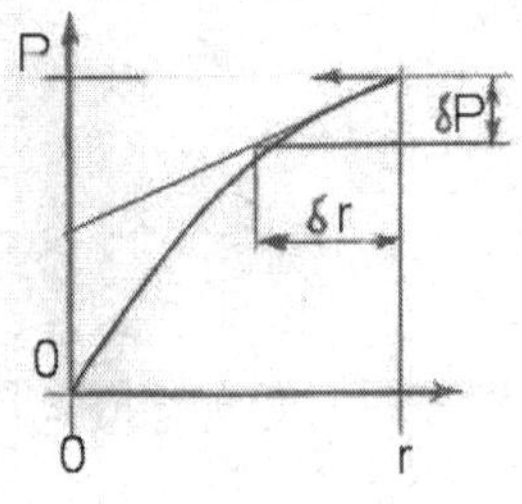

FIG8.1 CURVED PATH LINE ON P/r PLANE

Just in case the changes affect the relation between *dP/P* and *dr/r* we introduce γ a ratio illustrated in FIG.8.1. It is possible that path lines on the plot of *P* to a base of *r* will be curves so that tangents to these do not pass through the origin. So to allow for this possibility we assume:

$$\frac{dP}{P} = \gamma\frac{dr}{r} : \; So\ that : \; \frac{P}{P_1} = \left(\frac{r}{r_1}\right)^{\gamma} \qquad [8.1.5]$$

Then since, as shown in [7.3.16] p.178 and §7.5.1 p.184 for behaviour like a perfect gas:

$$\frac{dP}{P} + 3\frac{dr}{r} = C_{NT}\, dt \qquad [8.1.6]$$

Then substituting from [8.1.5] yields:

$$(3+\gamma)\int_{r1} dr/r = C_{NT}\int_{0} dt \qquad [8.1.7]$$

Which integrates to yield:

$$r = r_1 \exp\left[\left(\frac{C_{NT}}{3+\gamma}\right)t\right] \qquad [8.1.8]$$

And this differentiates to yield velocity *v* as :

$$\frac{dr}{dt} = v = r_1\left(\frac{C_{NT}}{3+\gamma}\right)\exp\left[\left(\frac{C_{NT}}{3+\gamma}\right)t\right] = \left(\frac{C_{NT}}{3+\gamma}\right)r$$

[8.1.9]

Which means that the Hubble constant H_0 in {8.1.4] has now become:

$$H_0 = \left(\frac{C_{NT}}{3+\gamma}\right) \qquad [8.1.10]$$

Equation [8.1.4] is, however, unaffected in any other way.

To reinforce the previous statement these equations give a perfect match with the Hubble law. It can be interpreted that a way has been discovered for determining part of the net creation curve as a function of pressure. The law previously obtained in §7.4 p.180 and illustrated in FIG.7.1 was only a first approximation. Now it is clear that for a considerable range of pressure the net creation rate is constant - showing that a large part of that first attempt was in fact close to an accurate prediction.

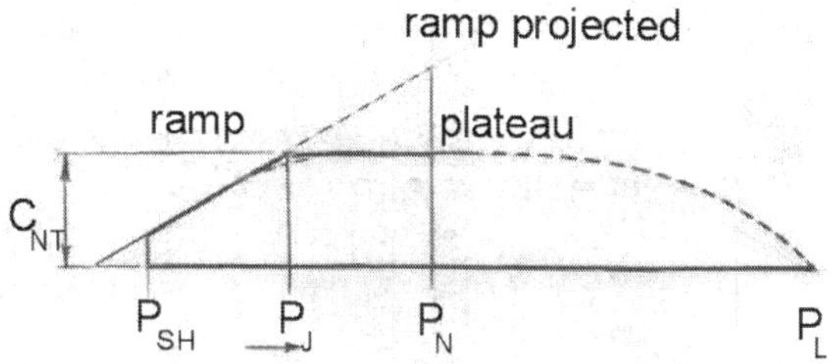

FIG.8.2 RAMP & PLATEAU CREATION RATE PROFILE

However, the entire pressure profile described by the right hand side of equation [8.1.4] cannot be correct. It will be correct as radius increases to a point where pressure has fallen to some limiting value P_J. Beyond that point a different relation between creation rate and pressure must obtain. This changed relation will also apply to all times prior to the transition to the creation rate plateau.

However, FIG.7.1 shows that a linear increase of creation rate with pressure prior to transition ought to be a reasonable assumption. These arguments led to the modified creation rate/pressure profile illustrated in FIG.8.2. No creation occurs until a sudden jump appears at pressure P_{SH}, corresponding to the initial shock of pure creation without any annihilation. This is followed by the 'ramp' stage in which creation increases in direct proportion to pressure until pressure P_J is reached. Then transition to the plateau region will be assumed to occur after pressure P_J has been reached. From then to the present era, where the pressure P_N is reached, the previously derived equations [8.1.4] to [8.1.10] will apply.

Ultimately as pressure increases to the liquidus value P_L the creation rate must fall to zero and then go negative. This part of the creation /pressure profile is shown dashed to indicate uncertainty since it is impossible to check by any experiment. Furthermore the sudden change from ramp to plateau is likely to be more of a gentle transition as also shown dashed but this refinement will be ignored.

Clearly the pressure gradients needed to provide accelerating expansion of the universe are modified by the profile shown in FIG.8.2. Part of the pressure/radius profile from zero to a radius r_J will have uniform creation rate and for greater radius the ramp will apply until the advancing edge is reached. There a shock front causes pressures to jump instantly to P_{SH}.

As we shall see this is going to lead to a maths inconsistency.

The next section will derive the pressure radius profiles for the initial phase when nct creation rate increases linearly with pressure.

8.2 The 'Ramp' phase: A solution with Creation Rate rising linearly with Pressure

This is to be an analytical solution for the case where $P<<P_L$ and $V_C = 0$ so that equation [7.3.16] p.178 is simplified to:

$$\frac{dP}{P} + \frac{dV}{V} = \frac{P}{P_J} C_{NT}\, dt \qquad [8.2.1]$$

The creation rate is $C_{NT}(P/P_J)$ when pressure is less than P_J and then remains constant at C_{NT} after the pressure P has exceeded the value P_J. The appropriate creation rate constant C_{NT} retains the same units as the Hubble constant. The combination $(P/P_J)C_{NT}$ makes the overall creation rate directly proportional to pressure instead of the uniform rate considered in nearly all previous derivations.

Path lines on the pressure/radius profile plot are assumed to be curves that satisfy equation [8.1.5] p.196.

Also, as shown in §7.5, [7.5.2] p.184 $dV/V = 3\,dr/r$ and from [8.1.5]:

$$\frac{dr}{r} = \frac{1}{\gamma}\frac{dP}{P} \qquad [8.2.2]$$

Where r_l and P_l are the radius and pressure of some arbitrary spherical shell element at the point where an arbitrarily chosen starting time t is chosen as zero. These factors allow [8.2.1] to be written as:

$$\left(1+\frac{3}{\gamma}\right) P_J \int_{P1} \frac{dP}{P^2} = C_{NT} \int_0 dt \,.$$

Which gives after integration:

$$(1+3/\gamma)\left[P_J/P_1 - P_J/P\right] = C_{NT}\, t$$

After re-arrangement this yields;

$$\frac{P_J}{P} = \frac{P_J}{P_1} - \frac{\gamma C_{NT}}{(3+\gamma)}t \qquad [8.2.3]$$

This means the pressure of an element is P_1 at time zero and steadily increases to P at time t until at time t_J, then $P = P_J$.
Since the radius r of the element is also increasing, and since [8.1.5] p.196 gives the relation between r and P then substitution and re-arrangement yields:

$$: r = r_J\left[\left(\frac{r_J}{r_1}\right)^{\gamma} - \frac{\gamma C_{NT}}{(3+\gamma)}t\right]^{-\frac{1}{\gamma}} \qquad [8.2.4]$$

This gives the radius of any element starting at r_1 at time zero. Its velocity v at time t can be found by differentiation, giving:

$$v = \frac{dr}{dt} = r_J\frac{C_{NT}}{(3+\gamma)}\left(\left(\frac{r_J}{r_1}\right)^{\gamma} - \frac{\gamma C_{NT}}{(3+\gamma)}t\right)^{-\frac{\gamma+1}{\gamma}} \qquad [8.2.5]$$

Differentiating again yields the acceleration of the element:

$$\frac{dv}{dt} = \frac{d^2r}{dt^2} = r_1(\gamma+1)\left(\frac{C_{NT}}{3+\gamma}\right)^2\left(\left(\frac{r_J}{r_1}\right)^{\gamma} - \frac{\gamma C_{NT}}{3+\gamma}t\right)^{-\frac{2\gamma+1}{\gamma}}$$

[8.2.6]

Now [8.2.4] can be used to replace the last term in brackets to yield the simplifications needed for deriving pressure profiles:

$$v = r_J\left(\frac{C_{NT}}{3+\gamma}\right)\left(\frac{r}{r_J}\right)^{(\gamma+1)} \quad \textit{and}$$

$$\frac{dv}{dt} = r_J(\gamma+1)\left(\frac{C_{NT}}{3\gamma+1}\right)^2\left(\frac{r}{r_J}\right)^{(2\gamma+1)} \qquad [8.2.7]$$

These equations show that. for $\gamma = 1$, the velocity increases as r^2 instead of just r as for the plateau case and with acceleration increasing as r^4 instead of r^2. However, simple though these derivations are considerable difficulties arise as soon as one attempts to use them in computer codes.

Suffix $_1$ in equations [8.2.3] to [8.2.6] means that an element of i-ther can be selected having an arbitrary radius r_1 between 0 and r_{SH} and its history of position r_1, pressure P_1 and speed v_1 will be provided. Fortunately r_1 is eliminated in equation [8.2.7] but the selection of r_J still presents a difficulty.

Clearly the arbitrary selection of this unit should not affect either the velocity or the acceleration for any value of r, yet these equations [8.2.7] imply that it does. It is therefore not possible to start at time zero to find v from considerations of the ramp stage on its own.

The methodology selected to overcome this difficulty is to specify P_J and P_N and work out the values of R_J and v_J from the plateau equation [8.1.4] p.195 giving the plateau pressure/radius profile since this gives r_J as a function of P_0/P_J. Then [8.1.9] p.196 yields v_J the velocity at P_J and r_J joining the plateau to the ramp profile.

The Ramp Pressure/Radius Profile

Equation [8.1.3] p.195 remains valid for the ramp case as well as for the plateau and so with the value of v_J and r_j now made available we can write for the ramp pressure/radius profile:

$$-\int_{PJ}\frac{dP}{P} = \frac{3C_{PF}}{v_P^2}\int_{vJ} v\,dv = \frac{3\,C_{PF}}{2\,v_P^2}\left(v^2 - v_J^2\right)$$

Which after substitution from [8.2.7] yields::

$$\frac{P}{P_J} = \exp\left[-\frac{3\,C_{PF}}{2\,v_P^2}\left(r_J^2\left(\frac{C_{NT}}{3+\gamma}\right)^2\left(\frac{r}{r_J}\right)^{2\gamma+2} - v_J^2\right)\right] \qquad [8.2.8]$$

Since this part of the profile joins at P_J : r_J onto the previous one, given by equation [8.1.4], the value of v_J is found by putting $P = P_J$ and $r=r_J$ in that equation which can be presented as:

$$v_J^2 = \frac{2\,v_P^2}{3\,C_{PF}}\ln\left(\frac{P_0}{P_J}\right) \qquad [8.2.9]$$

In order to evaluate the edge radius at the shock front where pressure rises instantly to P_{SH} at radius r_{SH} it is necessary to re-arrange [8.2.8] to yield:

$$\frac{r_{SH}}{r_J} = \left[\frac{v_J^2 + \dfrac{2\,v_P^2}{3\,C_{PF}}\ln\left(\dfrac{P_J}{P_{SH}}\right)}{\left(r_J\dfrac{C_{NT}}{(3+\gamma)}\right)^2}\right]^{\frac{1}{(2+2\gamma)}} \qquad [8.2.10]$$

In FIG.8.3 a solution is presented for $H_0 = 9.4\times10^{-19}$ s^{-1} corresponding to 29 km/s/M_{PC}. The curve P shows the pressure profile. The curve marked v is the v/c profile. The assumptions are:

$P_{0N}/P_L = 0.2$:

$P_J/P_{0N} = 0.7$ and

$P_{SH}/P_{ON} = 0.05$.

$\gamma = 1$ for both ramp and plateau.

The kink in both curves corresponds to the junction point J where ramp and plateau meet.

A very interesting observation is the sudden acceleration following the junction. The radius of the universe for the present era and measured to the advancing shock front of radius R_{SH}. moving at v/c=2.2, is 43 billion light years. However, the data was taken from Wright (2005) giving red shifts plotted to a base of D_L the 'luminosity distance' that was defined as the distance the supernovae were from us when observed. It had not been appreciated that the values of D_L had been scaled to the expected distance to which they would have moved by now. This means the distances, but nothing else, are about twice the correct values.

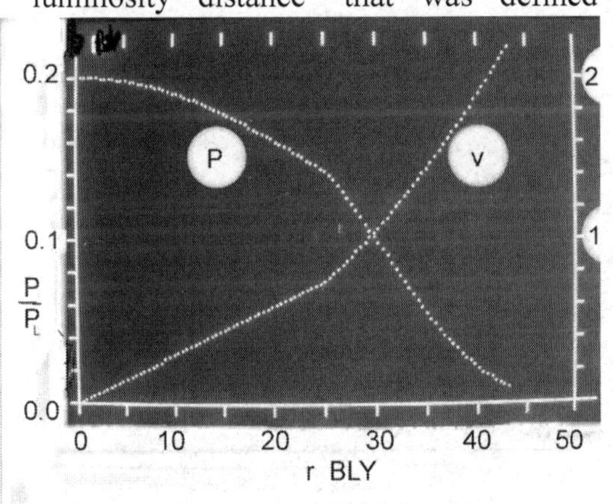

FIG.8.3 P/r & v/c - r PROFILES NOW

The pressure profile is the curve marked P and the kink marks the junction between ramp and plateau creation profiles.
The velocity profile is marked V is v/c using the right hand scale.
NOTE radius r in billions of light years BLY is only about half that shown owing to an error made in the interpretation of data.

At earlier times pressures everywhere will be lower but P_J will not have altered. Consequently at some past era the pressure at the origin i.e. r=0 was less than P_J then r_J became zero.

A problem arises since a divide by zero difficulty is presented. This difficulty is overcome by the procedure that will be considered in detail. But first a more serious problem arises.

A Problem for the Mathematician – inconsistency of the maths?

A first hint that something was wrong appeared when equation [8.1.4] p.195 was rearranged to give the radius of the shock front r_{SH}, where the value of pressure will be that of the shock front P_{SH}. Now the value of P_0/P becomes P_0/P_{SH} and can be revised to read $(P_0/P_L)(P_L/P_{SH})$ and P_{SH}/P_L will not change with time. The resulting equation is shown on the next page.

$$r_{SH} = \frac{1}{H_O}\sqrt{\frac{2v_P^{\;2}}{3C_{PF}}\ln\left(\frac{P_0}{P_L}\frac{P_L}{P_{SH}}\right)}$$

It follows from the above that when r_{SH} is specified by [7.5.7] p. 185 written as: $r_{SH} = r_{SH1}\exp(H_O t)$ then if H_O remains constant, as originally assumed, then the rate of increase in P_0/P_L is insufficient and requires H_O to change to obtain a match. Without changing H_O hopelessly wrong pressure/radius profiles are predicted using equations [8.1.4] p.195 or [8.2.8] for all eras other than the present.

Profiles thought to be correct were only obtained using these equations by progressively increasing C_{NT} and therefore $H_{O.}$ Only then were profiles obtained matching those derived by following path lines back in time using equations [8.1.8] p.196 and [8.2.4] p.199. Clearly this applied equally to both ramp and plateau. At the time this seemed to present a worrying inconsistency within the mathematics. However this was the view in 2005 but is resolved in Chapter 10.

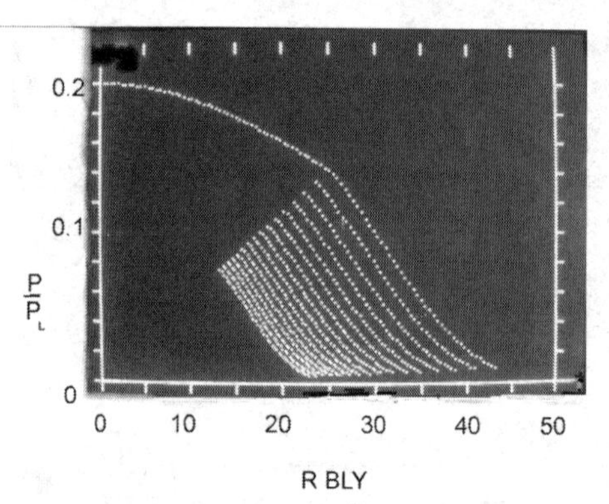

FIG.8.4 PROFILES GOING BACK IN 2 BY STEPS

We now return to the derivation as it appeared in 2005. Only for the present era is a satisfactory profile obtained – justified since here C_{NT} corresponds with the value used for the equations giving the evolution of i-ther with time.

For remaining profiles $H_0 = C_{NT}/(3 + \gamma)$ is no longer applicable. Only the ramp profiles, using [8.2.4] can be used at this stage to follow path lines backward in time. Then the creation rate is given by $C_{NT}\, P/P_J$ with C_{NT} constant as it should be. This step is illustrated in FIG.8.4 showing a set of 15 ramp profiles (the limit allowed by

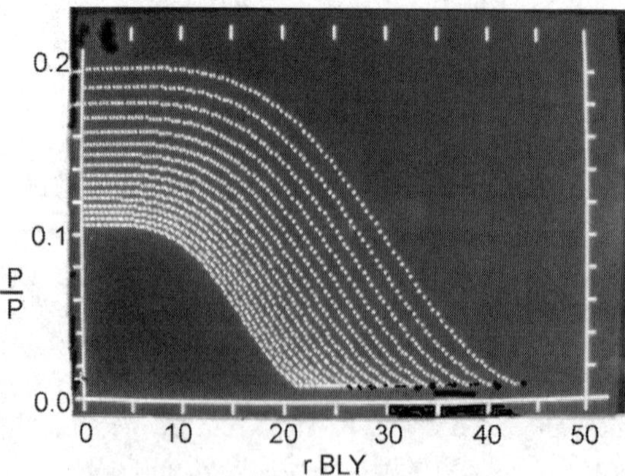

FIG.8.5 COMPLETED RAMP PROFILES in 2 billion year time steps

computer memory). The profiles are separated by intervals of two billion years and so cover evolution over a 30 billion year history. As can be seen, only the profile for the present era gives a complete ramp profile, and so it is necessary to use the remaining profiles as data input to complete them. These ramp profiles assumed $\gamma=1$ and show straight path lines converging toward the origin as required. This shows that this is the correct value for γ, for the ramp only, and simplifies the following analysis that is required.

With $\gamma = 1$ equation [8.2.7] p.199 for v can be re-arranged as:

$$v = \frac{C_{NT}/4}{r_J} r^2 = \frac{H_0}{r_J} r^2 \qquad [8.2.11]$$

Hence we can write:

$$\frac{H_0}{r_J} = \frac{v}{r^2} = \frac{v_J}{r_J^2} \qquad [8.2.12]$$

Clearly H_0/r_j is a constant for any given profile and since v and r are given by [8.2.4] and [8.2.5] by following path lines back in time both v and r can be taken from any element of any of the ramp profiles. This enables the profiles to be completed disregarding the plateau stage at this time. It is as if the ramp were projected along the dashed line indicated in FIG.8.2. The result is given in FIG.8.5 showing the completed ramp profiles. What is important now is that the value of points r_j, P_j for each profile are now established.

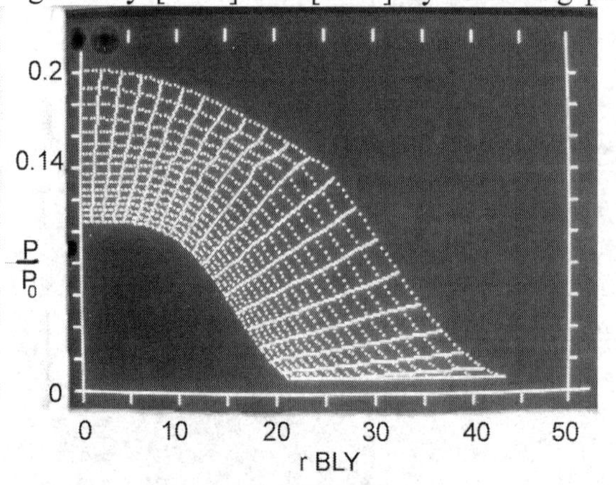

FIG.8.6 COMPLETE P/r PROFILE WITH PATH LINES ADDED

This makes possible the completion of the correct profiles since the plateau components meet the ramps at the points r_J,P_J. So now equation [8.1.4] for the plateau profiles at all epochs can be computed. This step is illustrated in FIG.8.6 showing both ramp and plateau profiles. Since P_{SH} is the same for all epochs all profiles are corrected to this common value.

Path lines have also been added as described in the following. Each profile is computed in GWBASIC with code name UN250409.

There are 100 radius, pressure and velocity elements in each array. Each array is two dimensional such as R(N,M) with N=0 to 100 and M=0 to 15. Selecting every fifth element N in each array, and connecting each corresponding element in each profile, defined by M, can establish path lines. Path lines obtained in this way are superimposed.

All path lines for the ramp stage appear straight and point toward the origin as required so proving the choice of $\gamma = 1$ as the correct choice. However, some of the path lines for the plateau component are not straight suggesting that the value of γ for the plateau should not be 1. Another value can be tried with the procedure repeated. Such a step will be made later but analysis of this set of assumptions is not yet complete.

Path lines on the time/ radius plot are required so that light propagation can be explored. Then red shifts need to be predicted.

8.3 Path lines in the Time/Radius Plane & Light Propagation

All arrays needed for plotting the path lines of particles of the i-ther on a radius/time plot are already available from the set of Pressure/radius profiles. The same information can now be used to find how light from distant galaxies moves toward the Earth.

The cosmic background radiation suggests that we are moving at 400 km/s relative to the origin centre of the universe. The computations already presented suggest the shock front is moving at 2.2 times the speed of light measured on Earth. It follows that our galaxy has to be situated at about 400/(2.2*300,000) = .06% of the present radius of the universe from its centre. It will also be shown, from the present analysis, that our astronomers are potentially able to see right to the edge of the universe. Since it appears symmetrical as viewed in all directions, this again suggests we are not far from the centre point. In consequence it is only necessary to consider the propagation of light in an inward radial direction.

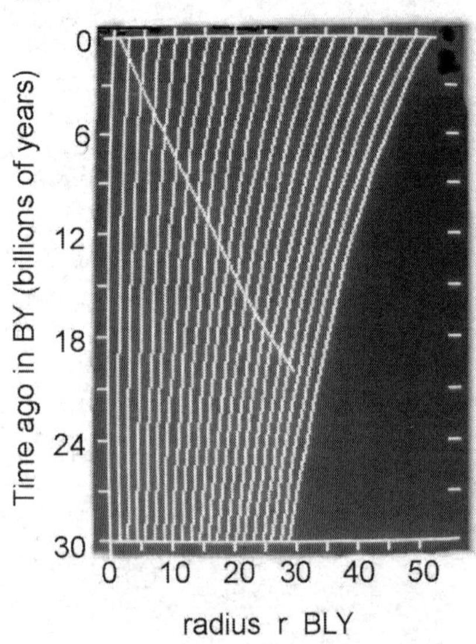

FIG.8.7 PATH LINES AND LIGHT PATH IN TIME/RADIUS PLANE

Two factors will interact to determine the speed of approach as viewed from Earth.

1 The 'observed' speed of light c_T governed by pressure of i-ther and propagating relative to the i-ther.

2 The speed of light relative to Earth c_O that is less than c_T owing to the speed of recession v. So $c_O = c_T - v$

Since the pressure is reducing with radius c_T is increasing and so these factors tend to cancel each other. Suffix $_0$ means $r = 0$ and the datum speed of light c_D will be negligibly different from c_0. The distance ΔS of travel in time Δt will be given by:

$$\Delta S = (c_T - v)\Delta t \qquad [8.3.1]$$

Evaluation of red-shifts, denoted by symbol Z is covered in detail in Chapter 9 and so will not be dealt with here. However, the maximum value of $Z = 1.755$ has been observed by astronomers for a remote supernova at a distance between 30 and 45 billion light years (BLY) according to the cosmologist Ned Wright (2005). This distance is calculated by extrapolating from the measured distance some 10 billion years ago using a special relativity basis. The values given here are computed from the Big Breed theory using equation [8.1.4] p.195 with assumed values of P_0 / P when $P = P_{SH}$, the pressure at the shock front, and r_{SH} is the radius at that front.

FIG.8.9 shows the red shifts computed from the data illustrated in FIGs 8.3 to 8.7. The component due to the Doppler effect caused by speed of recession is Z_D and the total due to the latter plus that caused by light needing to speed up as it approaches is Z. So Z-Z_D due to light speed-up is dominant. A blue shift due to photons falling in the effective gravitational field produced by i-theric pressure gradients associated with acceleration reduced the shift that is observed to Z_G. So Z-Z_G is that blue shift.

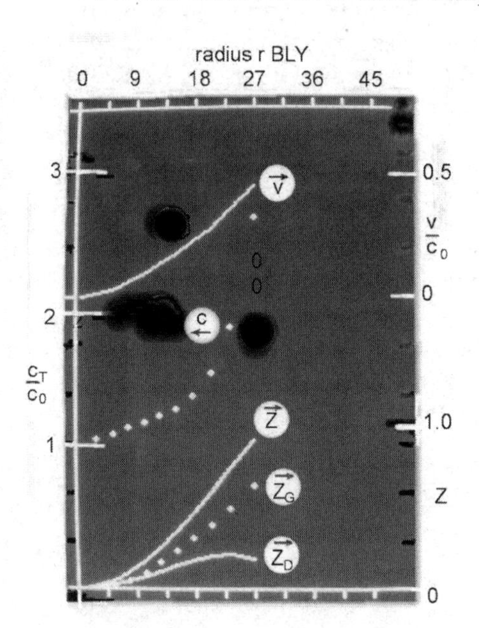

FIG. 8.9 RED SHIFTS Z, with wave and particle speeds

The curve marked v gives the computed values of v/c_O and that marked c gives c_T/c_O.. The very rapid increase of the latter when nearing the shock front explains why Z_D shows a maxima and falls as the shock front is approached.

Clearly different starting data needs to be used to determine the possible range of values the theory predicts.

It was thought necessary to first determine the effect of changing γ and the results of many runs were detailed in the original edition of this book. However, as recorded in Chapter 10 'Eureka Day' was 1st July 2010 and this changed everything. In consequence the rest of this chapter has been deleted.

8.4 CONCLUSION TO CHAPTER 8

The first section of this chapter made an important advance. It showed that pressure/radius profiles could be derived in a simple way. This became possible when it was realised that the mass increase due to creation in any element of i-ther would not add to the pressure gradients needed to provide accelerating expansion.

Most of this chapter, however, has simply illustrated how wildly wrong the investigator can be after being led onto a false track. As stated earlier, equation [7.5.3] p.184 had been misinterpreted to mean that pressure/radius profiles at all epochs would have similar shapes. In Chapter 10 how wrong this deduction was will become clear. Chapter 8 has, by way of example, only included one of the false avenues of investigation that have been explored. The conceptual error just mentioned caused many years of delay in solving the influence of pressure on the solution.

Furthermore the ratio of shock-front pressure to pressure at the origin point could only be guessed. The new and exciting insight to be described in Chapter 10 also enabled this important ratio to be given an accurate theoretical value from the theory.

Before the final solution can be properly understood it is necessary to detail the way the various components of the red-shifts were derived. This is the subject of the next chapter.

CHAPTER 9

EVALUATING ASTRONOMICAL RED-SHIFT DATA

The new solution for energy creation requires a totally new approach for the evaluation of red-shift data from remote galaxies and their supernovae. At present cosmologists, like Wright (2005) evaluate astronomical data using special relativity as its basis. Since this denies the existence of any background medium it is inappropriate as basis for the Big Breed theory. This is because the latter is based on the existence of i-ther and the quantum vacuum it spawns. The new approach is fully detailed in this chapter and shows five factors need to be combined.

9.1 Introduction

Remote galaxies were first shown to be red-shifted by the famous astronomer, Edwin Hubble, in 1929. This meant the wavelengths of light had increased from the values when emitted. Hubble interpreted this as being due to a Doppler effect and deduced that remote galaxies were receding from us at speeds directly proportional to their distance.

According to the cosmologist, Wright (2005), more complex analyses are now carried out. These are based on Einstein's theory of special relativity, with an infinite string of hypothetical observers as the 'frames of reference'. These stretch from any remote galaxy to Earth each passing the light along the line to the next.

Clearly this type of analysis is inapplicable in the case of a universe based on a background medium.

In this case light has to propagate relative to the medium with the speed of observers regarded as irrelevant. An observer moving in the medium could not see light as having the same value in all directions as a basic assumption of special relativity specifies.

However, some cosmologists use general relativity and say remote galaxies move with their own space-time and not have a Doppler shift. Instead space-time is stretched by the expansion and this stretches the wavelengths to produce the observed red-shifts.

This is closer to the approach used for the Big Breed theory but something is clearly wrong since space-time is being considered as if it were a real medium which is inconsistent with special relativity.

9.2 Light Propagation in an Expanding Medium

Quantum theory demands the existence a fluid-like medium called the 'quantum vacuum' and so light will be regarded as propagating through that background medium. However, in the complete Big Breed theory this is regarded as emerging from i-ther and can have low speeds, compared to that of light, relative to i-ther. However, for galaxies remote from Earth both mediums will be regarded as moving together since the error involved will be negligible. All arrays needed for plotting the motion of a photon of light through the quantum vacuum or i-ther on cosmological scales are provided by sets of pressure/radius profiles, examples of which were provided in the previous chapter.

The same information can now be used to find how light from distant galaxies moves radially inward to the origin point. The question now requiring an answer is the relevance of such a light path to that observed from Earth.

Since the light has to travel against the speed of recession, which increases with distance, it meets ever reducing recession speed as it approaches the observer. Consequently the photons of light have to continually speed up. Since momentum must be conserved this means the energy of photons must reduce with consequent increase of wavelength. This means an extra red shift is added to the (modified) Doppler-shift. As will be shown later in this chapter, the result is to predict speeds of recession that are less than half the established values and not more than 60% of the speed of light. It follows that the 'horizon' cosmologists assume exists, beyond which galaxies recede faster than light, can no longer be assumed to exist. Then since astronomers report that the universe seems the same when viewed in all directions and we see right to the edge, it follows that our galaxy cannot be far from the origin centre. The analysis to follow ignores the off-centre distance.

Unless our solar system is near the origin point of the universe the whole would appear lop-sided if we can see right to its edge.

It appears so uniform in all directions that cosmologists consider that we occupy no special place in the universe. They deduce it must be far larger than the volume we can observe. In consequence the analysis to follow is at variance with that of contemporary cosmology. This is most unfortunate but the reader is requested not to dismiss the approach to be considered here on grounds of this contentious aspect. The request is made for any judgement on this issue to be withheld, and the mind kept open, until the case for the change of methodology has been studied.

The universe appears symmetrical as viewed in all directions and our theory shows we are seeing nearly to its edge. In consequence it is only necessary to consider the propagation of light in an inward radial direction to an astronomer who is so close to the centre of the universe that the difference can be ignored.

Two factors will interact to determine the speed of approaching light as viewed from Earth. They are:

1 Light has to speed up as it approaches from a remote galaxy. Suffix $_0$ means $r = 0$ and the datum speed of light c_D will be negligibly different from c_O when Earth is so close to the origin point and so the speed as observed will be taken as $c_0 = c_D$. The speed of incoming light c_i, at great distance but relative to Earth, is less than what we will define as the 'observed speed of light' c_T at the distant galaxy, owing propagation against the recession speed v.

$c_i = c_T - v$: So using the Hubble law: $c_i = c_T - H_0 r$. [9.2.1]

2 The observed speed of light c_T increases as the pressure of i-ther falls as specified by the next equation. There is also a 'flat space' speed of light c that also increases as pressure falls.

In APPENDIX IV page 69 in PART I of this book a brief summary of the alternative to relativity is given, which was derived in conjunction with the Big Breed theory and yields quantum gravity. This is not a relativity-based theory and all kinetic energies are absolute being measured from 'local frames' that exist independently of observers. Therefore, unlike any relativity theory, precise values of kinetic energy can be allocated to any kind of particle. This allows 'sum energy' E to be defined as $E_0 + E_K$ where E_0 is the 'rest energy' and E_K the kinetic energy for any kind of particle. Equation [3.4.5] from *QUANTUM GRAVITY via Exact Classical Mechanics* (QGECM) relates c to sum energy E, c_T and i-theric pressure P. Since a photon has no rest mass it can now be treated as an object made only of kinetic energy. This means it is able to be treated as any massive sub-atomic particle or primary consisting of E_0 and E_K. With E_{K0} as the kinetic energy of a photon at the origin point, the equation adapted for the photon, becomes:

$$\frac{c}{c_0} = \exp\left[\frac{1}{5}\left(1-\left(\frac{P}{P_0}\right)^{\frac{5}{3}}\right)\right] = \frac{E_{K0}}{E_K}: \quad \frac{c_T}{c_0} = \frac{c}{c_0}\left(\frac{P}{P_0}\right)^{-\frac{1}{3}} \qquad [9.2.2]$$

For intergalactic distances the local frame coincides with the i-ther that is receding with the galaxy under observation. Light is therefore assumed to propagate relative to the i-ther. Confidence in this approach is justified since all experimental checks are matched just as well as achieved by general relativity.

I-theric density is not uniform and varies in the same ratio as its pressure. To explain why two speeds have to be considered a 'jump and dwell' model for photon motion needs to be imagined that fits in well with quantum theory. The photon is considered as if dwelling without motion and then making an instantaneous jump to the next position. As i-theric pressure falls dwell reduces, so explaining the increase of c without taking any change in the jump distance into account. However, the jump distance L has also increased, in the proportion $L/L_0 = (P/P_0)^{-1/3}$ so causing an extra speed increase to yield an observed speed c_T.

Both c and c_T appear in the solution.

Since the pressure reduces with radius increase, c_T also increases with radius. Consequently this tends cancel the speed up due to H_0 being constant with r reducing as shown by equation [9.2.1]

Space Gravity

Equations [9.2.1] and [9.2.2] provide the starting point for computing the path of light going backward in time right to the shock front defining the radius of the universe. These equations provided the basis of the computation plotted in FIG.8.7 using the same data as the previous figures, FIG.8.3 to FIG.8.6. To exaggerate the displacement of Earth from the origin centre Earth radius is taken as 5% of r_{SH}. Clearly the light path is almost a straight line and contrasts with the plot shown in FIG.7.2 of *QGECM* and reproduced as FIG.9 in this book on page 48. In that plot the light is a curve showing it speeding up. However, at that stage no account of the speed of light increasing with fall of pressure was taken into account.

Now the data needs to be used to evaluate the red shift of light from distant galaxies as observed from Earth.

9.3 Evaluating the Red Shift of Distant Galaxies

Five factors need to be taken into account:

1 The Doppler shift at the point of emission

2 An energy drop caused by light ever moving into regions of lower recession velocity so that photons need to keep speeding up.

3 An energy gain due to the 'space-gravitational field' caused by pressure gradients. This has nothing to do with the mutual attraction of matter. However, according to the theory of gravity

covered in *QGECM*, the force of gravity is one of 'negative buoyancy' caused by pressure gradients of the i-ther. Consequently photons will fall and gain energy as they move toward higher-pressure regions.

4 The effect of 'matter-gravity' caused by matter attracting light as well as other matter.

5 The absorption of energy by the electron population of space as shown to occur by Aspden (1984).

The problem is illustrated in FIG.9.1 p.213. Light from a remote supernova at distance D is emitted where the velocity of recession is v_D and light is emitted in a flash lasting time t. The star therefore moves a distance Δr in which $t = \Delta r \ / \ v_D$. The light from the tail of the flash, moving at speed c_{TD} relative to i-ther and against recession speed takes time $t_2 = \Delta r/(c_{TD} - v_D)$ to return to the distance the head of the flash started from. Consequently the observed time of the flash increases in the ratio $(t + t_2)/t = 1 + v_D/(c_{TD} - v_D)$. This ratio is also the ratio by which the wavelength of light is stretched and is defined as $1 + Z_D$ giving a slightly modified version of the 'Doppler shift' Z_D. The result is:

$$1 + Z_D = \frac{c_{TD}}{C_{TD} - v_D} \qquad [9.3.1]$$

However, this is not the major red shift that applies. The light now has to travel against an ever-reducing speed c of recession. As will be shown next a very large extra red shift is involved.

The Light Speed-Up Red-Shift

As illustrated in FIG.9.1 p.213 the light is now considered to move a distance x from its point of emission at distance D from the origin centred observer, so that its radius from origin O becomes r_1. At this point the i-ther is receding at speed v_1 so that photons are moving at speed $(c_{T1} - v_1)$ relative to the origin at O.

After a further travel of distance δx, the velocity of recession has reduced to v_2 and the 'observed' light speed has reduced from c_{T1} to c_{T2}. Simultaneously pressure has increased from P_1 to P_2 with the energy of photons changing from E_1 to E_2.

All parameters need to be expressed as ratios of those at the observer O assumed to be at the origin point of the universe

The philosophy needed to evaluate the energy change from E_1 to E_2 is quite challenging.

At first sight the effect of light speeding up seems to depend on the difference between $c_{T1} - v_1$ and $c_{T2} - v_2$ where c_T is the 'observed speed' of light.

However, a deeper consideration shows this to be an incorrect assumption as can be explained by inspection of equation [9.2.2] p.209 that shows E and c to be intimately related. For the photon E_K varies inversely as c.

Its momentum is $m_K c = (E_K / c^2)c = E_K / c$

So although the apparent speed is c_T the momentum carried is not $m_K c_T$. It is as if the photon makes a series of instantaneous jumps but during each dwell carries momentum as if moving at speed c.

If this is still not clear then let me try to explain this by recourse to analogy.

It is best to think of the analogy of the molecules of a gas exerting pressure on the containing walls: as in the kinetic theory of gases. This pressure is the combined result of bombardment of the surface by all the molecules that keep bouncing back after hitting it. Pressure is the change in momentum of each molecule, due to bounce-back, multiplied by the rate of arrival of molecules at the surface.

When a gas has its pressure increased, with temperature maintained constant, the gas density increases in proportion. The increase in pressure is due to the increased rate at which molecules arrive on unit area of surface and nothing else. The speed and momentum of each molecule of that gas has not changed. The increases of both density and pressure are due solely to increased packing density - as is also described by a reduction in the average separating distance between the molecules.

In a similar way each photon traversing the i-ther will maintain its momentum based on the change in speed c: not c_T since the latter appears only from change in the average separating distance of primaries.

The photon penetrates distance δx, without change of speed or momentum, prior to jumping back into equilibrium speed c_2. So momentum effects are evaluated from equation [9.2.2] based on the values of c it yields: not c_T. However, the latter is required for assessing the distance light travels in a given time i.e.

$$\delta x = (c_T - v) \times \delta t \qquad [9.3.2]$$

Two solutions will be considered in which E_1 and E_2 refer to a case with the space-gravity effect ignored. With the latter included these energies will be denoted E_{G1} and E_{G2} respectively.

In the latter case the photons will first gain energy δE_G, by falling in the space-gravity field from an initial energy E_{G1} to E_{G2}, so that from equation [9.2.2]:

$$\frac{\delta E_G}{E_0} = \frac{E_{G2}}{E_0} - \frac{E_{G1}}{E_0} = \frac{c_0}{c_2} - \frac{c_0}{c_1} \qquad [9.3.3]$$

Now light needs to speed up due to v_2 being less than v_1. This is the next item that needs to be considered. A photon, the carrier of light, is best considered, on entry to the element δx, able to maintain its initial velocity c_1 over the distance δx. At this point, owing to the change in recession speed, the photon is not moving at speed c_2 relative to the local i-ther but is moving at a speed that is less by the amount $v_1 - v_2$ also with change in c taken into account. In consequence the photon must now equilibriate.

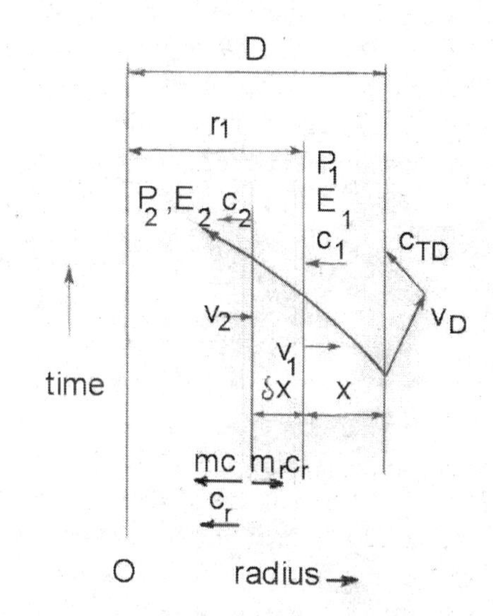

FIG.9.1 PHOTONS TRAVERSING SPACE ELEMENT δx

Momentum must be conserved and the possibility of partial reflection needs to be considered. The partially reflected photon of momentum $m_r c$, (not shown in FIG.9.1) needs to reduce energy of the transmitted photon mc to compensate for speeding up. Consequently for the reflected photon both the energy and the momentum $m_r c$ must be negative. As shown in Chapter 1 this means the partially reflected photon, made of negative energy, must travel opposite the direction of its negative momentum. This makes it propagate in direction c_r: the same direction as the transmitted photon. Consequently both photons move together and simply partially cancel. The effect is quite different from the partial reflection of light travelling from low to a high-density substance: like glass that is not in differential motion.

The state at $x + \delta x$ has flat space light speed c_2 that has changed from c_1 and a recession velocity v_2 different from v_1. Both changes need to be taken into account simultaneously in order to maintain conservation of momentum so that we can write:

$$m_2(c_2 - v_2) = m_1(c_1 - v_1)$$

Rearranging yields:

$$\frac{m_2}{m_1} = \frac{(c_1 - v_1)}{(c_2 - v_2)}$$

But $E = mc^2$ (derived from ECM without reference to relativity) and it is the energy change that is required. Then based on ratios of energy E_K to E_{K0}, the energy at the origin, we can write:

$$\frac{E_{K2}}{E_{K0}} = \frac{E_{K1}}{E_{K0}}\left[\frac{c_1 - v_1}{c_2 - v_2}\right]\left(\frac{c_2}{c_1}\right)^2 \qquad [9.3.4]$$

This equation does not allow for the energy gain due to photons falling in the space-gravity field given by [9.3.2]. With this addition at state 1 the photon energy will be E_{G1} instead of E_1 at the start of entry to element δx. Then with this substitution the pseudo-gravity gain has to be added to revise equation [9.3.4] to yield:

$$\frac{E_{G2}}{E_{K0}} = \frac{E_{G1}}{E_{K0}}\left[\frac{c_1/c_0 - (v_1/c_0)}{c_2/c_0 - v_2/c_0}\right]\left(\frac{c_2/c_0}{c_1/c_0}\right)^2 + \left(\frac{c_0}{c_2} - \frac{c_0}{c_1}\right) \qquad 9.3.5]$$

Where c_1/c_0 and c_2/c_0 are given by [9.2.2] using $P = P_1$ *or* P_2 as appropriate.

So far time has been considered moving forward and this is permissible but not expedient. This is because of the difficulty in arriving at Earth at the correct epoch. It is better therefore to start at Earth and work back in time. At each step the value of E_{G2} is known and E_{G1} needs to be found. This is readily achieved by the re-arrangement of [9.3.5] to provide the more useful equation:

$$\frac{E_{G1}}{E_{K0}} = \left[\frac{E_{G2}}{E_{K0}} - \frac{c_0}{c_2} + \frac{c_0}{c_1}\right] \Big/ \left[\left(\frac{c_1/c_0 - v_1/c_0}{c_2/c_0 - v_1/c_0}\right)\left(\frac{c_2/c_0}{c_1/c_0}\right)^2\right] \qquad 9.3.6]$$

Going back in time means δx is replaced by δr. Then after each time step δt seconds at observed light speed $c_T = (c_{T1} + c_{T2})/2$ the value of δr becomes $c_T \times \delta t$ metres. But in more convenient units of time in billions of years (BY) and distance in billions of light years (BLY):

$$\Delta r = \Delta t \times c_T / c_O \qquad [9.3.7]$$

After each step E_{G1} becomes E_{G2} for the next step until the emitting star or galaxy is reached to yield a final value of E_{G1}/E_O. This is equal to the wavelength ratio $1 + Z$ of observed divided by emitted light, before the Doppler ratio $1 + Z_D$ given by [9.3.1] is considered. The overall red shift Z_G, with E_{G1}/E_O given by repeated application of [9.3.6] therefore becomes:

$$Z_G = (1 + Z_D)E_{G1}/E_{KO} - 1 \qquad [9.3.8]$$

9.4 MATTER GRAVITY (Added July 2010)

It will be assumed that on the largest scales matter is uniformly distributed throughout the universe at an average mass density ρ. A blue shift will result that will be considered in isolation from the other shifts and then added to the value given by [9.3.8] in the form of a final correction.

Light is emitted from a radius R_E and travels radially inward to an observer placed at the origin point. It travels inward by distance x to radius r and then through an element of distance dx. As is known the gravitational forces from matter outside radius r cancel to zero and that inside acts as though concentrated at a central singularity.

The mass m_S equivalent to energy E_S inside radius r then becomes:

$$\frac{E_S}{c^2} = m_S = \rho_S \frac{4}{3}\pi r^3 = \rho_S \frac{4}{3}\pi (R_E - x)^3 \qquad [9.4.1]$$

The ECM equation for gravity derived in *QGECM* is energy based and not matter based as in Newton's equation. Furthermore this energy is called 'sum energy' being the sum of 'rest energy' (alias mass-energy) and kinetic energy. Since the photons of light have no rest mass they are considered made from kinetic energy alone, the new equation therefore applies equally to light as well as matter.

(Strictly speaking Newton's equation cannot be applied to light as it is restricted to the coupling of mass alone with its kinetic energy disregarded. With this inclusion, together with the non-uniform density of space, the result is a set of predictions that match observation just as well as both special and general relativity)

With the new gravitational constant $G_C = G/c^4$ the ECM equation for the force of gravity F_x is given by:

$$F_x = \frac{G_C\, E_S\, E}{r^2} : \quad \textit{and so is also} \quad \frac{G_C\, E_S\, E}{(R_E - x)^2} \qquad [9.4.2]$$

Substituting for E_S from [9.4.1] the result is:

$$F_x = \frac{4\pi}{3} \frac{G_C\, \rho_S\, c^2 E (R_E - x)^3}{(R_E - x)^2}$$

And the energy gain dE of the photon in falling distance dx with F_x acting on it is $F_x dx$: i.e. $dE = F_x dx$. So cancelling terms, multiplying both sides by dx and re-arranging in integral form yields:

$$\int_{EE}^{E0} \frac{dE}{E} = \frac{4\pi}{3} G_C\, \rho_S\, c^2 \int_0^{RE} (R_E - x)\, dx$$

Now $G_C c^2 = G/c^2$ and so after integration the result becomes:

$$\log e\left(\frac{E_{K0}}{E_{KE}}\right) = \frac{2\pi}{3} G\, \rho_S \left(\frac{R_E}{c}\right)^2 \qquad [9.4.3]$$

Then since, from quantum theory, the wavelength λ of photons varies inversely with the energy of the photon the above, after representing in its alternative form becomes:

$$\frac{E_{K0}}{E_{KE}} = \exp\left(\frac{2\pi}{3} G \rho_S \left(\frac{R_E}{c}\right)^2\right) = \frac{\lambda_E}{\lambda_0} \qquad [9.4.4]$$

Finally we will define Z_{MG} as: $Z_{MG} = \frac{\lambda_E}{\lambda_0} - 1$

This negative value needs adding to the result of Z_G given by equation [9.3.8] to yield the red-shift observed by astronomers.

Evaluating Z_{MG}

An internet search gave the average density of stars in the universe as equivalent to10^{-28} kg/m^3. However, astronomers say that at least ten times as much must be present as 'dark matter'. A very extreme case will be assumed in which dark matter is 100 times as great as the mass of visible stars. A value of 10^{-26} kg/m^3 will therefore be taken as this extreme case.

Then substituting $G = 6.673\times10^{-11}$ Nm^2/kg^2, c=2.9979×10^8 m/s with 3.156×10^7 s/year in [9.4.4] so that R_E can be expressed in billions of light-years (BLY) the result becomes:

$$Z_{MG} = \exp\left(-0.001392 R_E^{\ 2}\right) - 1 : \quad R_E \ \ in \ \ BLY \qquad [9.4.5]$$

As will be shown in Chapter 10 this is somewhat smaller than the effect of space gravity. Therefore with a more reasonable value of dark matter 10 times that of visible stars, this correction can be seen as negligible.

9.5 The Aspden Effect

This is a red-shift due to the electron population of space. Aspden, after deriving his theory showing that interaction with photons would cause the latter to lose energy, argued for a return to the concept of a static universe.

Cosmologists ignore the possibility of such interaction and no data appears to exist for assessing its magnitude. Therefore the red-shifts given in both Chapters 8 and 10 assume the Aspden effect to be negligibly small.

My own interpretation, however, would be that each photon/electron interaction could cause a non-negligible photon energy loss without deflection to produce a red shift. However, those colliding with dust would be scattered. These would produce no red shift but might produce reddening due to blue being scattered more than red. The difference is because reddening will not produce a shift in the spectral lines that identifies a red-shift. Both the Aspden effect and reddening will reduce intensity so making luminosity distance estimates too high. However, any incompatibility between red-shift and reddening observation might lead to the possibility of a partial correction being made.

9.6 APPLICATION TO CHAPTER 8

The result applied to the previous data is given by the dotted line marked Z_G in FIG.8.9. Its maximum value is 0.636 at a radius of 23.9 BLY and 20 BY ago. But this is not at the shock front. The latter has a radius of 28.6 BLY.

Also shown is the Doppler shift $Z_D = 0.216$ at the maximum value and contributes considerably to the total.

In addition a shift achieved by ignoring the energy gain by photons falling in the space-gravity field is included. This simply ignores the last term in [9.3.5]. This gives $Z = 0.992$.

So the difference between Z and Z_G shows the effect of this force.

It may seem confusing to the reader to see the light paths shown on Radius/Time plots, such as FIG.8.7 p.204 looking so straight when light speed-up is responsible for the major part of the red-shift. The reason is made clear in FIG.13.4 p.269 where a plot is provided that also shows the way the flat-space speed of light, which is responsible for this shift, actually has to speed up.

No further discussion regarding Chapter 8 will be made since, as shown in the next chapter an exciting new insight appeared in July 2010. This provided a major breakthrough so that previous derivations are now rendered obsolete and wrong.

9.7 CONCLUSION TO CHAPTER 9

A complete set of equations has been presented that allow the net red shifts of remote galaxies and supernovae to be predicted from sets of computed pressure/radius profiles. As will be shown in the next chapter accurate computations of these profiles appear from the input of net creation rates. What has really been presented therefore is a new way of interpreting observed red-shift data for evaluating the Hubble constant, the net creation rates of the i-ther, and its size as measured to the shock-fronted outer radius.

The next chapter makes it possible to utilise the methodology provided in this one so that astronomical observations can be used for evaluating the growth of i-ther in both size and pressure.

CHAPTER 10

EUREKA DAY – 1ST JULY 2010

A BREAKTHROUGH SOLUTION APPEARS

All suddenly became clear on 1st July 2010 as this chapter unveils. The new insight appeared when a change was made to investigate growth of the i-ther in forward time.

10.1 Introduction

The previous solutions described in Chapters 7 and 8 began with an estimate of the present size of the universe and worked backward in time. It was argued that the expanding i-ther would exist as a sphere having a shock fronted outer edge. However, the value of this shock pressure could not be quantified and so its value could only be guessed at. Furthermore the way pressure affects creation rate, and therefore pressure profile, caused difficulties that the mathematics was unable to properly handle. The solutions assumed a similarity of pressure profiles to be applicable. Then an alarming mathematical inconsistency appeared that seemed to invalidate those analyses.

In this chapter a switch to analysis in forward time is made starting from the arbitrarily small radius of ten kilometres. No inconsistency now arises, the mathematical difficulties are resolved and pressure/radius profiles of good accuracy are evaluated. At least a high confidence in their accuracy is achieved from the logic that emerges. By a major breakthrough, the magnitude of the shock front pressure also appears.

10.2 Pressure Development at the Origin Point with Time

A start will be made by repeating the derivation that led to equation [8.1.4] p.195. This is now presented with new equation numbering as an introduction. Then the new insights follow.

Only the existing mass within any given element of volume requires a pressure differential to provide acceleration. This is because, as the two phases gain energy in balanced amounts, their momentum gains also appear in balance and so cancel to zero. The two phases of the mixture, in bulk, do not couple strongly. If they did, meaning they had a 'coupling factor' $C_{PF} = 0$, then no pressure differential would be needed at all. The problem is actually simplified since each phase of the mixture is accelerated by its own pressure differential that is reduced in the proportion C_{PF}, and its most probable value has been assessed as 0.9. So now a simpler equation describing the pressure/radius profile can be derived. Equation [7.6.2] p.187 can be revised for the force dF acting on an element of i-ther of thickness dr to the form:

$$dF = -\pi(r\theta)^2 dP = C_{PF} dm\, dv/dt \qquad [10.2.1]$$

And $dm = \rho\pi(r\theta)^2 dr \quad where \quad \rho = 3P/v_P{}^2$

Showing that for the i-ther, treated as a perfect gas, the density ρ is related to pressure P and the average speed v_P of the primaries. Combining with [10.2.1] and putting in integral form yields:

$$-\int_{P0}\frac{dP}{P} = \frac{3C_{PF}}{v_P{}^2}\int_0 dr\frac{dv}{dt} \quad \& \quad \frac{dr}{dt} = v \qquad [10.2.2]$$

$$So \quad -\int_{P0}\frac{dP}{P} = \frac{3C_{PF}}{v_p{}^2}\int_0 v\,dv \qquad [10.2.3]$$

Integration of [10.2.3] and using Hubble's law yields:

$$\ln\left(\frac{P_0}{P}\right) = \frac{3C_{PF}}{v_P{}^2}\frac{v^2}{2} \;:\; \therefore \frac{P_0}{P} = \exp\left(\frac{3}{2}\frac{C_{PF}}{v_P{}^2}(H_0 r)^2\right) \qquad [10.2.4]$$

This suggests that at an early time when r, representing size of i-ther, was small, the pressure differences needed for acceleration would be negligible since $P \sim P_0$ is predicted. Then the term dP/P in equation [7.5.1] p.183 could be ignored. This also means that during this early period a shock front had to exist raising the pressure suddenly from zero to P_{SH} with all the i-ther following behind at a negligibly different pressure. The aim of this exercise is to determine how long such a rectangular pressure/radius profile could be assumed without incurring unacceptable error.

Assuming the i-ther can be treated as a perfect gas whose mass is growing at a rate C_{NA} proportional to the existing mass and comprising primaries of unchanging average speed we can write:

$$\frac{dm}{m} = \frac{d(PV)}{PV} = C_{NA}\,dt$$ Differentiation by parts yields:

$$\frac{dP}{P} + \frac{dV}{V} = C_{NA}\,dt$$

However, for the case where dP/P is negligible the first term can be ignored. Then since the volume $V = r^3 4\pi/3$ increases by $dV = r^2 4\pi dr$ it follows that $dV/V = 3dr/r$ and the equation in integral form becomes:

$$3\int_{r1} \frac{dr}{r} = \int_{V1} \frac{dV}{V} = C_{NA}\int_0 dt$$

Integration yields:

$$\therefore \quad r = r_1 \exp\left(\frac{C_{NA}}{3} t\right) \qquad [10.2.5]$$

Differentiating with respect to time yields the velocity, hence:

$$v = \frac{dr}{dt} = r_1 \frac{C_{NA}}{3} \exp\left(\frac{C_{NA}}{3} t\right) = \frac{C_{NA}}{3} r = H_0 r \qquad [10.2.6]$$

Which means that the Hubble constant H_0 has been changed from the value deduced of $C_{NA}/4$ in §7.5.1 p.184 [7.5.10] p.185 to $C_{NA}/3$.

Combining [10.2.5] and [10.2.6] applied to the shock-fronted edge of the i-ther, starting from r_{SH1} at time zero to radius r_{SH} at time t, where $P = P_{SH}$ and multiplying both sides by H_0 yields:

$$H_0 r_{SH} = H_0 r_{SH1} \exp(H_0 t) \qquad [10.2.7]$$

This can then be substituted in [10.2.4] to yield:

$$\frac{P_0}{P_{SH}} = \exp\left[\frac{3}{2}\frac{C_{PF}}{v_P^{\,2}}\left(H_0 r_{SH1} \exp(H_0 t)\right)^2\right] \qquad [10.2.8]$$

The result is an exponential inside an exponential and this means that P_0/P_{SH} will not start to rise noticeably above 1.0 for ages, thereafter the central pressure P_0 will increase very rapidly indeed.

Now the shock front is caused by sudden creation followed almost immediately by a rapidly evolving and almost balancing annihilation as cores develop. It is quite unlike shock waves in air that increase in pressure with speed increase. In that case it is the speed relative to the air in front that causes the shock pressure rise. There is only void, meaning nothing, in front of the i-theric shock. Relative speed cannot relate to nothing and the latter cannot have a flow. Indeed the pressure rise at the shock will depend only on the i-theric density needed to cause collapse of the fluid phase into cells each containing one annihilation core. So the pressure at the shock front will have a constant value for all values of time.

The speed of the growing edge has to be measured from the origin point of i-ther because that is the only place stationary as compared to the average velocity of all primaries in the whole i-ther. However, this speed will not affect the value of P_{SH}.

It follows that pressure at the origin point can be calculated from

[10.2.8] in units of P_{SH}, since the latter will be an unchanging constant. The following table shows how the i-ther can be expected to grow from an assumed radius of 10 km at time zero based on Hubble's constants of 71 km/s/M_{PC} and 29 km/s /M_{PC}, which convert to 2.3×10^{-18} s^{-1} and 9.4×10^{-19} s^{-1} respectively.

It is first necessary to point out that only growth of i-ther: not the universe of matter, is considered. In this theory, unlike the big bang, everything including space, time and matter, do not appear at the same instant. In the Big Breed theory matter only emerges from i-ther after an adequate time has elapsed to allow it evolve the intelligence needed to organise the creation of matter.

TABLE 10.I ORIGIN PRESSURE RISE WITH TIME

H_0 = 71 km/s/M_{PC}: R_{SH} in BLY				H_0 = 29 km/s/M_{PC}			
t BY	R_{SH}	P_0/P_{SH}	*v/c*	*t* BY	R_{SH}	P_0/P_{SH}	*v/c*
0	1E-16	1	7E-18	0	1E-16	1	3E-18
100	1.3E-13	1	1E-14	1000	7E-4	1	2.2E-5
200	1.9E-10	1	1E-11	1200	.283	1.00004	8.4E-3
300	2.7E-7	1	2E-8	1240	.926	1.00048	.0275
400	3.9E-4	1	28E-6	1260	1.68	1.0156	.0498
500	.554	1.001	.0403	1280	3.04	1.0051	.0901
520	2.37	1.019	.172	1290	4.08	1.0093	.1212
530	4.90	1.083	.356	1300	5.49	1.0169	.163
535	7.05	1.180	.512	1310	7.39	1.0308	.219
540	10.13	1.407	.736	1320	9.95	1.0564	.295
545	14.57	2.025	1.058	1330	13.38	1.1044	.397
550	20.95	4.301	1.522	1340	18.0	1.197	.534
555	30.12	20.42	2.188	1350	24.2	1.385	.719
560	43.32	511	3.15	1360	32.6	1.802	.967

One important aspect of this table appears when comparable values of P_0 /P_{SH} are chosen for the two Hubble constants used. The theoretical radius of the i-ther is smaller for a larger Hubble constant. This means that if the latter is accurately measured then the theory provides a measure of the size of i-ther and therefore the radius of the universe of matter emerging from i-ther.

The calculations use SI units so distances had to be converted from M_{PC} to metres. There are 3.26×10^6 light years per M_{PC} and $3.156\times10^{16}\times c$ metres per light year. However, the radial distance of the shock front from the origin point is given in billions (10^9) of light years (BLY) and time is given in billions of years (BY).

The starting shock front radius at time zero corresponds to about ten kilometres but will have grown to about 13,000 km at t = 100 billion years for $H_0 = 71$ km/s/M$_{PC}$ but takes about 280 BY to reach this size when $H_0 = 29$ km/s/M$_{PC}$. The age of i-ther depends on the choice of starting radius that cannot yet be calculated but could be anywhere between a few centimetres and perhaps 10 kilometres.

What is most important, however, is the many billions of years that need to elapse before a significant pressure increment from shock to origin point appears – about 500 BY and 1253 BY respectively for the same 0.1% pressure elevation. Then P_0 grows rapidly being an exponential as function of an exponential according to equation [10.2.8].

It is interesting to compare the two cases at 535 BY and 1338.7 BY respectively since P_0 / P_{SH} are 1.18 in both cases. For the higher value of H_0, however, $R_{SH} = 7.05$ billion light years as compared to 17.3 BLY for the lower H_0 value. Furthermore the shock speed ratios v/c are .512 and .514 respectively – in other words identical.

10.3 The Extra Growth due to Diffusion

When equation [10.2.6] is applied to calculate the speed of the shock front at an early time a ridiculously low value is returned. This cannot possibly be realistic since primaries at the shock front will spread out in front of it thinning until the density is too low to produce any collision breeding. Since no annihilation will be occurring at this stage, then a very rapid build up of pressure will occur at the point where breeding starts until collapse to form annihilation cores happens.

What this means is that, added to the speeds of growth previously considered will be a 'diffusion growth rate': the shock front will have a speed $E_X v_P$ added. Where E_X could be up to about 0.1 meaning an extra shock front speed 10% of the average speed v_P of the primaries. At present this is only a guess but the theory is potentially able to be developed to provide an accurate value for E_X.

With this refinement equation [10.2.6] is modified to:

$$v_{SH} = \frac{dr}{dt} = r_1 \frac{C_{NA}}{3} \exp\left(\frac{C_{NA}}{3} t\right) + E_X v_P \qquad [10.3.1]$$

Where r_1 is the radius of the shock front where t is arbitrarily chosen to be zero and v_{SH} is the shock front velocity at time t. And on integration the radius of the shock front r_{SH} becomes:

$$r_{SH} = r_1 \exp\left(\frac{C_{NA}}{3} t\right) + E_X v_P t \qquad [10.3.2]$$

Equation [10.2.7] is now extended to yield:

$$H_0 r_{SH} = H_0 \left[r_{SH1} \exp\{H_0 t\} + E_X v_P t \right]$$

When substituted in [10.2.4] the result for origin pressure development with time becomes:

$$\frac{P_0}{P_{SH}} = \exp\left[\frac{3}{2}\frac{C_{PF}}{v_P^{\,2}}\{H_0(r_{SH1}\exp(H_0\,t) + E_X v_P\,t)\}^2\right] \quad [10.3.3]$$

The result is a drastic reduction in the estimated age of the i-ther.

As soon as P_0 / P_{SH} has increased to about 1.01 it becomes necessary to consider the effect of pressure on the rate of creation. This matter will be considered next and this is where the new breakthrough appeared.

10.4 Net Creation Rate as a Function of Pressure

The rate of creation without any annihilation was derived as a function of the pressure ratio P/P_L in equation [7.3.16] p.178. It was argued that $V_C = 0$ is reasonable since this was a factor allowing for the volume taken up be primaries, but when annihilation is included this has an opposing effect. This equation can then be re-written as for a perfect gas to yield:

$$\frac{d(PV)}{PV} = C_{BT}\frac{P/P_L}{(3 + P/P_L)^2}dt \quad [10.4.1]$$

Where breed collision creation rate C_{BT} was shown by [7.3.15] to be a constant provided no annihilation was taking place. Annihilation will reduce this by about 42 orders of magnitude at the shock front and then as pressures rise the rate of annihilation will steadily increase until at the i-theric liquidus pressure P_L annihilation cancels creation exactly to produce an equilibrium state. It will be assumed reasonable to replace C_{BT} by a many times smaller net shock front creation rate C_{SH} and then multiply the remainder of [10.4.1] by a factor giving a linear reduction to yield zero at the liquidus. The result is:

$$\frac{d(PV)}{PV} = C_{SH}\frac{P/P_L}{(3 + P/P_L)^2}\left[\frac{1 - P/P_L}{1 - P_{SH}/P_L}\right]dt \quad [10.4.3]$$

So now if the net creation rate at pressure P is C_{NT} and the ratio C_{NR} is defined as $C_{NR} = C_{NT}/C_{SH}$ then [10.4.3] reduces to:

$$C_{NR} = \frac{C_{NT}}{C_{SH}} = \frac{P/P_L}{(3 + P/P_L)^2}\left[\frac{1 - P/P_L}{1 - P_{SH}/P_L}\right] \quad [10.4.4]$$

This is plotted in FIG.10.1 and shows a curve very similar to the dotted curve obtained years earlier and presented in FIG.7.2p.182. The new curve looks like an inverted parabola with the maximum shifted from the position where $P/P_L = 0.5$ to 0.43 except that the position of the shock front has not yet been established.

The first guesses for the shock front placed it at $P_{SH}/P_L = 0.01$ and computation using [10.3.1] to [10.3.3] run until P_0/P_{SH} rose to about 1.05: a 'pressure elevation' of .05. The intention was then to apply [10.4.4] using a Hubble constant changed to $C_{NR}H_0$ with H_0 being the originally assumed value.

Now wait for the breakthrough!

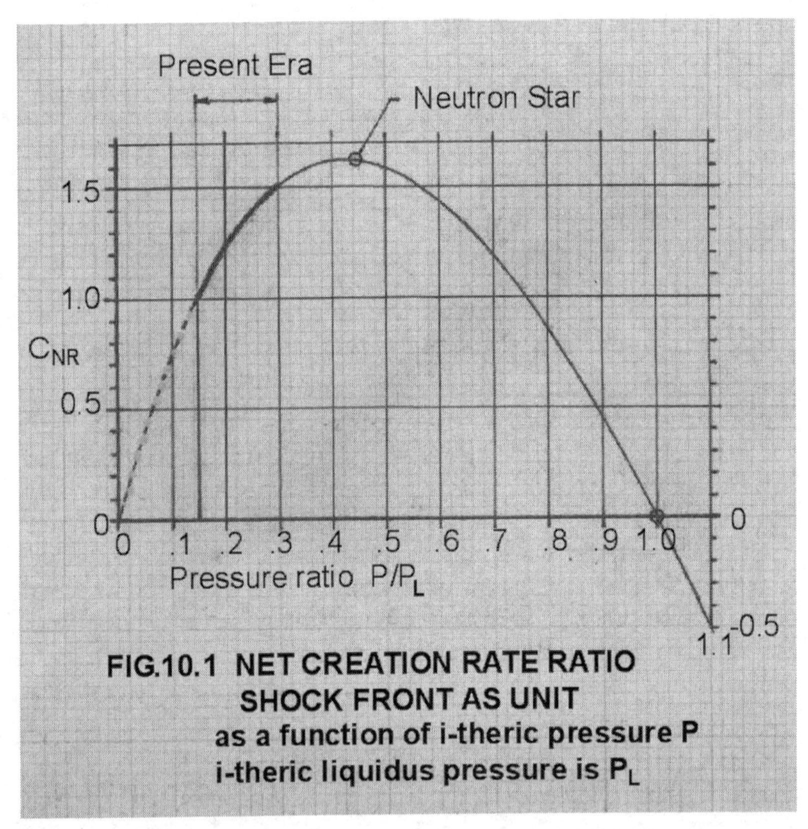

FIG.10.1 NET CREATION RATE RATIO SHOCK FRONT AS UNIT as a function of i-theric pressure P i-theric liquidus pressure is P_L

The new creation rate almost doubled the pressure elevation making the computation go unstable. The pressure rise kept on increasing iteration after iteration. It was then realised that the shock pressure ratio P_{SH}/P_L had been chosen too low. When chosen too high the creation rate caused the pressure gain to drop. So in reality the shock front has to jump to a value where stability is achieved. This value turned out to be:

$$P_{SH}/P_L = 0.15$$

So at last we have a firm value for the shock front pressure!

10.5 More exciting findings were soon to emerge

For the next experiment a test was made to see what error would be involved using the existing theory to a time that would yield red shifts matching existing astronomical data. This meant extending to cover a range of much larger pressure elevations. These would certainly involve considerable change of C_{NR} from the unity value assumed in the computation of TABLE 10.I to some other value given by equation [10.4.4].

The computer program, now referenced as "Un150710", was arranged to produce the preliminary growth from r_{SHI} = 10 km to a specified radius r_{SH}. A value of 12 BLY was chosen for the first test with: H_0 = 45 km/s/M$_{PC}$, P_{SH}/P_L = .15 and E_X = .05.
From this point a full set of 15 pressure/radius profiles was computed at one billion year intervals with results stored in a two dimensional array. In this N represented a number between 0 and 100 proportional to r/r_{SH}. The number M represented the profile number and so also the time in billions of years from the new starting point of time zero at r_{SHI}.

This meant the storage of $R1$(N,M), in BLY; $P1$(N,M), meaning P/P_L values and $U1$(N,M), meaning u/c_0 where u is the radial speed of primaries and c_0 the datum speed of light fixed at the central origin point of the universe.

The shock front, however, has an extra diffusion speed. To take this into account an extra value of N is incorporated making the total 101. This specifies the shock front. At time zero this coincides with the N=100 array values. At all other times the difference between N=100 and N=101 is the extra caused by diffusion. All other N values show path lines of particles and galaxies after these have formed and so even for element N=100 its pressure increases.

Pressures for any profile M and any element of radius N are then calculated from [10.2.4] p.219 and presented in the form:

$$\frac{P_0}{P_L} = \frac{P_{SH}}{P_L} \exp\left(\frac{3}{2} \frac{C_{PF}}{v_P^{\,2}} (H_0\, r_{SH})^2 \right) : \qquad [10.5.1]$$

$$\frac{P_N}{P_L} = \frac{P_0}{P_L} \exp\left(-\frac{3}{2} \frac{C_{PF}}{v_P^{\,2}} (H_0\, r_N)^2 \right) \qquad [10.5.2]$$

10.6 The required net creation rate: the energy demand

What has been specified by equations [10.2.4] and [10.2.5] is a pressure increase dP/dt at pressure P and a growth rate of dV/dt of all elements, each of volume V. The values of P and V need to be averages during the time Δt between profiles. The combined effect of these specified increases of pressure and volume is an energy demand that can be expected to differ from one element to another. The energy supplied will also change with pressure and volume. The problem is to compare the energy required with that which can be supplied.

Derivation of Demand

Let K be any profile number from which the next, specified as M, is to develop. All such arrays are split into 100 elements denoted N. So in code "Un150710" two dimensional arrays, R(N,M), P(N,M) and U(N,M) specify the radius, pressure, and radial velocity respectively of every element from N=1 to N=100. N=0 specifies the origin point and so the radial distance between N=0 to N=1 defines the first element as a small sphere. All the rest are spherical shells. A final element is N=100 to N=101 and this represents the extra caused by diffusion growth.

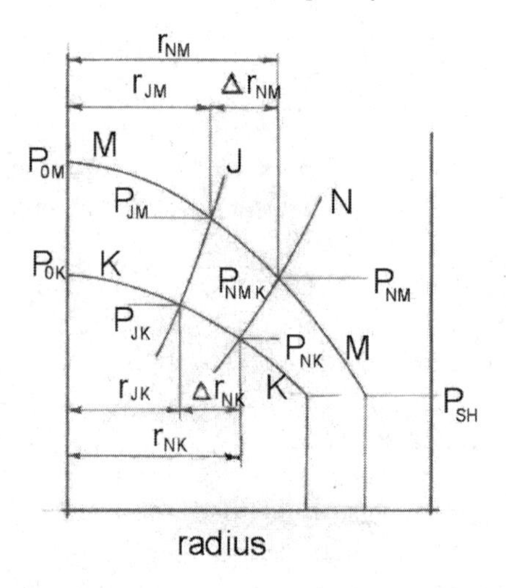

FIG.10.2 NOMENCLATURE for PRESSURE/RADIUS PROFILE COMPUTATION
M is the P/r profile 1 BY after the K profile & N is the path line of element N that has radial width Δr so this is distance from path line J to N
P_{NMK} is used to find C_{NR}

Now all three variables are specified with radius and pressure given by equations [10.2.5] and [10.2.4] respectively. In consequence the gain in *PV* i.e. $\delta(PV)$ can be calculated for every element in an interval of 1 BY (billion years). The average *PV* can also be obtained from these two profiles. From this the ratio $\delta(PV)/(PV)$ can be calculated and this determines the <u>required</u> creation rate C_{RC} . This is specified as: $C_{RC} = E_{demand}/C_{SH}$.

This can then be compared with the creation rate ratio C_{NR} since this is $E_{supplied}/C_{SH}$ as provided by equation [10.4.4] p.223. This is defined as: $C_{NR} = C_{NT}/C_{SH}$. These two ratios can be evaluated and compared to see how good a match is provided.

The average pressure P_{NMK} is needed for use with [10.4.4] to give the value of C_{NR}. This pressure can be seen from FIG.10.2 to be found as follows. Path line J, moving as time flows, marks the inner radius of an element and path line N its outer radius. With J = N-1 and K = M-1 the previous profile 1 billion years (BY) earlier is specified. The average pressure for element N at profile K is:

$$P_K = (P_{NK} + P_{JK})/2$$

and at profile M is $P_M = (P_{NM} + P_{JM})/2$.

Then $P_{NMK} = (P_K + P_M)/2$.

The average volume of the element is required on which to evaluate the energy creation during time Δt. This is based on the volume at K, which is: $V_{NJK} = (R_{NK}^3 - R_{JK}^3) \times 4\pi/3$
and the volume at M, which is $V_{NJM} = (R_{NM}^3 - R_{JM}^3) \times 4\pi/3$.

So the average volume is $(V_{NJM} + V_{NJK})/2$ and the volume gain is $V_{NJM} - V_{NJK}$. Then the equation for the energy demand becomes:

$$\frac{\delta(PV)}{PV} = \frac{P_M\left(R_{NM}^3 - R_{JM}^3\right) - P_K\left(R_{NK}^3 - R_{JK}^3\right)}{P_{NMK}\left(R_{NM}^3 - R_{JM}^3 + R_{NK}^3 - R_{JK}^3\right)/2} \qquad [10.6.1]$$

If C_{RC} is the ratio of energy demand to that provided at shock front pressure P_{SH} then this can be expressed in the alternative form:

$$\frac{\delta(PV)}{PV} = C_{RC}\, C_{SH}\, \Delta t : \quad \textit{Noting that} \quad C_{SH} = 3H_0$$

From this combination the value of C_{RC}, the required creation ratio, can be found to compare with the target ratio C_{NR} for the same value of N from [10.4.4]. The value of C_{NR} for N = 0 is C_{OR}. This means the error is based on the target ratio given by the maximum pressure P_{0M} at profile M. It should be noted here that this pressure and all the others shown in FIG.10.2 are given in units of P_L.

The percentage error is then given by:

$$Error\% = 100\frac{\left(C_{RC} - C_{NR}\right)}{C_{0R}} \qquad [10.6.2]$$

There are 16 arrays altogether counting M=0 as the first. So the gains for profile M=1 are from K=0 to M=1. Consequently only 15 out of the 16 arrays provide gains and errors.

We are now ready to present the results of some computations in which *Err %* is the percentage error given in the bottom rows of TABLE 10.II.

10.7 Results of Computation

Input data:	H_0	P_{SH}/P_L	E_X	R_{SH1}
	45 km/s/M_{PC}	0.15	0.05	12 BLY

M = 1 **TABLE 10.II COMPRISES M=1 TO 15**

N	**1**	**20**	**40**	**60**	**80**	**100**	**101**
R	**.125**	**2.51**	**5.03**	**7.54**	**10.05**	**12.56**	**12.64**
u/c_0	**.006**	**.116**	**.231**	**.347**	**.463**	**.579**	**.652**
P/P_L	**.186**	**.184**	**.180**	**.172**	**.162**	**.15**	**.15**
C_{NR}	**1.159**	**1.153**	**1.134**	**1.102**	**1.058**	**1.002**	**1**
Err%	**-.63**	**-.497**	**.186**	**.312**	**.995**	**1.88**	**0**

M = 5

N	**1**	**20**	**40**	**60**	**80**	**100**	**101**
R	**.151**	**3.02**	**6.04**	**9.06**	**12.09**	**15.1**	**15.51**
u/c_0	**.007**	**.139**	**.278**	**.417**	**.556**	**.696**	**.769**
P/P_L	**.207**	**.204**	**.197**	**.185**	**.170**	**.152**	**.15**
C_{NR}	**1.241**	**1.232**	**1.204**	**1.158**	**1.093**	**1.012**	**1**
Err%	**-1.36**	**-1.19**	**-.78**	**-.010**	**.901**	**2.21**	**0**

M = 10

N	**1**	**20**	**40**	**60**	**80**	**100**	**101**
R	**.190**	**3.80**	**7.61**	**11.41**	**15.21**	**19.02**	**19.93**
u/c_0	**.009**	**.175**	**.350**	**.525**	**.701**	**.876**	**.949**
P/P_L	**.255**	**.250**	**.236**	**.214**	**.187**	**.157**	**.15**
C_{NR}	**1.395**	**1.382**	**1.34**	**1.268**	**1.165**	**1.035**	**1**
Err%	**-2.27**	**-2.12**	**-1.73**	**-.92**	**.519**	**2.61**	**0**

M = 15

N	**1**	**20**	**40**	**60**	**80**	**100**	**101**
R	**.239**	**4.79**	**9.58**	**14.36**	**19.15**	**23.94**	**25.49**
u/c_0	**.011**	**.220**	**.441**	**.661**	**.882**	**1.102**	**1.176**
P/P_L	**.357**	**.346**	**.316**	**.271**	**.219**	**.166**	**.15**
C_{NR}	**1.585**	**1.573**	**1.530**	**1.437**	**1.284**	**1.075**	**1**
Err%	**3E-4**	**-.359**	**-1.21**	**-1.50**	**-.216**	**2.98**	**0**

10.8 The breakthrough is demonstrated by this set of tables!

Note how the errors in percentages shown in the last row of each set are never more than a few percent and so are acceptably small. It is strange that in every case the greatest error is at N = 100.

This is the major breakthrough! What this table has shown is that the simple theory first used and presented in §7.7 p.56 was accurate. It ignored the effect of pressure and for many years triggered a search for a refined approach that would correct the anticipated errors. It had not been appreciated that the net creation rate profile would cancel the pressure effect to yield acceptable accuracy.

The thickened part of the net creation rate curve given in FIG.10.1 shows the range covered by our universe in the present era. This shows the central region is close to the maxima of net creation rate. In consequence the fortuitous match represented by the small values of error given in TABLE 10.II cannot be expected to apply for the distant future.

In TABLE 10.III below red shifts are presented for the same set of input data.

An important finding is that in order to roughly match the maximum red shift of 1.755 reported by astronomers for type 1a supernovae, the Hubble constant assumed had to be drastically reduced. A value had to be selected of 45 km/s/M_{PC} as compared to the value of 71 km/s/M_{PC} that appears from the very crude method of linear extrapolation conventionally applied. In effect this obsolete method assumes $H_0\, r$ to be a constant with respect to time so that galaxies are moving with constant speed. Clearly with an accelerating expansion this assumption is invalid. Then as the Big Breed theory has demonstrated, when it is the net creation that is causing expansion, then Hubble's constant H_0 is invariant with both time and space.

TABLE 10.III COMPUTED RED SHIFTS & DISTANCES

Input data: H_0 45 km/s/M_{PC} P_{SH}/P_L 0.15 E_X 0.05 R_{SH1} 12 BLY

(1)	(2)	(3)	(4)	(5)	(6)	(7)	(8)	(9)
M N	Z_D	Z	Z_G	Z_{MG}	P/P_0	***R***	u/c_0	c_T/c_0
14 5	.053	.111	.084	.083	.9207	1	.0526	1.055
13 10	.10	.221	.169	.164	.8515	2.004	.1005	1.106
12 15	.143	.330	.254	.241	,7911	3.014	.1440	1.153
11 21	.191	,463	.363	.341	.7379	4.030	.1926	1.199
10 27	.235	.596	.472	.437	.6891	5.045	.2364	1.242
9 34	.284	.757	.608	.558	.6451	6.061	.2843	1.284
8 41	.329	.919	.744	.677	.6062	7.074	.3274	1.323
7 49	.378	1.112	.911	.826	.5698	8.08	.3737	1.362
6 58	.432	1.346	1.112	1.003	.5355	9.089	.4225	1.401
5 67	.479	1.581	1.315	1.183	.5051	10.09	.466	1.439
4 77	.530	1.862	1.559	1.402	.4761	11.08	.5115	1.476
3 88	.584	2.196	1.850	1.666	.4486	12.07	.5583	1.514
2 100	.640	2.593	2.197	1.986	.4223	13.05	.6058	1.552

As TABLE 10.II shows, the theory predicts the radius of the universe when the Hubble constant has been chosen. In this case for M=15 at N=101 (the shock front) it is 25.5 billion light years. Then as TABLE 10.III shows in column (5) the theory predicts the red shift that astronomers should observe reaching almost 2.

So runs have to be made with different radius at M= 0, 12 BLY in this case, and different values of H_0 until acceptable values of red shift, such as this, are returned.

This input value of H_0 is really a statement of the net creation rate at the shock front given by $C_{SH} = 3\ H_0$. For this case it works out to $C_{SH} = 4.277\times10^{-18}\ s^{-1}$. It does not need to correspond with the value obtained from the predicted recession velocity and radius.

Therefore from the end to end points of the radius/time plot a new value for the Hubble constant can be found. This means using the speed and radius at N = 100 from columns (8) and (7) respectively. The result is 45.38 km/s/M_{PC} which is little different from the assumed starting value of 45.

Red shift analysis, based on derivations given in Chapter 9, shows some surprising features. The Doppler shift given in column (2) is only about 1/3 of that caused by light having to speed up due to meeting lower and lower speeds of recession and this despite the light speed c_T, shown in column (9), slowing down. This slowing is due to i-theric pressures rising as shown in column (6).

The blue shift due to space gravity is shown in column (4) to produce a small reduction and this is reduced still further by the mass gravity effect shown in column (5). However, the latter is based on an average mass density of matter that could be about three times too high and so the predicted red shift astronomers would record is probably nearer the values given in column (4).

With the value of diffusion extra growth coefficient E_X =.05 assumed. what has been called the pre-history age of the universe has been drastically reduced. This is the age to the start of detailed computation. This was assumed as the time needed to increase its radius from 10 km to 12 BLY. This is now reduced to 47.04 BY from the mind boggling values given in TABLE 10.I. So age to the present era becomes:

47 + 15 = 62 billion years.

Column (1) shows that for this case the maximum value of N reached by the time-reversed starlight is 100 meaning it is almost at the shock front and M=2 means the starlight was emitted 15 - 2 = 13 BY ago. This, however, is not the time from creation of matter by the supposed big bang. The star first had to travel out to the distance R of 13.05 BLY given in column (7). The speed of recession is given in column (8) and just after the big bang would be nearly equal to c_0; so a rough average speed of about $0.8c_0$ could be assumed. So the star would take about 13/0.8 = 16.2 BY to reach the position given in the last row. So the age of matter then becomes: 13 + 16.2 = 29.2 BY.

This is more than twice the accepted age of 13.7 BY.

10.9 Summarising a few other computations

The predicted 'observed' red shift is given by Z_{MG} that includes the blue shift caused by mass gravity. This was based on an average mass density of 10^{-26} kg/m^3. This value is the assumed average mass density of stars and dark matter and, as already stated, could be several times too large. If this is the case then the effect is negligible and the values of Z_G can then be taken as predicting the observed red shifts. The maximum so far observed is 1.755 and so predictions are probably on the high side. However, the initial assumptions can be adjusted to give lower values. This is permissible since what has been derived is a new way of determining the Hubble constant, together with the size of the universe, from red-shift data.

The effects of changing R_{SH1} and H_0

Brief summaries are now presented of runs made with different values of input data to give some idea of the way variables interact. Input data is chosen to give red shifts not too far away from the maximum values so far observed. The pre-history phase is first calculated from an assumed value of Hubble constant H_0, quoted in units of km/s/M$_{PC}$ to accord with convention, and the end of this phase shock front radius R_{MO} in BLY. The diffusion excess growth factor E_X has also to be supplied. This data is supplied in column (1) of TABLE 10.VI together with case letters A to F.

Column (2) gives the end state of what has been called the 'pre-history phase' in which, starting from 10km radius, T_B is the age of this phase in BY. The shock front speed, as a ratio of light speed at the origin point, u_{SH}/c_0 is included together with P_0/P_{SH}, which shows how the origin pressure has developed up to time T_B. The value H_{O2} is a Hubble constant evaluated from the final computation of the light path from a star close to the edge of the universe. There is no reason for this to coincide with the value originally assumed since that is really $C_{SH}/3$ a statement of the rate of creation at the shock front. In the second phase when 15 pressure/radius profiles are explored the creation rate varies.

These all assume $P_{SH}/P_L = 0.15$.

The first surprise is given by comparing case E and F since the result may seem counter-intuitive. Case E had no diffusion to increase growth and in Case F, $E_X = 0.1$. It hardly seems to affect anything except the pre-history age T_B in BY (billions of years) and the red shifts which seem to change in the wrong direction. Cases A and D also only differ in the value of E_X. Again the red shifts are lower when there is a diffusion excess growth. Of course this excess growth is also incorporated in the selected end of pre-history growth so perhaps the effects just considered are not so unreasonable after-all.

Since the net creation rate is about 42 orders of magnitude smaller than the breeding rate it is impossible to derive a theoretical method of evaluation. Values can only be assumed to give predicted growth that matches observed values. What the table does demonstrate in column (3), by cases A, B and C, is the way the creation rate C_{SH} , represented in column (1) by H_0 , ($C_{SH} = 3H_0$) has to be reduced to compensate for the increased value of R_{MO} . The latter is the end of pre-history shock front radius assumed as data input to provide the starting radius for computation with M=0.

Symbol M (column (4)) is the pressure/radius profile number that increases in units of ΔT_B=1BY. The three cases, A,B and C also show how the pressure P_0/P_L at the origin at the present era given by M=15 (meaning 15 BY after M=0) is highest for the highest creation rate. Yet red shifts Z, shown in column (4) are little affected. It is unfortunate that all red shifts are not evaluated at equal distances from the shock front. These distances are given by N which is the element number out of 100. So the radius represented is $R_{N,M}/R_{100,M}$. So these distances are only measured from the shock front when $E_X = 0$ since N = 101 at the shpck. The distances of the emitting star therefore correspond to element number N, in Column (4), and are given by R_{NB} in BLY (billions of light years) in column (5). The time ago of the light emission is given by 15 – M. This is in BY since each profile is separated in time by a billion years.

Also in column (5) is P/P_0 . This is the pressure of i-ther at the star divided by pressure at the origin centre of i-ther that is assumed negligibly different from that of the Earth based observer. In consequence the speed of light c_T is greater at the star than at a central observer, in ratio c_T/c_0 as shown in column (6). Also shown in that column is the speed of recession of the star given as u_{NF}/c_0. The significance of this is that the speed of recession is always less than half that of light propagation. This means that the entire universe is potentially within range of observation from Earth.

These distances and speeds of recession are of course computed values based on the Big Breed Theory just as are the red shifts. In this way the input data needs to be chosen so that the computed red shifts match those observed. In this way the new theory provides a way of deriving the Hubble constant from observed red shifts and reads out the distance of the star simultaneously.

No other theory predicts the radius of the universe. Consequently the Big Breed theory enables a new check to be made by comparing this theoretical distance with the luminosity distance that is observed. The latter is likely to be a gross overestimate, however, owing to scattering of some of the light by dust and by the Aspden effect of absorption by free electrons.

Now we return to column (4) to look at the red shift values in more detail. The Doppler shift Z_D is always only about 1/3rd of the shift $Z - Z_D$ caused by c, what might be termed the 'momentum

TABLE 10.IV RESULTS USING N=0 FOR SHOCK FRONT EXPLORING FROM 15 BY AGO UNTIL THE PRESENT

(1) (2) (3) (4) (5) (6)

Data | At M = 0 | At M = 15 | Values at remote galaxy

Case	E_X	T_B	u_{SH}/c_0	R_{SH}	P_0/P_L	u_{SH}/c_0	N	Z_D	Z_G	P/P_0	U_{NF}/c_0
H_0	R_{M0}	P_0/P_{SH}	H_{02}	R_1		u_1/c_0	M	Z	Z_{MG}	R_{NB}	c_T/c_0
A	.05	37.2	.507	22.6	.416	1.25	88	.585	1.97	.392	.591
55	9	1.175	50.4	20.9		1.18	5	2.36	1.83	10.4	1.60
B	.05	47.0	.553	25.5	.357	1.16	99	.64	2.20	.422	.606
45	12	1.212	45.4	23.9		1.102	2	2.59	1.99	13.1	1.55
C	.05	55.2	.614	29.2	.369	1.29	98	.603	1.99	.426	.582
40	15	1.268	37.6	27.7		1.134	1	2.36	1.742	14.2	1.574
D	0	1061	.507	20.9	.360	1.18	98	.667	2.36	.421	.622
55	9	1.175	50.9	20.9		1.178	4	2.77	2.21	11.0	1.554
E	0	1303	.553	23.9	.322	1.10	97	.657	2.18	.472	.588
45	12	1.212	42.4	23.93		1.102	2	2.55	1.983	12.6	1.482
F	.10	34.3	.553	27.0	.398	1.25	91	.572	1.85	.400	.577
45	12	1.212	41.8	23.94		1.102	3	2.25	1.68	12.5	1.586

component' of light, having to speed up as it approaches. This is a very large extra red-shift that cosmologists do not take into account when basing evaluation on Einstein's theory of special relativity. In consequence much lower values of Hubble constant are read out than the 71 km/s/M_{PC} that conventional analysis returns based on the same data.

The red shifts observed will be smaller than Z due to the blue shifts due to space gravity, given by $Z - Z_G$, and that due to mass gravity $Z_G - Z_{GM}$. The latter is likely to be over-predicted since it is based on an average mass density of matter of 10^{-26} kg/m^3 that could be several times too high. Consequently it is probably safer to assume Z_G as the value observed by Earth bound astronomers than Z_{MG} , but the true value lies somewhere between the two.

10.10 Depicting overall Results

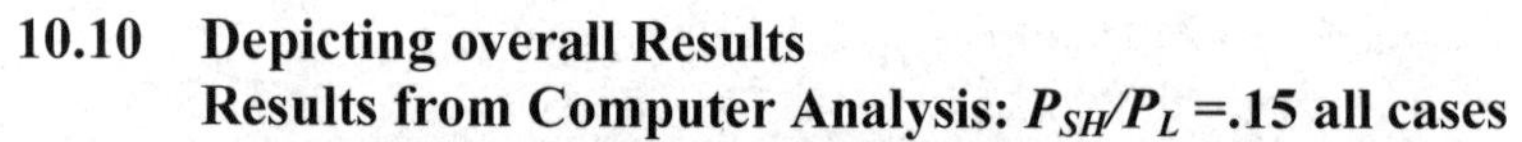

Results from Computer Analysis: P_{SH}/P_L =.15 all cases

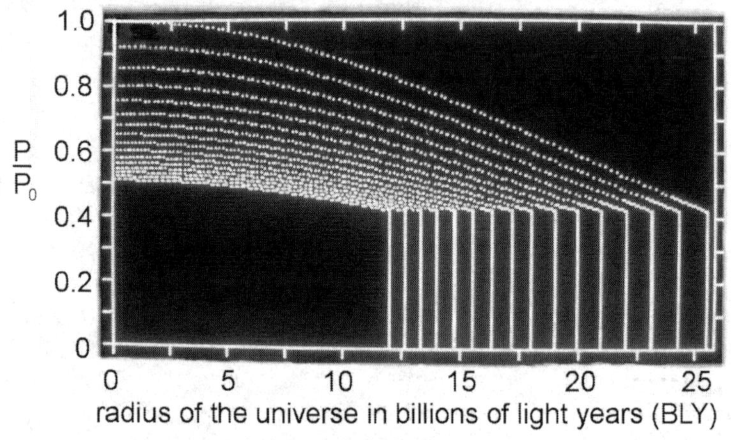

FIG.10.3 PRESSURE/RADIUS PROFILES STARTING 15 BILLION YEARS AGO
CASE H_0= 45: E_X= .05: R_{m0}=12 BLY.
P in units of P_0, the pressure at the origin point now

Each profile is described by arrays of 101 elements, symbol N. Each curve is one of 16 profiles from 0 to 15, symbol M, at one billion year intervals. These limits are all that can be permitted by memory considerations. An example of photographs taken of plots on the computer screen is presented in FIG.10.3 for the case given in TABLES 10.II and 10.III having a data input of H_0 = 45 km/s/M$_{PC}$, E_X = 0.05 and starting radius R_{M0} =12 BLY. Of course P_{SH}/P_L = 0.15. Pressures are presented as P/P_0 and for this case the origin pressure P_0/P_L = 0.346 so P_{SH}/P_0 = 0.434.

In FIG.10.3 the 16 pressure/radius profiles are shown each bounded by a shock front shown by the vertical lines. Short inclined lines from N = 101 join these to the first element N = 100 of the main array shown as dots each representing the position of an element N. The rapid rise of pressure in regions close to the origin point is particularly evident. What gives the impression of a slightly increasing shock front pressure is an optical illusion.

FIG.10.4 is the same plot as FIG.10.3 but path lines representing primaries of the i-ther, or co-moving galaxies, are superimposed by joining up each fifth element N to the same numbered element in all other profiles. These show that, as primaries recede from the origin point, each experiences a pressure increase that accelerates faster than the radius.

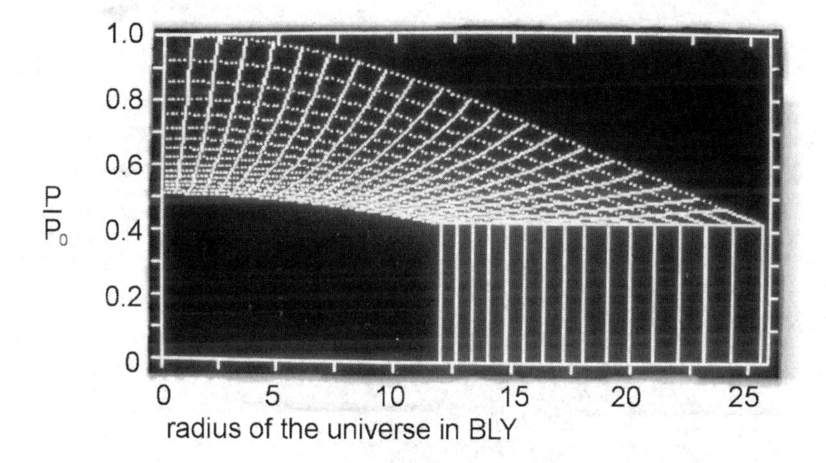

FIG.10.4 PRESSURE/RADIUS PROFILES WITH PATH LINES ADDED
Same case as FIG.10.3

The next plot, FIG.10.5 uses the same data but now time is the ordinate plotted against radius. The fanning out of path lines clearly illustrates the accelerating nature of the expansion. All path lines remain equally spaced except for one: element 101. This gains on element 100 and this is due to the extra growth caused by diffusion at the shock front.

The line of negative slope shows the light path from a distant supernova close to the advancing shock front as it propagates to the origin point. Chapter 9 showed that this theory implies that our galaxy cannot be far from the origin centre so that this plot can be taken to apply to an Earth-based astronomer with little error.

This line was actually calculated with time reversal. Surprisingly it is almost a straight line despite the need for light to speed up due to its meeting reduced recession speed. This is a feature consistently replicated in all the computations made with different input data. This was a surprise because a curvature due to light needing to speed up as it approaches had been expected. This is explained by the propagation speed c_T, which has been termed 'observed', reducing due to meeting a medium of increasing pressure as light comes in toward the observer. Clearly the two effects almost cancel each other. The linearity permits a new value of the Hubble constant to be calculated. The reason a red-shift appears due to light speeding up is because the flat space speed of light c, governing momentum, is speeding up as shown later in FIG.13.4 p.269.

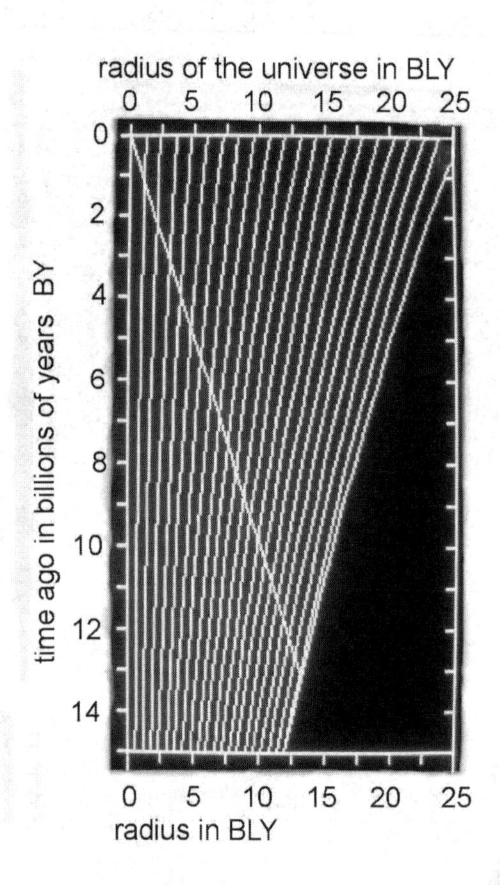

FIG.10.5 TIME /RADIUS PLOT showing path lines and light path same case as FIG.10.3

Finally we come to FIG.10.6 giving the predicted red shifts. Curve Z_D is the Doppler shift to which is added $Z–Z_D$, which is due to the momentum of photons having to be conserved as they move into space receding at ever slower speed. The total is Z and clearly the Doppler shift is the minor component. Then a blue shift is caused by photons falling in the space-gravitational field. This reduces the net value to Z_{SG}. A further blue shift to curve Z_{MG} is caused by mass gravity. This final value is the red shift measured by astronomers. However, this is based on an average mass density, inclusive of 'dark matter', of 10^{-26} kg/m^3. This is a value often adopted by cosmologists but it could be several times too high.

Two other curves are presented in FIG.10.6. That marked c_T/c_0 is the observed speed of light divided by that at the origin c_0. This is higher than the speed of propagation c in flat space by an amount caused by change of i-theric pressure. The top curve is the corresponding speed of recession v also divided by c_0. Even close to the shock front the latter has not exceeded 0.6 as compared to c_T/c_0 = 1.55. This means there is no 'horizon' from beyond which light cannot reach Earth. In consequence astronomers are potentially able to see right to the very shock fronted edge of the universe.

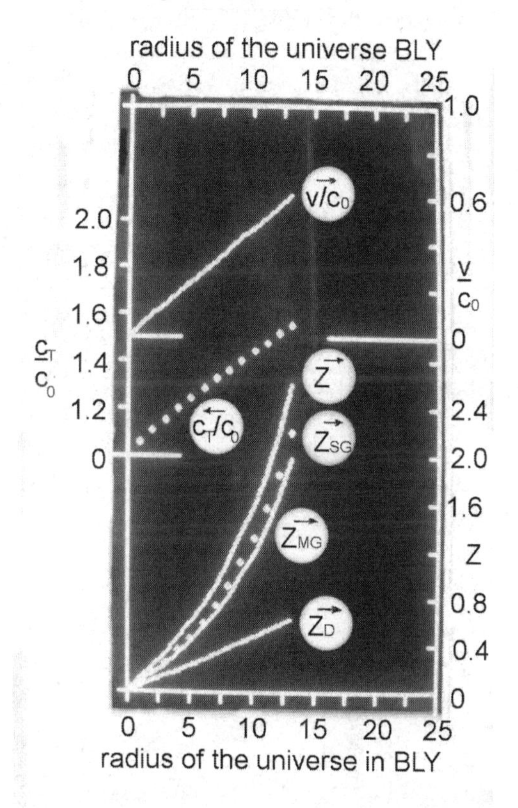

FIG.10.6 RED SHIFTS Z LIGHT SPEED RATIO c_T/c_0 & REGRESSION SPEED RATIO v/c_0

A recap on a previous conclusion seems now to be required. It is unfortunate that Wright (2005) and other cosmologists approached, all refused to divulge the raw data of distances of remote supernovae as determined from their luminosity. All they would provide were the distances estimated for their positions in the present era. Since these estimates used Einstein's relativity theories, which are inapplicable for a theory in which light propagates through a background medium, all the data except red shifts were rendered valueless.

Therefore data input was chosen that led to the predictions of observed red shift roughly equal to the maximum of 1.755 reported by Wright (2005). So the distances shown are those computed from the Big Breed theory.

As is evident from a study of TABLE 10.VI the value of the assumed Hubble constant has to be reduced when the radius of the universe, which appears from the theory, is increased. Only when the latter is known from observation can the Big Breed theory provide a firm value for the Hubble constant. However, the study has already shown this value is far smaller than that derived by conventional analysis based on special relativity.

10.11 Exploring Higher H_0 values and the Far Future

The limit of applicability of the computer code used for

previous analyses was applicable up to $P_0/P_L = 0.38$. Beyond that errors became unacceptable. To extend the range a simple iterative loop was therefore added. This reduced the pressure of each element until the error fell below 1%. Until then it had not been possible to attempt to explore the accepted value of Hubble constant of 71 km/s/Mpc since pressures were found to exceed the limit.

With this refinement it is not yet possible to include diffusion growth since, for reasons not yet discovered, this upsets the iteration to yield false pressures close to the shock front.

The following result may be of interest. To yield a desired end result of $H_0 = 71$ km/s/Mpc an input figure of 81 was found to be required. Then also $P_0/P_L = 0.545$ was predicted, which is well above the value permissible without resort to error correction.

The following table gives remaining figures for the end result:

TABLE 10.V For Hubble Constant $H_0 = 71$ km/s/Mpc

(1) Data	(2) At M = 0	(3) At M = 15	(4) Values at remote galaxy	(5)	(6)
E_X H_0 R_{M0}	T_B u_{SH}/c_0 P_0/P_{SH} H_{02}	R_{SH} P_0/P_L u_{SH}/c_0 R_1 u_1/c_0	N Z_D Z_G M Z Z_{MG}	P/P_0 R_{NB}	U_{NF}/c_0 c_T/c_0
0 81 4.5	712 1.29 3.63 71.0	15.6 .545 1.29 15.6 1.29	86 .552 2.08 8 2.51 2.00	.303 7.59	.630 1.77

This table is set out as in TABLE 10.IV p.233 to make comparison easy. What is most noticeable is the drastic reduction in radius for the present era given in column (3). Since the universe is unlikely to be as small as this in the present era the need to reduce H_0 is justified. The reason for the large disparity between the high input value of H_0 and the end value has not yet been studied. Such large differences do not appear for the cases given in TABLE 10 IV.

The theory also now enables the future growth of i-ther to be explored. This has no practical value since by definition no possibility of checking the result by experiment could possibly be countenanced. However, some readers may be interested to see what is predicted to happen and so the following table presents a prediction for the next 15 billion years.

A run started from the radius R_{SH} = 25 BLY given by computation for the present era with $E_X = 0$. A successful result appeared at the first attempt. It explored the pressure and radius development for the next 15 billion years.

As shown in TABLE VI column (3) the pressure P_0 at the origin has moved over the peak of the net energy creation curve, shown in FIG.10.1 p.224 at P/P_L = .43, by the next 4 billion years. In 15

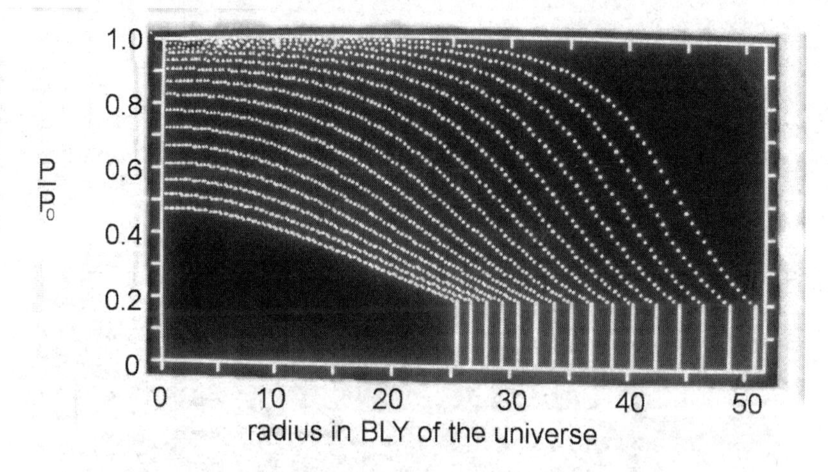

FIG.10.7 EXPLORING THE FUTURE FOR 15 BY (billion years)
16 pressure/radius profiles at 1 BY intervals
Pressure P_0 at origin point as the unit of pressure
the lowest curve is the profile for the present era

billion years, as shown in column (7), it has reached the value .75, which is ¾ of the way to the i-theric liquidus condition where $P/P_L = 1.0$. In consequence the rate of pressure rise has flattened off and the pressure/radius profile has become very flat topped. This is shown more dramatically by the photos of plots on the computer screen given in FIGs.10.7 and 10.8.

TABLE 10.VI THE NEXT 15 BILLION YEARS

Data H_0=45: E_X=0: R_{M0}=25.5: ΔT_B=2

(1) T BY	(2) R_{SH} BLY	(3) $\frac{P_0}{P_L}$	(4) $\frac{v_{SH}}{c_0}$	(5) T BY	(6) R_{SH} BLY	(7) $\frac{P_0}{P_L}$	(8) $\frac{v_{SH}}{c_0}$
1	25	.346	1.151	9	38.6	.671	1.779
2	26.3	.375	1.209	10	40.5	.695	1.866
3	27.6	.409	1.269	11	42.5	.713	1.957
4	28.9	.446	1.332	12	44.6	.728	2.053
5	31.9	.486	1.468	13	46.8	.738	2.153
6	33.5	.527	1.540	14	49.0	.746	2.258
7	35.1	.607	1.616	15	51.4	.752	2.368
8	36.8	.641	1.696				

FIG,10.7 shows the 16 pressure/radius profiles, The shock front still has the same pressure P_{SH}/P_L =.15 but now at P/P_L =.75 hardly any pressure increase with time occurs. It would take infinite time to reach the liquidus state.

In FIG.10.8 path lines have been added and make an even more interesting display. Path lines all now appear showing a pronounced falling-off in the rate of pressure rise experienced by every particle carried along with the i-ther.

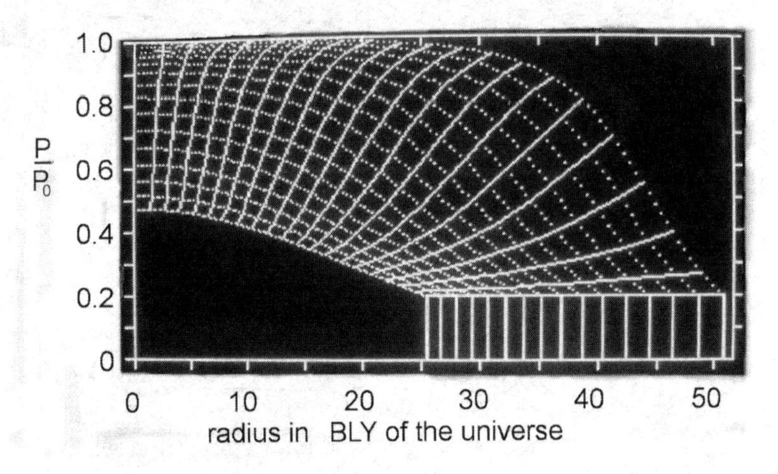

FIG.10.8 EXPLORING THE FUTURE FOR 15 BY
16 pressure/radius profiles at 1 BY intervals
with path lines added

To explore further the time interval between profiles was doubled to 2 BY. Computation has to start from the present era since the first profile is the starting one that cannot have error correction. So now we go 30 billion years into the future (but will ignore up to M=6).

TABLE 10.VII THE NEXT 30 BILLION YEARS: ΔT_B=4

(1)	(2)	(3)	(4)		(5)	(6)	(7)	(8)
T BY	R_{SH} *BLY*	$\frac{P_0}{P_L}$	$\frac{v_{SH}}{c_0}$		*T BY* *N*	R_{SH} *BLY*	$\frac{P_0}{P_L}$	$\frac{v_{SH}}{c_0}$
14	48.6	.7483	2.237		24	77.0	.7656	3.545
16	53.3	.7560	2.453		26	84.4	.7679	3.887
18	58.4	.7598	2.689		28	92.7	.7711	4.262
20	64.0	.7617	2.949		30 81	20.3	.7749	0.935
22	70.22	.7647	3.233		30	101	.7694	4.673

Columns (2), (4), (6) and (8) are at the shock front where N=101.
Columns (3) and (7) are at the origin point where N=0
Except where N= 81 is given. This shows the point where the pressure is a maximum – instead of at the origin. This, however,

appears to be simply computational error.

What is very interesting is that the maximum pressure is reached at about 22 BY. After that the variations shown could, again, be interpreted as computational error. This is supported by the value of C_{NR} that was also tabulated. This progressively fell from the peak value of 1.6 to 1. Then this remained constant within error limits. This means that the creation rate never falls below that of the shock front so that $P_0/P_L = .766$ from that point on.

It means that the ultimately stable liquidus state can never be reached except locally where black holes arise and of course in the ultimate filamentous structure of the i-ther itself.

This seemed an extraordinary prediction needing a check. The suspicion was that path lines would not keep diverging in the way specified by the Hubble's law. If H_0 started to reduce then pressures would be allowed to keep on increasing. This seemed reasonable since this constant is directly proportional to the creation rate and this is falling off when P/P_L exceeds 0.43.

Consequently in the next section a further revision is attempted in order to resolve this issue.

10.12 Further Refinement for theorists wishing to make further contribution (otherwise jump to page 244)

The way the maximum error always appears at N=100, as shown in TABLE 10.II p.228, presented one tantalizing puzzle. This element is the one closest to the shock front before the addition due to diffusion is made.

The simplest remedy replaced equation [10.3.1] that adds $E_X v_P$ to the velocity of element 100 by simply putting $v_{SH} = H_0 R_{SH}$. This is strictly not mathematically consistent but eliminates the problem. The relatively high error at this element 100 was entirely due to the extra growth by diffusion and vanishes when this is omitted.

The main problem, however, concerns the puzzle mentioned in the previous section. The creation rate has been shown directly proportional to the Hubble constant. I therefore thought the latter must vary since creation rate varies with pressure. Then path lines would not always remain equidistant from one another as the theory developed so far has predicted. This meant that to achieve perfection further refinement would be required.

When the creation rate starts to fall as pressures increase, as shown by FIG.10.1 p224, then H_0 should reduce progressively as radius reduces beyond the value where P/P_L exceeds about 0.4. Each element in this region will not grow as fast as elements near the edge. However, the existing formulation makes path lines near the edge remain unaltered as if the deficit in growth near the centre made no difference. This means that, near the centre, elements are not growing as fast as the space suggested has allowed. Unfilled gaps would be left in the central regions and of course this is disallowed. To fill these gaps the following procedure was adopted.

Two extra one dimensional arrays were added, each representing a spherical shell element of space. One was H_{0EN} representing the effective Hubble constant of element N. The other was ΔR_N, the radial width of element N. This allowed the radial growth of each element to be evaluated individually by adapting the equation [10.2.5] to give: (and noting suffix M is understood to apply)

$$\Delta R_{NM} = \Delta R_{NK} \exp(H_{0EN}\,\Delta t) \;\; \& \;\; R_{NM} = \sum_0^N \Delta R_{NM} \qquad [10.11.1]$$

This meant using the value of H_{0EN} from previous profile K to find the values of ΔR_N after time Δt for profile M and then adding from origin point radially outward until the shock front is reached.

Then assuming a constant acceleration f and using two simple equations of motion we can write for the radius gain in any element such as N from profile K to profile M in time Δt:

$$R_{NM} - R_{NK} = v_{NK}\Delta t + \tfrac{1}{2} f\,\Delta t^2 \;\; \& \;\; f\Delta t = v_{NM} - v_{NK}$$

Combining these yields:

$$R_{NM} - R_{NK} = v_{NK}\Delta t + \tfrac{1}{2}(v_{NM} - v_{NK})\Delta t = \tfrac{1}{2}(v_{NK} + v_{NM})\Delta t$$

Re-arranging yields the speed v_{NM} for each element as:

$$v_{NM} = 2(R_{NM} - R_{NK})/\Delta t - v_{NK} \qquad [10.11.2]$$

The revised method of computation then allowed a simple way of matching to the net energy supply. The latter is provided by the curve shown in FIG.10.1 that provides the energy supply ratio C_{NR}. For the energy demand equation [10.6.1] specifying the energy demand ratio C_{RC} was used. Note that element number J = N-1, for the profile number M, being evaluated. Calculation starts from the same element in the previous profile K, so K = M-1. The Hubble constant will be increased in proportion to C_{NR}, that of the energy supply available. However, it will be reduced as the energy demand ratio C_{RC} increases. Consequently both ratios need inclusion. The method combines equations in [10.5.1] and [10.5.2].

Then the combination can be presented in the form:

$$\frac{P_{NM}}{P_L} = \frac{P_{JM}}{P_L} \exp\left(\frac{3}{2}\frac{C_{PF}}{v_P^{\,2}}\left(v_{NM}^{\,2} - v_{JM}^{\,2}\right)\right) \qquad [10.11.3]$$

The ratio of C_{NR}/C_{RC} multiplied by H_0, provides the effective Hubble constant H_{OE} for element N i.e . H_{OEN}

This meant starting from the known pressure of P_{SH}/P_L at the shock front and working inwards ultimately to the origin point.

At each step the error correction routine was originally applied to match energy supply to demand in the way that seemed to satisfactorily give the solution illustrated in FIG.10.8 and the previous tables.

When exploring the future where $P/P_L>.4$ results were now catastrophic! Pressures rose far too fast and soon $P/P_L>1$ appeared before total failure occurred.

The reason was that use of the error correction routine was no longer justified. This made $C_{NR} = C_{RC}$ at all steps so that the Hubble constant never changed. A mathematical inconsistency had been introduced by assuming energy supply must match demand. The supply has to be less than demand in the future. The deficit in each element has to come from the energy already existing in that element.

Equations [10.6.1] and [10.6.1A] p.227 are still valid for obtaining C_{RC} and [10.4.4] p.223 still applies for obtaining C_{NR} but equating them is disallowed. The pressure ratio P/P_L gives C_{NR} (for supply) and the value of C_{RC} (demand) has to use the value of P_{NM}/P_L obtained from [10.11.3]. This means that the pressure gradient needed for acceleration has to be satisfied. This has to match the value given by [10.6.1] governing energy demand. This requires elaboration.

A value of P_{NM}/P_L has to be assumed and used in [10.6.0] to get the average pressures needed to insert in [10.6.1]. The latter provides $\delta(PV)/(PV)$ to be combined with [10.6.1] to yield:

$$C_{RC} = \frac{\delta(PV)}{PV}\frac{1}{3H_0\,\Delta t} \qquad [10.11.4]$$

The average pressure P_{NMK}/P_L from [10.5.0] is the value of P/P_L that needs to be used to trim the value of C_{NR} using [10.4.4].
Then P_{NM}/P_L has to be found from [10.11.3]. The values of H_{OEN} Obtained this way will differ from the initial values taken from the previous profile K and alters all radii, velocities and pressures. Iteration has to continue, at every value of N, until the effective Hubble constants agree with the values corresponding to the computed pressures for each value of N. The final values of both H_{OEN} and ΔR_N can then be made those for use in computing the next step. No need exists to make these two dimensional arrays.

This will make the Hubble constant H_{0E} reduce as pressures increase beyond values applying in the present era. The previous solution can therefore only be regarded as a first approximation. An almost flat pressure will tend to develop as illustrated in FIG.10.8 but path lines will not diverge so rapidly as that solution implies. The new refinement has not yet been fully developed. Iteration problems have proved formidable.

10.13 How Astronomy can provide a check on the Big Breed Theory

The theory is based on a value of the average speed of primaries v_P of 1.464 times the speed of light as measured on Earth i.e. 4.392×10^8 m/s. This was arrived at in §7.2 p.174 on the basis of pressure waves in the i-ther moving at the speed of light. This was justified by the experiments of Taylor and Hulse who measured the loss of energy from a binary pulsar in close though highly elliptical orbit. The measured loss agreed with that predicted by general relativity based on gravity wave generation. The new ECM mechanics matches this but the gravity waves are now pressure waves.

Hence good agreement of the predicted distance of remote supernovae and observed values could provide useful confirmation of the theory. This radius is also dependent on the phase coupling factor considered in Chapter 6. The value of 0.9, used in the computations needs to be refined, but there is potential in the theory for achieving this.

Another factor that is potentially computable is the diffusion excess growth factor E_X that so far has only been guessed. However, as TABLE 10.VI shows, this can vary over a considerable range without influencing the results appreciably. This mostly affects the age of the universe prior to the time before it reached a radius of about 12 billion light years (what has been termed its 'pre-history').

10.14 Neutron Stars and Black Holes

Also shown in FIG.10.1 is the point applying to the surface of a neutron star. As shown on p.70 of QGECM the i-theric energy density at the surface of a neutron star, of 1.4 times solar mass, is only 1.476 times that at Earth – this despite the enormous difference in mass density. In consequence the surface of a neutron star exists where the net creation rate is about at its maximum value.

On further collapse to produce a black hole the i-theric liquidus state will be reached where net creation is zero. This sets a limit to the density within the black hole and solves the problem of the impossible 'singularity' that Einstein's theory predicts will exist. All physicists admit singularities – matter existing at infinite density in a point of zero size – is an impossibility.

Of course this is not the ultimate density. The ultimate value will be about four times greater than that at the i-theric liquidus and is the value reached at the annihilation cores that have to form at the centres of all those minute cells of which the i-ther is composed. At this density enormous annihilation occurs.

10.15 Conclusion to Chapter 10

A major breakthrough that occurred on 1st July 2010 enabled a major difficulty to be resolved. This was the effect of changing pressure on the creation rate and the consequent effect on the Hubble constant. It had already been established that the growing i-ther would have a shock-fronted spherical boundary but previously the magnitude of its pressure jump had not been established. Now it became clear that below 15% of the liquidus pressure P_L, solutions were divergent. The pressure at the front would grow until it reached a stable value of 0.15 of P_L.

Then it transpired that the greater energy needed for the pressure increases associated with acceleration was cancelled very nearly by increased rates of energy supply. Consequently the Hubble constant did not vary with time right up to the present era.

Since an extra red-shift had been identified in Chapter 9 the Hubble constant was greatly reduced to only about 60% of that established by analysis based on special relativity. In consequence speeds of recession reduced so much that the entire universe became observable. With the accelerating expansion explained as due to net creation, for the first time, the size of the universe could now be predicted from red shifts and luminosity distance data.

Until about 20 billion years ago the pressure difference from the origin point to the shock-fronted edge was negligibly small but then started to rise at a doubly exponential rate up to the present era.

An attempt to explore the future for the next 15 billion years showed that pressures around the origin point would rise at a steadily reducing rate instead of the rate increasing as in the past.

For the history of the universe up to the present era, a fully consistent theory has been made available. It provides a new way of analysing astronomical data that we consider renders existing methods obsolete. We hope theorists will consider the implications with extreme care.

There is no longer a mystery surrounding Dark Energy since the accelerating nature of the universe is now understood to be a consequence of solving the problem of the cosmological constant.

Angular-size-distance and more consideration of red-shifts are provided in the **ADDENDUM** on page 292 and needs a study.

The next steps for further advance are suggested in Chapter 11.

CHAPTER 11

WHAT NEXT? – SOMETHING IS NEEDED FROM A COMPUTER SPECIALIST – A COLLISION BREEDING CODE

What is needed next is a full computer analysis of collision breeding leading to spontaneous collapse into cells with cores. This might prove self-organisation could actually happen.

11.1 INTRODUCTION

What still invalidates the big bang theory is the problem of the cosmological constant. The Big Breed Theory has provided a solution, showing that a tangle of filaments must exist at the ultimate level of reality: the i-ther. This has been deduced from logic but total proof requires the annihilation cores to be seen forming up to create these filaments. This could only be achieved by a computer simulation. The details of opposed energy collision dynamics have been covered in Chapters 2 to 6 and the mathematics involved needs to be applied in a fully three dimensional simulation. It needs to start with two positive and two negative primaries, which then breed by repeated collision. When the number density has reached some critical value a spontaneous collapse to form one or two annihilation cores needs to be demonstrated. Later with many cores forming a tangle, the hope would be to see some further self-organisation. This would be due to the known organising power of energy-fed chaos.

Such organisation has been known to occur spontaneously by mathematicians who study chaotic systems. Many interesting articles and books on the subject have appeared. I found a book by John Gribbin (2005) captured the theme admirably. He shows how different kinds of chaotic systems, starting off as completely random arrangements, will self-organise if subjected to a constant energy input. The resulting patterns always have an astonishing degree of dynamic organisation of a kind that could never have been predicted from the simple rules or properties of the individual components. The i-ther has all the ingredients required for such self-organisation since a huge energy supply, by collision breeding, is continuously available everywhere. Indeed, the collapse of the initial field of breeding primaries to produce the filamentous structure would be the first stage of such self-organisation.

If this could be demonstrated by computer simulation it would prove that i-ther really is able to evolve. Certainly the filaments will be continually buffeted by the random bombardment of primaries. This will cause them to vibrate and produce waves travelling great distances in all directions. Some will be pressure waves moving through the gas like breeding annulus of cells, others will be longitudinal waves of far greater propagation speed. The latter will travel along the filaments themselves. Interference patterns will appear and so will local resonances. The questions are:

"Could such wave action be the underlying cause of the quantum level of reality being based on wave mechanics?"

And, "Could non-locality be explained?"

The first person to give some leads toward answering these questions might well be a candidate for a Nobel Prize!

Such an ambitious project would demand the use of a parallel processing super-computer. A less demanding project could just try to see if a tangle of filaments with random forced excitation could lead to some form of organised wave propagation.

That such wave action alone could be the cause of our world of matter appearing as it does was a proposal made by one of our supporters, the engineer Alan Middleton. It seems to me a proposal worthy of further study. My original concept was that the filaments would lead to the formation of something like a neural net that could evolve the learning and memory capacity of proven artificial neural nets. Both ideas need testing and only a super-computer simulation has the potential of achieving this.

The reason for the second question is that no theory yet exists that can explain non-locality. A pair of particles, like photons, can become 'entangled' which means that when one is perturbed the other instantly reacts regardless of separating distance. This appears to require some form of information travel at an almost infinite speed. Fortunately the filaments have the potential to provide an answer. They are the ultimate solids having no pores at all and so can be expected to be very stiff – a high Young's modulus. Also their net mass density is close to zero since their positive and negative mass components almost cancel. These are the characteristics that predict almost infinite speeds of wave propagation. And information can be carried by waves.

In the next section some further detail of the way such codes could be set up is considered. Some results of my own early attempts to make a start on such projects are also briefly described.

11.2 Two Dimensional Simulation as the Starting Point

My own effort began in 2005 using a rectangular boundary filling the computer screen. Primaries were represented as circles, with positives shown on the screen in magenta whilst negative primaries appeared in cyan. Areas enclosed by the circles were made proportional to their energies. Two positive and two negative primaries emerged together from a central point. They moved away from each other and then made reflections from the boundary until collisions occurred. Reflections assumed a cosine rule random angle distribution using the RND function. Only linear motions were considered for simplicity, with the spinning motions of primaries ignored. This worked quite well until numbers had increased so that about eight to twelve primaries were jostling in the field.

Then unforeseen difficulties appeared. Two would tangle in a corner and orbit one another in a manner that was impossible. It was impossible because the code contained no force at a distance routine that could cause motion in a circle. The difficulty arose when, by chance, one primary was bouncing back from a boundary as another approached near the same place.

A change to an oval boundary improved matters but the effect was not eliminated. Reflection was replaced by having a primary moving out of the field replaced by one coming in at a randomly selected location and angle. It helped but did not eliminate the difficulty. When the trouble was overcome the two primaries that would have tangled, to move in orbits about each other, now started bouncing away and coming back as if clapping hands. Again this was a false simulation.

After several months of frustrating effort I finally abandoned this project. I mention this experience to show that even a simple case written in two dimensions does not lead to an easy task.

11.3 Three Dimensional Simulation

Until the difficulties of two-dimensional simulation have been overcome it would be unwise to attempt the required three-dimensional simulation. However, it will be assumed the difficulties have been overcome. An oval boundary is recommended and now a slice of space needs representation. The oval boundary now represents a surface, as if formed from a strip of paper of specified width with ends glued together. Extra virtual boundaries need adding as flat sheets to form a closed box.

Then fully three dimensional collisions between primaries can be simulated with motions in all x,y and z directions considered. In addition to the oval wall, from which new primaries need injection to replace those leaving, the flat boundaries limiting the width of the slice of space need to be considered. For the best simulation replacements really need insertion as close as possible to the places at which the outgoing primary made its exit.

Perhaps a safer way would be to specify a spherical boundary in order to avoid corners of any kind.

Ultimately it should be possible to see filamentous and blob type annihilation cores building. This requires the simulation of several thousand primaries and at this stage they would so fill the screen as to block out the view of cores. At this stage, therefore, the primaries need to be either represented by dots or not plotted at all. Then only the cores would appear. To provide a three dimensional impression, however, it would seem best to use the stereoscopic type of imagery. The surfaces of cores are now represented by red and green coloured dots and viewed through red and green filters so that each eye can only see one kind of dot. By such means the cores will then jump into fully three dimensional view.

The hope is that the vibrating tangle would show some meaningful self-organisation.

Some tricks might need to be tried to secure this kind of action. For example cores might be given a breaking strength so that when vibration forced by random buffeting exceeded a specified amplitude, breakage of a filament would occur with re-connection to another filament or blob at some other place. This might be necessary in order to provide some dynamic freedom for self-organisation to succeed. Then again each core could be permitted a liquid surface so that flow could occur.

Of course one could add all kinds of properties. The objection that would be raised is that nobody knows what the properties of the i-ther really are so idle speculation is being applied. What matters, however, is that if any combination of properties shows self-organisation then proof that it is possible will have been achieved.

The main point of such an exercise is to try and find a way of explaining how matter could arise from the energies and organising power of i-ther. In the next chapter the way this could be applied to a big bang type creation scenario is considered.

CHAPTER 12

THE CREATION OF MATTER

The energies needed for creation have been provided by our i-theric theory based on opposed energy dynamics. Quantum theory tells us that matter is organised by waves or is even totally comprised of waves and waves are produced by the i-ther

Although quite a large number of different matter-creation scenarios could be supported, this chapter concentrates mainly on application to the big bang theory.

12.1 Introduction

The existing big bang theory has all the matter ever to exist produced in a blinding flash by a so-called 'inflation'. This concept can be retained but generated in a totally different way in order to eliminate the false prediction it produces: the excessive rate of expansion of the universe said to be 10^{120} times too high.

I hope the reader will not be put off by the speculative part of the theory that will be considered first. This is introduced to make the link with the Big Breed theory. After that the mathematics will refer to established details of the big bang approach.

12.2 Now to consider the speculative component

To prepare for what follows it first needs to be pointed out that the i-ther had to emerge from the void many billions of years before it could organise its energies to create what we see as matter. It will be assumed that some form of routine is generated by the self-organising power of i-theric waves. This specifies the properties of elementary particles such as quarks, electrons and photons. These routines have a replicating capacity like the viruses that plague our computers. Unlike viruses, however, a replication cut-off feature is provided so that the creation switch-off feature, lacking in the existing theory, is included. The number of replications permitted is specified with the count recorded at each generation so that all replication ceases at the same instant. In this way a philosophy is provided that takes into account the weird features of wave-particle duality on which the quantum level is based.

There are now two creative events. The first produced the sub-quantum background medium of i-ther. This had to evolve by the self-organising power of chaos for many billions of years. Only then could the required computational power of i-ther emerge. At this stage the second creative event of the big bang could happen.

Galaxies could eventually condense from the primordial gas cloud of abstract matter, due to the abstract forces also incorporated in the computational routine. Initially they would travel radially outward under their own inertia at high speeds relative to i-ther. They would continually slow by mutual gravity until they equilibrated with i-ther. Then they would continue moving with the i-ther with its slowly accelerating expansion.

The reader may have spotted a problem here and this will be considered later.

Many people will say this is excessively speculative. However, to us it seems less speculative than other proposals such as those requiring an infinite number of universes all splitting continually to create new ones. It has to be remembered that quantum mechanics is weird in the extreme and shows a lack of reality. Its description is abstract and virtual suggesting some form of computer generation.

Readers who reject this proposition are invited to contribute a better suggestion that makes use of the energies the i-ther has provided.

What follows next, however, does not depend on the model outlined being correct or the only explanation possible.

12.3 THE COSMIC BACKGROUND RADIATION

The big bang theory has it that after inflation a gas cloud resulted having an extreme temperature so that it existed as plasma. The photons of light were being repeatedly absorbed and re-emitted by the freely moving electrons so that at this stage the universe was opaque and thermal equilibrium existed. This means all different species of particle had the same average kinetic energies. During isentropic expansion temperature fell, according to known thermodynamic principles, until it reached 3,000°K. Then a gradual transition to a normal gas occurred as electrons attached themselves to protons to produce hydrogen gas. Then the whole became transparent so that photons could now travel freely and unhindered.

It is from this point that the present analysis will start. It will consider details of light propagation within this transparent medium relying to some extent on the theory provided in one of our books called, *QUANTUM GRAVITY via Exact Classical Mechanics*.

According to cosmologists photons took 115,000 years to decouple from the initial plasma ball left after the big bang and decoupling was complete by an age of 487,000 years. We were somewhere within this ball which was expanding fast with photons coming at us from all directions. So if the gas expanded at a speed close to that of light the radius of this ball of gas would be something less than a radius of about 487,000 light years. We will interpret this as the PVF radius: the radius at which photon decoupling is complete.

They give average density of mass including dark matter for the present era as 10^{-26} kg/m^3. At the PVF radius the same mass existed as it does now, according to big bang theory, and from the Big Breed theory the present radius of the universe can be taken as 25 billion light years. From these figures and with volume changing as the cube of radius, the density is increased 1.35×10^{14} times. This converts to the low density of only 1.35×10^{-12} kg/m^3.

Even so the mass works out to the huge value of M=5.54×10^{53} kg.

12.4 Mathematical Analysis of Light Bending

We will start analysis by finding the radius r_0 at which, according to Newtonian mechanics, a photon of light will orbit a point mass M in a perfect circle. The centripetal acceleration is force/mass = GM/r_0^2 given by Newton's theory of gravity and this acceleration is also v^2/r. So putting $r_0 = r$ and $c = v$. results in:

$$r_0 = \frac{GM}{c^2} \qquad [12.4.1]$$

This is what is known as the 'gravitational radius' r_0.

For the mass M deduced in §12.3 r_0 works out at 43.8×10^9 LY – many times the PVF radius and more than twice the radius of the universe now.

But the radius of curvature R_N at any other radius r, measured from the origin point of the universe, with R_S as the surface radius of the attracting sphere of uniform density, and with photons moving at angle β to the radial direction, [12.4.1] then yields:

$$R_N = \frac{R_S^{\ 2}}{r_0 X \sin\beta} : X = \frac{r}{R_S} \ \textit{when}\ r < R_S\ \textit{otherwise}\ X = 1 \qquad [12.4.2]$$

This is readily deduced since when $r<R_S$ the effective mass is that enclosed within a sphere of that radius r since the effect of the spherical shells outside cancel to zero.

According to page 62 in our book *Quantum Gravity*, however,

Newtonian mechanics is revised to make it an exact theory. Then keeping the same definition for r_0 as given in [12.4.1] the radius of curvature is modified owing to the variation of the energy density of space ε from the datum value ε_D where both G and c are measured. (i.e. $c=c_D$ for evaluating r_0 at Earth's surface)
The non-uniformity of space density has an almost identical effect to Einstein's curved space-time. The exact equation for the radius of curvature R_{ECM} for both light and matter at a radius r from the centre of a point mass M and moving at angle β to the radial direction is shown to be given by the ECM derivation as:

$$\frac{1}{R_{ECM}}=\left\{\frac{\varepsilon_D}{\varepsilon}+\left(\frac{c_T}{w}\right)^2\left(\frac{\varepsilon}{\varepsilon_D}\right)^{\frac{2}{3}}\right\}\frac{r_0}{r^2}\sin\beta \qquad [12.4.3]$$

For matter the speed of the object is w but for light $w = c_T$. (See [9.2.2] p.209 for clarification.) Also for anything inside the universe the mass M is only the mass within the sphere of radius r since that outside this radius produces forces that cancel to zero. Consequently the value of r_0/r^2 for an average density of matter in the universe ρ now becomes.

$$\frac{r_0}{r^2}=\frac{G}{c^2}\rho\frac{4\pi}{3}r \qquad [12.4.4]$$

A final component is the curvature $1/R_{SP}$ of a photon trajectory caused by the acceleration of expansion of i-ther. This is derived in §7.5.1 p.184 and by equating to $c^2/(R_{SP}\sin\beta)$ yields:

$$\frac{c^2}{R_{SP}\sin\beta}=H_0^{\,2}r: \quad Hence: \ \frac{1}{R_{SP}}=\left(\frac{H_0}{c}\right)^2 r\sin\beta \qquad [12.4.5]$$

This curvature can be added arithmetically to that of [12.4.4] to yield the combined radius of curvature R_C.

The following table, calculated from these figures, shows something rather extraordinary.

12.5 The Gravitational Focussing of Starlight

EXPLANATION of TABLE 12 I

Columns (4) to (9) refer to the present era with N in column (4) being the element number in the array with 101 referring to the shock front. Columns (5) and (6) use the pressure/radius data provided by §10.7 TABLE 10.II M=15 p.228 for the present era. P/P_0 is pressure P of i-ther as a proportion of that at the origin point P_0. Then $X=r/R_S$ as in [12.4.2] and corresponds to column (5).

For comparison with the radii of curvature at the PVF photon decoupling radius the same values of X are used. For this case i-theric pressure is constant so $P/P_0 = 1$ all cases. Also the effect of space acceleration is negligible and so is not recorded

TABLE 12 I Radii of curvature for photons at $\beta = 90^{\circ}$

(1)	(2)	(3)	(4)	(5)	(6)	(7)	(8)	(9)
R_N	R_{ECM}	X	N	r_{BLY}	P/P_0	R_N	R_{ECM}	R_C
∞	∞	0	1	0	1.0	∞	∞	∞
29.08	14.54	.1879	20	4.79	.969	75.00	37.29	27.05
14.54	7.269	.3758	40	9.58	.885	37.50	18.28	13.33
9.698	4.849	.5634	60	14.36	.759	25.01	11.64	8.592
7.272	3.636	.7513	80	19.15	.613	18.76	7.972	6.022
5.817	2.909	.9392	100	23.94	.465	15.00	5.456	4.272
5.464	2.732	1.0	101	25.49	.420	14.09	4.790	3.805

$R_S = 4.87 \times 10^5$ LY at PVF radius $R_S = 25.49 \times 10^9$ LY present era
Columns (5), (7), (8) & (9) are in 10^9 light years

At Photon Decoupling Time

Considering the case of the PVF radius first, columns (1) & (2) are radii of curvature in light years, with $\beta = 90^0$. $X = r/R_S = .19$ means at 19% of the surface radius, the radius of curvature is only $R_{ECM}/R_S = 0.003\%$ of the surface radius based on the ECM theory.

Of course these figures only give the maximum radii for photons moving in any tangential direction. Moving radially, either out or in, there is zero curvature. However, the slightest deviation from the radial direction when photons are moving substantially outward and the angle β will rapidly grow. The photon will soon have $\beta = 90^{\circ}$ and still continue to increase. It will therefore turn round and travel toward the origin point becoming ever more radially directed. Gravity is focussing almost all the light released toward the centre. It will yo-yo continually re-crossing the centre point to make orbits shaped rather like a figure 8 and very few photons will escape.

Much effort is being expended at the present time in the study of the CMB that is claimed to contain information concerning the seeds of non-uniformity that grew to form the galaxies we see today. It is the state at the time of photon decoupling that is being observed. It is difficult to see how this information can be disentangled after being focussed in the manner this study predicts.

At the Present Era

The curvature of light paths reduces faster than the rate at which the universe expands as shown by the greatly increased radii of curvature in the present era shown in columns (7) to (9). The

Newtonian value is much larger than that given by ECM mainly owing to the effect of non-uniformity of energy density of space associated with gravity. This is additional to that caused by the acceleration of expanding space. This addition reduces the radius of curvature even more as shown by column (9).

At the same 19% of shock front radius R_S the radius of curvature is now slightly greater than R_S. Even now, however, a strong tendency exists for starlight, and particularly scattered starlight, to focus at the centre. But now it will be a rather soft focus. From that focus it should re-emerge moving radially outward. This suggests the need for a full computer study that would not be difficult to program. It would trace the paths of photons starting from random positions and with totally random initial direction in all three dimensions. The concentration and isotropy at selected positions would be the output.

12.6 Yo-Yo Light: the Universe as a Hall of Mirrors: a Caution

At $X = 1$ the radius of curvature is only 15% of the shock front radius and so light moving outward but at an angle β greater than about 15 to 30 degrees when X is about .8 will rapidly increase this angle until it turns round and starts to travel inward again. It will become ever-more radial until crossing near the origin point to perform figure of 8 shaped orbits similar to those predicted at the PVF condition. This is Yo-Yo light.

What this also means is that cosmologists could be making estimates of size of the observable universe much larger than it really is. Light can leave the opposite side of a nearby galaxy, travel out to great distance whilst turning round and then be observed. It would appear as a galaxy at great distance. It would also be seen at a much earlier epoch than the same nearby galaxy - also recorded from a direct straight line photon trajectory. Imagining in three dimensions suggests every galaxy could produce multiple images appearing to exist in many places all over the sky.

Furthermore if one considers an observer near the shock front the light moving out will seem to be speeding up as it approaches - just as happens when the observer is situated near the origin. So whether going in or out the photons will lose energy and suffer a red shift. Energy will not be conserved for the photon as it will continually lose energy as it yo-yos.

Cosmologists will have no way of knowing light which has started to go away and then turned round will not have been going away all the time and so will be liable to greatly overestimate the distance away as measured in a straight line. Therefore the safest policy, since now the size of the universe appears from our derivations, is to use data from red–shifts not greater than about 1.0. Then the radius of the shock front can be found from our computer program using just the red-shifts and luminosity distances as input data over this limited range.

Yo-Yo-ing light can never escape but will ultimately add to the CMB until fading away altogether. The tendency to become ever-more radially directed when coming in means that it will tend to focus toward the origin point as already mentioned.

Visible light would be affected equally. It is just possible therefore that the computer study previously mentioned would predict the existence of a powerful star-like object at the centre of the universe. This would be no ordinary star since matter would not necessarily exist there. The light would contain no spectral emission or absorption lines and so this might be the tell-tail signal for which astronomers might search.

Could this be a quasar? If only one had been found this would seem a real possibility. However, even today TABLE 12 I suggests the universe could be acting like a hall of mirrors. We see some galaxies directly but light from the same one, emitted in a different direction, could turn round to be received by the same telescope. It would appear a more distant and older galaxy of greater red shift and would appear at a totally different point in the sky. Therefore a central quasar might be seen in several different places scattered about the universe.

12.7 AN ALTERNATVE MODEL OF CREATION AND RESOLVING A DIFFICULTY (27 August 2010)

Another model is permitted by the filamentous structure of i-ther. These filaments have zero net mass and so can have infinite speed of wave propagation to carry information instantly to any place. In this case matter could be instantly created as a diffuse cloud everywhere. Matter creation, as mathematical organisation, would be arranged mathematically to cut off at some chosen instant.

It has been shown by Spaniol and Sutton (1994) that the gravitational constant G can be derived from those of magnetism and electricity. This proves that a unification of all these three forces exists. However, electromagnetism is shown by Feynman (1985) to be quantum based and so gravitation as well as electromagnetism needs to be regarded as abstract: being mathematically contrived. Furthermore Lerner (1992) following Alfven shows a plasma state could have existed that moulded primordial matter into proto-galaxies by electromagnetic forces. This now appears as an optional extra but there seems no reason to bar its incorporation.

These unification theories enable the difficulty mentioned early in this chapter to be resolved. The solution presented by derivation of Hubble's law, which predicts an ever-accelerating expansion, strictly applies only to i-ther. Application to matter depends upon galaxies having only low speeds relative to surrounding local space – i-ther. However, mutual gravitational attraction together with the Hubble acceleration, which produces the effect of space gravity, ought to cause matter to fall toward the centre of the universe.

It is also possible that gravity has a long but finite range so that super-clusters do not attract each other. Cosmologists only assume the range of gravity is infinite: there is no proof of this.

Space gravity, however, was based on the assumption that mass gravity is a real force of negative buoyancy caused by an asymmetry of rest and kinetic energies of i-ther. This is the model explored in our Quantum Gravity book but now it looks as though this could be wrong. If instead mass gravity is considered purely abstract, like electromagnetism, then no asymmetry of i-ther is required. In this case space gravity will not exist because the positive and negative accelerating forces cancel to zero.

With these two assumptions accepted remote galaxies have no tendency to fall toward the origin centre and so will accelerate with i-ther. In this way the theory matches observation.

The big bang theory will still be one valid solution since at small sizes all matter will be slowed by gravity and the effects on light will remain as evaluated in TABLE 12 I.

12.8 A dilemma

With mass gravity of long but finite range the effects on photons travelling from near the shock front to the origin centre will cancel to zero. Photons gain energy toward the centre of a super-cluster and then lose the same amount as they travel out again.

Another issue then arises. For total consistency it can be argued that the two blue shifts are disallowed in which case an even smaller Hubble constant needs to be input to bring the predicted red shifts down a little. A run of "Un150710" with input $H_0 = 33$, $r_1 = 16$ BLY & $E_x = .05$ yielded $r_{SH} = 28$ BLY and with $P_0 /P_L = .263$ for the present era. Then $Z = 1.82$ and $Z_D = .561$ at .88 of the distance to the final path line nearest the shock front. This point is at $r = 14.1$ BLY and 15 BY ago. This is the limit: the first profile computed. So the code cannot give values nearer the shock front. The pressure here is $P/P_L = .156$ and so is not much greater than at the shock front. Recession speed $v/c_D = .481$ and $c_T/c_D = 1.337$ at this point.

This has resulted in a larger estimate for the radius of the universe than the 25.5 billion light years BLY predicted by the input data giving FIGs.10.3 to 10.6.

In connection with the size of the universe it is worth noting that Wright (2005) gives 1.755 as the maximum red shift observed so far. He gives this a luminosity distance, the distance based on light intensity with no absorption assumed. This distance converts to between 30 and 45 BLY after processing by relativity theory to give the distance it would be now. However, Aspden (1984) has shown that free electrons in space will reduce light intensity. This suggests Wright's distances are likely to be too great. Even higher red shifts of about 6 have been recorded for very remote galaxies. This also suggests considerable light absorption must be occurring.

12.9 Conclusion to Chapter 12

Outward travelling starlight has been shown to have the tendency to turn round and travel radially inward especially during early phases of evolution of the universe As a consequence one would expect to see a higher intensity of the CMB at opposite sides of, what appears to us, as the 'celestial sphere' if Earth is far from the origin point of the universe. This suggests our galaxy is near the origin centre, so giving support to previous deductions. The computer study proposed ought to predict this expected departure from isotropy.

And new probes are being sent to study the CMB to try and find clues that could lead theorists to an understanding of what Dark Energy is. A European probe is being specially designed for this to add to the mix of those already in orbit – all studying the CMB! According to the study made in this chapter there is no way this expensive space program could yield anything more than misleading information! So this is a matter needing urgent attention.

However, it was finally realised that since the force of gravity has already been shown unified with electromagnetism, and since the latter has to be regarded as abstract, so gravity must be an abstract force of long but finite range. Space gravity then also vanished and so both blue shifts evaluated in Chapter 9 had to be discounted.

A further consequence was that the Hubble constant could possibly be reduced to as little as 33 km/s/M_{PC} with the radius of the universe increased to 28 BLY and its age greatly increased.
Some debate on these issues seems to be required.

To complete this PART II a final effort will be made, in the next chapter, to assess the ages of the universe, of matter and i-ther.

CHAPTER 13

AGES OF I-THER AND THE UNIVERSE

In Chapter 10 the shock front was shown to travel faster than particles by an amount that could only be guessed. This chapter explores the errors this could cause in estimating the ages of the i-ther and the universe of matter. Furthermore the error this could cause in evaluation of the Hubble constant and the present size of the universe are explored.

13.1 Age of the Universe of Matter

The age of the universe needs a lot of explaining. If we use the cosmologists' estimates the universe, it is 13.7 billion years (BY) old. We will analyse this value on the basis of the big bang theory that yielded this figure. In a big bang scenario matter will be thrown out initially moving at speeds varying between zero and almost the speed of light. We will concentrate on a star that formed, say 1.7 billion years later, from matter not moving at all. If light travels at a constant speed c then light from that star will have travelled a distance of 13.7 – 1.7 billion light years (BLY).

Consequently simple logic suggests our planet is 12 BLY from the central origin point of the universe. Such a deduction is unfortunately totally at variance with the current beliefs of established cosmologists.

They say we have no special location in the universe and the previous logic defines a special location. Worse they state that the universe has no centre and is finite yet unbounded: a statement that arises from their sophisticated maths based on geometries that, without resort to analogy, defy imagination. To logic confined to the Euclidean geometry of common sense this appears an internally contradictory statement.

They also say we observe only an undefined portion of the universe. This is because they estimate, from red-shifts, that some stars are moving away ay over 0.8 of the speed of light and deduce that further away some could be receding faster than light and could not be seen. So they posit a 'horizon' where galaxies have a speed of recession equal to that of light. Consequently from this perspective astronomers can only see the 'observable universe'. Some cosmologists say the universe could be infinitely large.

However, by their own admission they have no idea of any way to solve the problem of the cosmological constant and to them Dark Energy still poses an unsolved mystery.

So what is the truth?
Readers of this book now have the advantage of understanding solutions to these two problems that are highly relevant to the issue. Chapter 9 introduced an extra red-shift, caused by light having to counter recession velocity. This factor cannot be reached by any relativity theory and has a crucial effect: it drastically reduces the speeds of recession. Furthermore, with a theory able to predict an accelerating expansion, together with this reduction of speed, the size of the universe also became predictable. In Chapter 10 it was deduced that we must be located close to a true origin centre of the universe and that we are potentially able to see right to its edge.

If we were at the centre, then looking back to the edge we might observe a star whose light has taken 12 BY, to reach us. So the light started out 12 BLY away and so was at a radius of 12 BLY from the origin centre. But this does not mean the age of the universe is 12 BY since the star had to travel out at an average speed about 0.6 times the speed of light from zero radius to 12 BLY initially to get to that position, and this would take 12/0.6 = 20 BY. So this makes the age of the universe from the big bang 12+20 = 32 BY.

Some cosmologists complain that 13.7 BY does not allow enough time for galaxies to form and some astronomers claim that some stars appear to be much older than this. Perhaps our increased estimate could be of considerable help in solving this dilemma.

To be more general an observer could be regarded as located at some arbitrary radius, say R_O, from the origin centre. A star is seen at some smaller radius R_S from that centre. Then the time for light to travel distance $R_O - R_S$ is $(R_O - R_S)/c$. However, the time to get to R_S has to be added and this is R_S / v_{av} where v_{av} is the average speed from big bang to the point where the light seen was emitted. This makes the age of the universe T_{AGE} equal to:

$$T_{AGE} = \left(R_O - R_S + R_S / (v_{av}/c)\right)/c \qquad [13.1.1]$$

When the star is at a greater radius than the observer the result is:

$$T_{AGE} = \left(-R_O + R_S + R_S / (v_{av}/c)\right)/c \qquad [13.1.2]$$

In either case whatever the value of R_O only when R_S is zero is it valid to ignore the time for the star to reach its observed position.

Why do cosmologists ignore this addition? Surely all cosmologists cannot have missed so simple a matter unless they are all operating on a logic that makes no concession to common sense.

13.2 A more Rigorous Evaluation

We will now make a better estimate for the average speed of matter arising at a big bang and decelerating due to mutual gravitational attraction until it reaches the speed of recession computed by our code 'Un150810'. An exact mathematical solution will not be derived here since it requires a finite difference solution and the computer memory available will not permit this addition to the code. Instead it will be assumed that matter starts out at speed v_0 and decelerates at a linear rate with respect to time. The deceleration is assumed zero at the time the light to be observed was emitted from a star at recession speed v_e at radial distance R_e. This makes a nice little problem for the student to take on and check the following result. The time T_D from big bang to cover distance R_e becomes:

$$T_D = R_e 3/(v_0/c + 2v_e/c) \qquad [13.2.1]$$

The time T_D is in BY when R_e is in BLY

The acceleration f at time t with T_D now in seconds is given by:

$$f = \frac{2c}{T_D}\left(\frac{v_0}{c} - \frac{v_e}{c}\right)\left(\frac{t}{T_D} - 1\right) \qquad [13.2.2]$$

The radius r reached at time t becomes:

$$\frac{r}{cT_D} = \left\{\frac{v_0}{c} - \left(\frac{v_0}{c} - \frac{v_e}{c}\right)\left[\frac{t}{T_D} - \frac{1}{3}\left(\frac{t}{T_D}\right)^2\right]\right\}\frac{t}{T_D} \qquad [13.2.3]$$

13.3 The Ages of the Universe and I-ther

First the age of the universe

A big bang type of creation starting from a point size is to be assumed. Then the age will be the time T_D, for a star to travel to the edge of the universe and reach the position from which it released the light to be observed, plus the time T_{STAR} that the light took to reach Earth.

Since code "Un150910" could not trace light right back to the shock-fronted edge it was modified by extrapolation to that edge and re-named "Un280910". This involved making the time interval dT_B between pressure/radius profiles adjustable. However, dT_B was maintained equal for all intervals separating the pressure/radius profiles numbered M = 0 to 16, but adjusted so that the edge lay between M=0 and M=1. Therefore with 15 time intervals the time T_{STAR} became (15 - ΔM) dT_B where ΔM is the fraction of the time interval between M=0 and M=1 at which the light had to start from the edge. So the age of the universe of matter becomes:

$$AGE = 15dT_B - T_e + T_D : \quad T_e = \Delta M \, dT_B \qquad [13.3.1]$$

Second the Age of the i-ther

The i-ther is undergoing ever-accelerating growth and provides the shock-fronted edge. This is assumed to start from a sphere one metre in radius and grow according to equation [10.3.2] p.222, which involves an exponential term plus the term $E_X v_p$. The latter provides the excess shock-front speed due to primaries diffusing forward to cause collision breeding prior to arrival of the shock itself. The proportion E_X has not yet been evaluated and could lie anywhere between 0.0 and 0.3. Consequently this range will now be explored to determine what errors could be involved.

The age to M=0 has been termed its pre-history: symbol T_{BM0}. However, as illustrated in FIG.13.1 the big bang starts sooner by an amount T_D - T_e. Consequently the age of i-ther to the time T_{BYI} of the big bang becomes:

$$T_{BYI} = T_{BM0} + T_e - T_D \qquad [13.3.2]$$

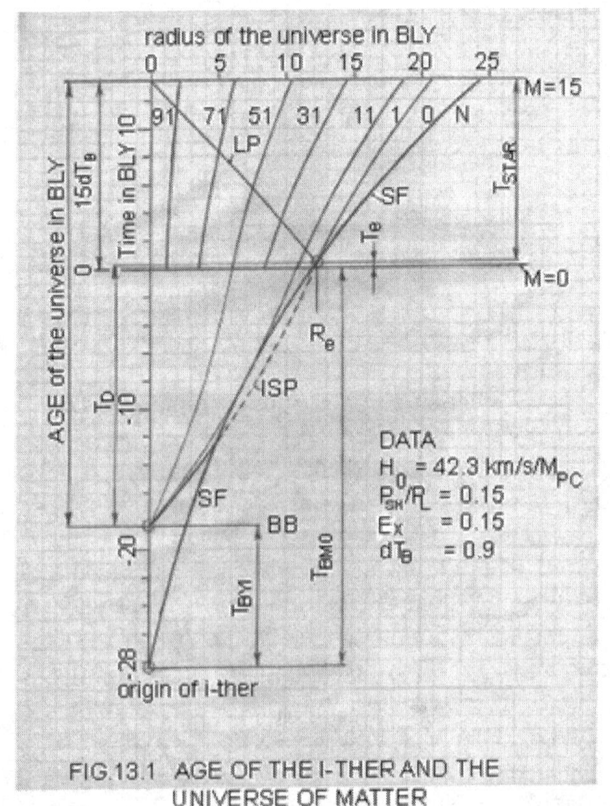

FIG.13.1 AGE OF THE I-THER AND THE UNIVERSE OF MATTER

FIG.13.1 gives a solution plotted to scale with time plotted against radius to illustrate the complete picture.

The shock front SF is shown starting from the 'origin of i-ther' and at a time T_{BM0} later the first detailed P/R profile is computed for profile number M=0 that specifies time zero.

This zero, with dT_B=0.9 is therefore set as: $15dT_B$

= 13.5 BY ago as measured from the present era.

After the origin of i-ther and T_{BYI} later the big bang, marked BB is initiated with matter at the outer edge of its growing fireball initially moving at the speed of light so $v_0 /c = 1$ in [13.2.1] to [13.2.3]. However, this reaches the shock front and so disappears. The dashed line ISP shows the path it would have followed to reach the point at radius R_e and time T_e after time zero.

However, light from a star, travelling along light path LP from the edge of the universe, would have to start from the point R_e, T_e, just to reach the origin centre. That star would then have followed the path marked N=1. The shock front having N=0 moves faster.

Clearly the only star that could have existed without colliding with the shock front would have element number N = 11 and so would have started out at a speed v_0 about 89% of light speed.

Another complete star path is hown at N = 51.

13.4 How Shock Speed Affects AGE & HUBBLE CONSTANT

TABLE 13 I shows how the ages of matter and i-ther are affected over the range $E_X = 0$ to 0.3.

TABLE 13 I All having R_{M0} = 12 BLY: E_X varies

(1)	(2)	(3)	(4)	(5)	(6)	(7)	(8)	(9)
E_X	N	R	P/P_L	v/c_0	R_e	T_{STAR}	v_{NF}	N_F
H_0					Z_e	T_{BYI}	R_{NF}	$(c-v)/c_0$
dT_B					T_{BM0}	AGE	Z_{NF}	Z_D
0	101	21.61	0.15	0.9421	12.19	13.15	0.5114	93
42.6	100	"	"	"	2.204	1358.2	11.72	0.6392
0.9	0	0	0.2624	0	1376.4	31.67	2.068	0.6124
0.05	101	22.87	0.15	0.9923	12.32	12.98	0.5251	96
42.4	100	21.56	0.1608	0.9352	2.217	30.53	11.98	0.6404
0.9	0	0	0.2789	0	48.65	31.67	2.108	0.6067
0.10	101	23.78	0.15	0.9850	12.34	13.00	0.5211	96
42.3	100	21.17	0.1702	0.8720	2.205	17.13	11.98	0.6450
0.9	0	0	0.2764	0	35.34	31.76	2.083	0.6041
0.15	101	24.61	0.15	0.9586	12.31	13.09	0.5125	95
42.3	100	20.75	0.1773	0.8083	2.207	10.06	11.88	0.6504
0.9	0	0	0.2676	0	28.35	31.83	2.049	0.6048
0.20	101	25.53	0.15	0.9437	12.28	13.25	0.5077	94
42.6	100	20.44	0.1835	0.7555	2.199	5.469	11.81	0.649
0.9	0	0	0.2629	0	23.80	31.87	2.033	0.6041
0.25	101	26.61	0.15	0.9161	12.38	13.36	0.5040	95
42.1	100	20.17	0.1879	0.6942	2.188	2.152	11.97	0.6498
0.92	0	0	0.2545	0	20.7	32.4	2.036	0.6129
0.30	101	27.46	0.15	0.8748	12.403	13.52	0.4962	96
41.3	100	19.74	0.1893	0.6288	2.203	-0.309	11.99	0.6562
0.93	0	0	0.2429	0	18.41	32.7	2.019	0.6187

The analysis uses equations [13.2.1] to [13.3.2]. It will be remembered that the shock-front moves at a speed $E_X v_P$ faster than the closest path line but the value of E_X has not yet been evaluated. The values shown in both TABLE 13 I and TABLE 13 II are guesses to cover the possible range in order to determine the magnitude of errors that wrong values might incur.

Furthermore, to provide a fair comparison, the radius R_{M0} at starting P/R profile M=0 has been fixed at 12 BLY in TABLE 13 I with the Hubble constant adjusted until the red-shift Z_e, extrapolated to the shock-front at R_e, T_e, is as close to 2.2 as possible. In this way the possible errors caused by wrong choice of E_X on the Hubble constant, deduced from red-shift data, can be assessed. Indeed this provides an example of the way astronomical data of any luminosity distance, symbol R, together with observed red-shifts Z need to be analysed by the Big Breed theory in order to evaluate the speed of recession, Hubble constant and radius of the universe.

Instead of fixing the starting radius $\boldsymbol{R_{M0}}$, that will correspond nearly to the 'luminosity distance' measured by astronomers, and has also to be guessed (since cosmologists refuse to yield raw data) the following table fixes E_X at 0.1 and allows three values for $\boldsymbol{R_{M0}}$.

TABLE 13 II All having $E_X = 0.1$: R_{M0} (in BLY) varies

(1)	(2)	(3)	(4)	(5)	(6)	(7)	(8)	(9)
R_{MO} H_0 dT_B	N	R	P/P_L	v/c_0	R_e Z_e T_B	T_{STAR} T_{BYI} AGE	v_{NF} R_{NF} Z_{NF}	N_F $(c-v)/c_0$ Z_D
16	101	31.15	0.15	0.9864	16.16	17.01	0.5082	92
32.3	100	27.85	0.170	0.8817	2.202	22.56	15.35	0.6578
1.15	0	0	0.2769	0	46.7	41.36	1.984	0.5859
12	101	23.78	0.15	0.9850	12.34	13.00	0.5211	96
42.3	100	21.17	0.1702	0.8720	2.205	17.13	11.98	0.6450
0.9	0	0	0.2764	0	35.34	31.76	2.083	0.6041
8	101	15.89	0.15	0.9921	8.233	8.656	0.5232	93
63.7	100	14.15	0.1709	0.8835	2.209	18.58	8.004	0.6442
0.6	0	0	0.2788	0	23.51	21.17	2.092	0.6042

The previous tables use code "Un280910".

TABLE 13 I shows how H_0 has to be changed to match a target Z_e = 2.2 when the starting P/R profile (M=0) has a shock-front radius $\boldsymbol{R_{MO}}$ of 12 BLY. What is most surprising is that H_0 , shown in column (1), only changes from 42.6 to 41.3 km/s/M_{PC} over the huge range of E_X from 0 to 0.3, when the red-shift for a star at the shock-fronted edge is close to the target value of 2.2 as given in column (6), row 2.

The major effect of E_X is on the pre-history age of i-ther. This is T_{BM0} given in column (6) row 3. This is over 1300 billion years when $E_X = 0$, meaning that the shock-front only moves at the same speed as the path lines of the expanding universe. This is based on a starting radius of one metre but the age would be infinite if this was zero. It only needs this shock to move 5% of the average speed of primaries faster than the closest path lines to reduce this age to less than $T_{BM0} = 50$ billion years, as shown by the next data set. Now the starting radius can be reduced to zero without noticeable effect.

As shown in column (7) the time for light to reach Earth from a star at the edge of the universe, given as T_{STAR}, hardly varies from 13 BLY over the entire range of E_X , nor does the life of the universe, *AGE*, measured from the big bang change very much from the value of 31 BY. However, the time available T_{BYI} (column (7) row 2) for the i-ther to self-organise by energy-fed chaos changes dramatically. If two billion years is long enough then E_X can be as high as 0.25 but at 0.3 a big bang scenario becomes impossible since it begins before the i-ther has originated as shown by the negative value of T_{BYI}.

Columns (8) and (9) give details of the star closest to the edge (suffix $_{NF}$) that could be given accurate analysis by the computer code. The number N_F would be 101 if it reached the edge and 0 represents the origin centre. So the ratio N_F /100 is the relative distance of the star from the centre to the path line 100.

The ratio $(c - v)/c_0$ gives the amount the flat space speed of light c has to speed up to reach a value of unity and is the cause of the major red shift $Z_{NF} - Z_D$ with these values shown in columns (8) and (9) respectively.

Other interesting results are given in columns (2) to (5) that show details for the present era. When N=101 values at the shock-front are represented whilst at N=100 the first path line, starting at the shock-front for the first P/R profile given by M=0, is represented.

When $E_X = 0$ the radius from the origin point, given in column (3) is the same for N=0 and 1 of course. For all other cases the difference increases as E_X increases as is to be expected. However, i-theric pressure P_0 at zero radius as a ratio of liquidus value P_L , which is P_0/P_L , changes very little. It rises slightly from .262 to .276 as E_X rises from 0 to 0.1 and then falls slightly to 0.243 as E_X increases to 0.3. Also the speed of particles and galaxies, corresponding to path lines and represented by v/c_0 at N=100, steadily falls from 0.94 to 0.63 although the speed of the shock front rises and then falls only a little.

TABLE 13 II shows how a change in starting radius R_{M0} affects H_0. About 15 BY ago it must lie in the range R_{M0} = 8 to 16 that yields values for the shock front radius at the present era of between 31 and 16 BLY respectively as shown in column (3). Simultaneously H_0 (giving C_{SH}) reduces from about 64 to 32 km/s/M_{PC} as R_{M0} is increased from 8 to 16 BLY.

The most important finding from the study is that the value of E_X makes very little difference to the value of H_0 determined from the red-shift data provided by astronomers. This value is a measure of the net i-theric creation rate given by $C_{SH} = 3\ H_0$. This is important since the value of E_X has not so far been established and only guesses can be used at the present time. However, the analysis shows values of E_X = 0.1 to 0.2 can be used with confidence.

13.5 The Hubble Constant falls below $C_{SH}/3$

The value of H_0 shown in column (1) of TABLE 13 I & II specify the creation rate, that is $C_{SH} = 3H_0$, at the shock front and evaluated from the input data of R and Z. This does not, however, correspond closely to the values of H_0 at all radii except for $E_X = 0$. For higher values it was found necessary to use an error correction for some arbitrary element and number 30 was adopted. This was necessary since for E_X greater than 0 the difference between energy demand and supply became excessive and had to be corrected, for each value of M, to reduce the difference generally to less than ±1%. The following tables show how H_0 falls off as M increases.
In each case R_{M0} = 12 BLY (billion light years)

TABLE 13 III Case $E_X = 0.1$, $C_{SH} = 3{\times}42.3\ M_{PC}$, $dT_B = 0.9$

M	1	3	5	7	9	11	13	15
H_0	42.09	41.71	41.37	41.12	40.90	40.73	40.57	40.48

TABLE 13 IV Case $E_X = 0.15$, $C_{SH} = 3{\times}42.3\ M_{PC}$, $dT_B = 0.9$

M	1	3	5	7	9	11	13	15
H_0	41.84	41.03	40.31	39.72	39.21	38.79	38.41	38.07

TABLE 13V Case $E_X = 0.25$, $C_{SH} = 3{\times}42.1\ M_{PC}$, $dT_B = 0.9$

M	1	3	5	7	9	11	13	15
H_0	41.13	39.45	38.06	36.92	35.95	35.07	34.31	33.64

13.6 Eliminating an Embarrassing Difficulty

Wright (2005) gives two equations on page 3 of his website, both derived from special relativity, purporting to connect the red-shift with speed of recession. One, he rejects as wrong, is the "Special relativity Doppler-shift law", he quotes as:

$$1 + Z = \sqrt{[(1 + v/c)/(1 - v/c)]} \qquad [13.6.1]$$

However, he says the correct red-shift-velocity law is:

$$1 + Z = \exp(v/c) \qquad [13.6.2]$$

The latter is based on the curious idea of having an infinite number of hypothetical observers connecting a remote type 1A supernova to the observer. Each passes the light signal on to the next.

The values from the Big Breed Theory use code "Un280910" and input data:

$H_0 = 42.3$, $P_{SH}/P_L = 0.15$: $E_X = 0.15$: $R_{MO} = 12$ BLY: $\Delta T_B = 0.9$ BY

$H_0 = 71$ km/s/M_{PC} is the cosmologists value for 'Hubble constant'

TABLE 13 VI RED SHIFTS {Wright *v/c* from [13.6.2]}

(1)	(2)	(3)	(4)	(5)	(6)
Z	Z_D	R BLY	v/c_0	$(c - v)/c_0$	Wright v/c
1.524	0.498	10.29	0.4350	0.714	0.9258
1.759	0.548	11.09	0.4721	0.6811	1.0149
2.049	0.6047	11.88	0.5125	0.6504	1.148
2.207		12.31	0.5309		1.165

Columns (1) to (5) are from the Big Breed Theory with values taken going back along the light path having a speed of light c increasing with radial distance. However, the bottom row is extrapolated to the shock front that specifies the radius of the i-ther. Symbol c_0 is the speed of light at the origin centre. The value of v/c given in column (6) from Wright's equation has c constant and equal to the value c_0.

The value of v/c given in column (6) uses Wright's equation [13.6.2] with column (1) as input data. Wright also states that the maximum red-shift so far measured is 1.755. It must be an embarrassment to both him and other cosmologists that this leads to a speed of recession greater than that of light. This puts it outside the observable universe and yet the value has been recorded!

Columns (1) and (2) show how relatively small, from the Big Breed Theory, is the contribution by the Doppler shift. Column (3) gives the distance of the light path from the origin centre in billions of light years and column (4) gives the value of recession speed as a ratio of the observed speed of light at the origin. This value, given by the Big Breed Theory, is little more than half the speed of light. This means the new approach shows we are potentially able to observe the entire universe. The way the light is speeding up, to cause the extra red-shift, can be judged from column (5).

The most important comparison to be made is between columns (4) and (6) showing how the Big Breed Theory yields such comparatively low speeds of recession – less than half of Wright's value!

13.7 Leading to a plot of Red-Shift with Distance

Photographs of the computer screen using code "Un280910" show the complete solution for what is thought to be the most probable case. FIG.13.2 shows the set of pressure radius profiles with path lines added and FIG.13.3 shows how he path lines appear on a base of time. The small circle at the end of the light path shows the extrapolation of the light path to the shock front to define the point R_e, T_e marked in FIG.13.1 p.262.

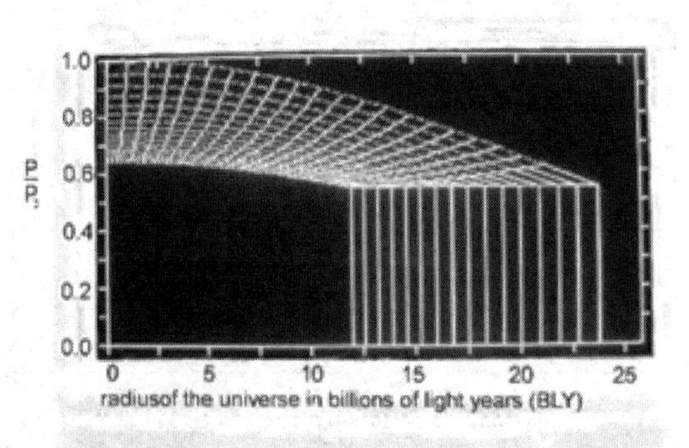

FIG.13.2 16 PRESSURE/RADIUS PROFILES AT 0.9 BILLION YEAR (BY) INTERVALS

CASE H_0 = 42.3, E_X = 0.1, R_{MO} = 12 BLY, P_0/P_L = 0.2764

FIG.13.4 shows how the red shifts vary with radius and again a small circular blob shows the extrapolation to the shock front. So this point corresponds to Z_e in column (6) TABLE 13 I whose target value was 2.2. The blue shifts Z_{SG} and Z_{MG} , caused by space and matter gravity that were derived in Chapter 9, would not be relevant according to the final argument presented in Chapter 12. Therefore these should be ignored and the curve marked Z should correspond with what astronomers observe.

The Doppler red-shift is Z_D and is clearly the minor

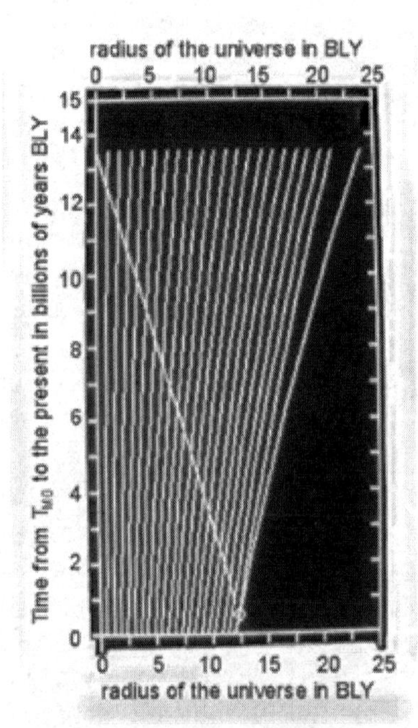

FIG.13.3 TIME/RADIUS PLOT showing path lines, shock front (right) and light path. Data as for FIG.13.2

component of the total, whereas it corresponds to the only red-shift according to contemporary cosmologists. This is one of the major features of the Big Breed Theory that leads to major changes in understanding of the nature of the universe.

The difference in red-shift $Z - Z_D$ is caused by the 'flat speed of light', at speed c relative to space, having its photons speed up due to falling speed of recession v that light encounters as it approaches the observer. This difference $c - v$ is plotted in FIG.13.4 as $(c - v)/c_0$ showing the amount of speed-up that occurs. The datum value c_0 is the light speed at the observer who is situated, in all these plots, at the central origin point of the universe.

On the other hand what was defined as the 'observed speed of light' c_T in our book '*QUANTUM GRAVITY etc.*', no longer corresponds with what the observer sees and so is now a confusing name. It is the speed inclusive of the effect of i-theric density change and is measured relative to that kind of space. As can be seen in the central graph, this speed, given as ratio c_T/c_0, increases with radius. However, the space is receding at the speed v shown in the uppermost curve as ratio v/c_0. The difference $c_T - v$ is nearly constant. This results in the almost straight line path of light that is shown in FIG.13.1 and FIG.13.3.

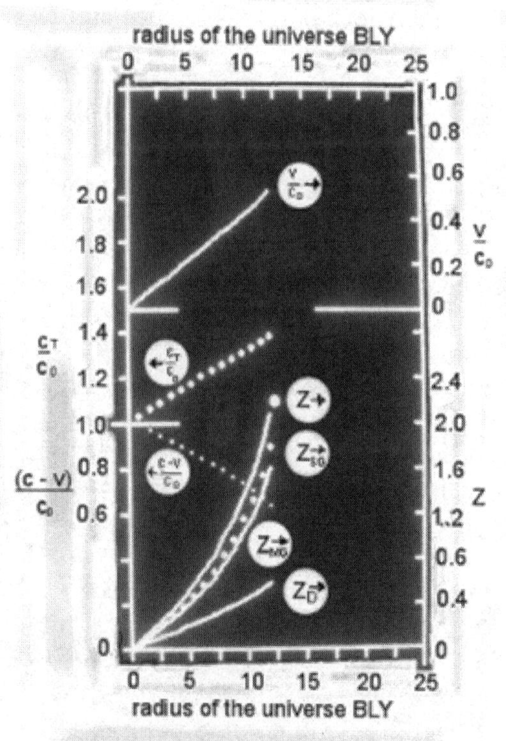

FIG.13.4 RED SHIFTS Z
FLAT LIGHT SPEED-UP RATIO $c/c_0 - v/c_0$
OBSERVED LIGHT SPEED RATIO c_T/c_0
& REGRESSION SPEED RATIO v/c_0

It may seem confusing that an almost constant 'observed' speed of light c_T exists and yet a large red-shift occurs caused by light speeding up as it approaches. The reason is fully explained in Chapter 9 where it is shown that this $Z - Z_D$ red shift is caused by the flat speed of light c speeding up: not c_T.

13.8 A NEW INSIGHT APPEARED 1st November 2010

I suddenly realised that the momentum equation used to evaluate red-shift in Chapter 9 p.211 will also cause the image of a remote galaxy to be magnified by refraction. This will partially counter the inverse reduction with distance that would normally be expected. An internet search soon showed this effect has been observed by astronomers.

The theory is given in the ADDENDUM on page 292.

13.9 CONCLUSION TO CHAPTER 13

We found in Chapters 10 and 12 that, contrary to established belief, our galaxy must be close to the central origin point of the universe. Then at the start of this chapter it followed, as a consequence, that the age of the universe of matter had to be more than twice the established figure of 13.7 billion years. This was because the star from which light was emitted had first to travel out from the origin point of creation to the observed position. This travel time is ignored by cosmologists.

Next a problem arose due to the speed of the i-theric shock front being greater than that of the nearest path line by an unknown amount. It was therefore necessary to determine the possible errors involved by finding the predictions arising from as wide a range of this speed increment as possible. This was covered by allowing the factor E_X to vary over the range 0 to 0.3. It was assumed the red shift Z at the shock fronted edge would be fixed at 2.2 for all cases as if the computer code "Un280910" was being used to evaluate a given set of astronomical data.

The effect on the computed life of i-ther was huge but the age of the universe of matter and the shock front creation rate C_{SH} , derived from the data, were hardly affected. The Hubble constant turned out to be rather smaller than specified by C_{SH}. However, the conclusion reached is that the result is a very encouraging because it means the theory can be used with confidence to evaluate astronomical data with acceptable accuracy if E_X is assumed to be from 0.1 to 0.2.

The radius of the universe is probably about 24 BLY but needs the input of luminosity-distance data to give a more accurate figure that can be quoted with confidence.

The analysis also suggests the desirability of making a search for a theoretical derivation to give a value for E_X for improvement of accuracy in the estimate of age of the i-ther.

CHAPTER 14

CONCLUSION

Collecting Six ideas for new research

Some projects could win anybody having maths or physics qualifications a NOBEL PRIZE!

These are highlighted in bold typeface.

Six areas of research unfold that should encourage mathematicians and computer specialists to continue where this study ends.

The main conclusion drawn is that, for the first time, a solution has been provided that resolves the mystery of what Dark Energy really is, as spin-off from elimination of the major flaw invalidating the big bang theory. The latter is known as 'the problem of the cosmological constant' that is 10^{120} *times too big.*

The energy required for creation of our universe had to originate billions of years before matter arose. More important still - the means for this energy to emerge from the void of zero energy has been provided. This resulted in the prediction of an accelerating expansion of the universe of about the correct order of magnitude – so solving the Dark Energy Mystery!.

A further development showed new principles need incorporation in the analysis of astronomical red shift data. These lead to a greatly reduced Hubble constant and, again for the first time, provide a prediction for the size of the universe.

This applies if our universe appeared by either a big bang or in any other way. It was the provision of a means for creation switch-off, lacking in the big bang theory, which led to these conclusions.

After studying PART II of this book the reader is recommended to return to PART I. This is really an easy to read summary of the mathematical derivations. This should help fix the main points in the mind of the reader.

PART I, a popularisation, began by showing the current big

bang model for creation of the universe to have serious flaws. The worst was its prediction of a rate of acceleration many billions of times greater than astronomers could permit. This is still known as 'The problem of the Cosmological Constant'. By ignoring this problem and assuming gravity to be the only major long-range force cosmologists concluded that the expansion of the universe had to be always slowing down. Then in 1998 astronomers studying remote type 1a supernovae found it was speeding up. Cosmologists resorted to the invention of the concept of 'Dark Energy' producing anti-gravity forces at long range that overcame the attraction of gravity. This has been claimed to patch up the big bang approach to make it match the data but still the major problem remains.

This study began, in 1987, from the realisation that the quantum level of reality was so weird that it could not be the ultimate reality. Its elementary particles had to exist as abstract waves in order that a single one could pass through a pair of narrow closely spaced slits in a mask and yet, after thousands of such particles had passed in this way, produce an interference pattern. This meant the particles had to be ephemeral with unreal wave-like qualities as they travelled only to collapse into particles when stopped. They only appeared as real when observed by a mind.

Such experimental evidence seemed to mean that an even deeper level of reality existed. This was to be called 'i-ther' from which the quantum level of reality emerged.

Since the i-ther needed to be made from truly real particles they could not operate on quantum principles. If they did an even deeper level would be required for its emergence - leading to an unsatisfactory infinite regression. Consequently for the mechanics of the i-ther a reversion to a revised Newtonian mechanics was derived that extended its range of applicability up to the speed of light and to the most extreme gravitational fields. This 'Exact Classical Mechanics' matched all experimental data just as well as Einstein's theories of relativity as shown in a companion volume. So this mechanics could be used at the i-ther level, penetrate the quantum level and be equally applicable at the macroscopic level of reality. This had been specified as a requirement by the author before derivation had begun in order to justify the approach.

The reader can judge whether or not the achievement of this objective does indeed justify the approach.

Furthermore, at i-ther level, energies were no longer limited to discreet values that needed quantum jumps from one energy level to another. Kinetic energy, whether of the linear or rotational kind, could be considered to change smoothly from one state to another, as it seems to do at the macroscopic level.

In order to solve the problem of the cosmological constant this deeper level had to consist of the ultimate real particles called 'primaries' that existed in two opposite and complementary forms. These constituted two intimately mixed 'phases': one made from positive energy with the other made from a negative kind. The latter was defined by interchanging the forces of 'action' and 'reaction' in Newton's laws of motion. It was explained in PART I of this book how there is nothing strange about the concept of primaries made from negative energy since if two made of this kind of energy were to collide responses would be identical to those made of positive energy: the negative effects would cancel out.

However, when primaries made of opposite energies collide a very different response was shown to occur. This is briefly explained in **PART I**. By simple non-mathematical logic it is shown that when opposites collide each gains energy of its own kind. This occurs because momentum has to be conserved and causes both kinds of energy to be forced out of the void of zero energy in equal and opposite amounts. Then energy is conserved as well as momentum. The rate of creation would be prohibitively huge. However, a creation switch-off mechanism is inherent in the approach. Mutual annihilation results when large numbers of primaries converge together from all possible directions. This not only provided a solution for the problem of the cosmological constant: it also predicted that the universe must be in a state of ever-accelerating expansion. In this way both major problems of the existing big bang model were resolved.

Then in **PART II** the new concept introduced is called 'opposed energy dynamics' and given a detailed mathematical analysis. This first considered only linear motions of primaries of opposite energy showing that a breeding effect is produced if no rest energy of colliding primaries is lost. Then it was found possible for some rest energy to be mutually annihilated and this was sufficient to almost cancel the gains. However, in **Chapter 6** the effect of spinning motions was also considered. In some collisions the kinetic energy of spinning motion was found, on detailed mathematical analysis, to be lost but at other times a gain in spin energy occurred. With full statistical analysis it was found that gains predominated. They more that offset possible losses of rest energy and so the breeding effect was fully supported.

In the same chapter a further difficulty was also resolved. This was called the 'phase coupling factor'. In the first publication of

1992 it had been assumed that, for bodily motions of i-ther, the two phases would couple to each other as strongly as to themselves. Then each would pull on the other to eliminate its inertia and so accelerating motion would arise without any pressure gradients being involved. This would be classified as a phase coupling factor $C_{PF} = 0$.

However, opposed energy dynamics showed that the deflections experienced when unlike primaries collided was far smaller than for collisions of like primaries. In a limiting case $C_{PF} = 1$ could arise. Then each phase would bc totally uncoupled from its partner as far as gross motions are concerned and for its acceleration the same pressure gradients would need to apply as if the other phase did not exist. The phase interaction would then be limited to breeding and mutual annihilation. However, $C_{PF} = 0.5$ seemed more likely.

A surprise discovery appeared as a result of programming the random motions of a mixture of positive and negative primaries. The random clumping, characteristic of chaos, was demonstrated but what had not been anticipated was that the clumping always appeared in different places for the two phases of the mixture. This meant that collisions between primaries of opposite energies would not be anything like as high as originally proposed. This showed C_{PF} must approach unity. A value of $C_{PF} = 0.9$ was decided upon for cosmological application. However, the influence of spinning motions has not yet been included in the estimates of $C_{PF.}$

Our first proposal for further research has now appeared: completing the theory of spinning motions in order to obtain a more accurate value for phase coupling factor C_{PF}.

Then in **Chapter 7** rates of collision were considered and shown to produce hyper-astronomical rates of energy gain by collision breeding of the order 10^{69} W/m^3. This corresponded to a creation constant C_{BT} of 9×10^{24} s^{-1} in the growth equation $dV=V\times C_{BT}\,dt$. The value required to match astronomical observations is, however, only about 2×10^{-18} s^{-1}. This meant that the formation of cores of mutual annihilation needed to cancel the breeding effect to about 1 part in 10^{42}.

No difficulty was presented, however, since at the 'i-theric liquidus' state of ultimate density total annihilation had to occur. At lower densities a net creation remained that was controlled by the degree of acceleration of the expansion of the i-ther and therefore of the universe. It was in this way that the problem of the cosmological constant was resolved. The ever-accelerating expansion of the universe emerged as a secondary consequence in the manner to be summarised next.

The creation theory allowed the rate of increase of mass in any element of volume to be proportional to the mass present. If the i-ther behaved as a perfect gas this meant $d(PV)/PV = C_{NA}dt$ where C_{NA} is the creation rate assumed constant. It meant that some energy was absorbed in producing a pressure increase with the rest causing an increase of volume. When a fixed proportion was allowed for the pressure term then integration by the calculus became possible and yielded Hubble's law, $v = H_0R$, in which $C_{NA} = 4H_0$ where H_0 is the Hubble constant. Differentiation showed the expansion existed in an ever-accelerating state.

A force F was needed to produce the acceleration. The derivation of this part of the solution began with the assumption that $F = d(mv)/dt$ which expands to $mdv/dt + vdm/dt$. In other words that the force needed to produce accelerated motion is equal to the rate of change of momentum. This meant the rate of change of mass dm/dt contributes to the force required.

This force could only be supplied from pressure gradients. This meant that even though there might be a sudden jump in pressure at the growing edge, the pressure at the origin point would be higher than at the edge. This seemed to complicate the mathematics to such extent as, at the time, to render the problem insoluble.

One possibly useful theorem emerged as a result of the presence of the term dm/dt but the main aim of solving the growth problem could only be regarded as very approximate at this stage.

Chapter 8 began by a simplification in the equation relating pressure gradients to the acceleration they produced. It had been suddenly realised that the term dm/dt should not be incorporated. This was due to the simultaneous creation of matching amounts of positive and negative energy and it followed that the associated momentum gains would cancel to zero as well.

However, the simplification involved did not solve the main difficulty of providing an accurate means of predicting the growth of both pressure and radius. The theory was far from exact at this stage owing to the mathematical difficulty arising from pressure and volume increasing simultaneously. The remainder of Chapter 8 explored several of the attempts, by way of example, to resolve the difficulty. None of these were really satisfactory and several years were to elapse before a major breakthrough resolved the difficulty.

One approach, that initially showed promise, used a finite difference method. This split the growing sphere of i-ther into 100 nested spherical shell elements. Each was treated individually, with a creation rate varying with pressure that caused growth of both pressure and volume, whilst simultaneously being accelerated radially outward by radial pressure gradients. This was only partially successful owing to the rapid development of instabilities and so still remains undeveloped.

This has now been relegated to APPENDIX V since it is no longer essential. However, it has not been abandoned since a fully developed computer code might be able to predict possible wave effects that cannot yet be simulated.

Our second proposal for further research is therefore to devise a way of developing a computer code that does not suffer from instability. It needs to have the potential of predicting any wave effects superimposed on the radial growth and rising pressures of a large array of nested radial shell elements of the i-ther.

At this time the derivation of a fully satisfactory solution of the growth problem was left unresolved for attention in the future.

Instead, in **Chapter 9** attention was paid to the interpretation of the red-shift data produced by astronomers. A new approach was demanded since established methods are based on Einstein's relativity theories that cannot be incorporated when the existence of a background medium is recognised.

This is a hugely important matter for cosmology as it leads to a very different picture of the universe from that given by current cosmology. Using relativity, for example, the cosmologist Ned Wright (2005) postulates a string of hypothetical observers stretching from a distant galaxy to the astronomer's telescope. The reason seems to be the need to consider the speed of light appearing the same for all observers. Each is assumed to pass the light signal along from one to another until the telescope is reached.

Such a model is untenable as soon as light is assumed to propagate through a medium. Consequently the Big Breed approach presented a totally different interpretation. Now light had to propagate through an expanding medium whose size could be estimated owing to the accelerating state. Contrary to contemporary physics, that postulates a universe that is finite yet unbounded, a universe finite and having a definite edge is predicted. This edge is like a shock fronted wave except that in this case the shock front bounces the void of nothingness into a mix of opposed primaries. Behind the shock front pressures are greater as required to produce an accelerating expansion.

The conclusion now is that we are potentially able to see right to the edge of the universe and so can determine its size. Then for two reasons we can now determine our position within the universe. Since we see uniformity on the largest scales when looking in any direction it was concluded we exist close to the centre of the universe. We cannot be at its exact centre as otherwise the CMB, the microwave background radiation, would seem exactly uniform. In fact anisotropy exists that can be interpreted as a speed of 400 kilometres per second. This is very small as compared to about 0.6 times the speed of light the theory predicts for motion of the edge. Consequently Earth was shown to be well within 1% of the centre as compared to the distance to the edge.

It follows that light needs to arrive in almost radial straight lines from galaxies close to the edge. Furthermore light now has to propagate through the medium: not a string of observers arranged in an arc. That medium, the i-ther, is expanding and so light starts off near the edge having to overcome a high radial speed of recession. The latter reduces with reduction in distance and so, from the astronomer's viewpoint; the light speeds up as it approaches.

An extra red shift is involved since the kinetic mass, from which the photons of light are made according to ECM theory, has to reduce as speeds increase in order to conserve momentum. The photon's energy reduces in proportion to its kinetic mass with consequent increase of wavelength. The resulting extra red shift is shown to be more than twice the Doppler shift that still applies. Two blue shifts, of small amount, reduce the total a little. These are due to mass gravity and space gravity. The latter is a consequence of the acceleration. The result is to predict a far lower Hubble constant and speeds of recession. Consequently another project appears.

Thirdly a project for future cosmologists has made its debut: Re-evaluating all existing astronomical data to refine our understanding of the cosmos with light propagating through space (instead of a string of hypothetical observers). We make available a computer code for downloading from the internet that will enable this study to be carried out cheaply and easily.

In **Chapter 10** a major breakthrough is described! This occurred on 1st July 2010. Up until then the pressure jump at the shock front, marking the advancing edge of the universe, could only be guessed. Furthermore no satisfactory solution had been found for calculating pressure and radius increases. All now fell neatly into place.

Computations had only used a guessed value of P_{SH}/P_L, the pressure at the shock front divided by that of the liquidus state. Values of around 0.001 had been chosen. The curve showing creation rate as a function of pressure ratio P/P_L was almost an inverted parabola - except that the point of maximum pressure was at $0.43P_L$ instead of 0.5. Computations of pressure increase due to acceleration were going divergent and the cause was found to be that the energy available due to pressure increase was larger than required. So pressures increased uncontrollably.

However, when a high value of P_{SH}/P_L of 0.2 was chosen the reverse occurred. The energy available was insufficient. A value of 0.15 was exactly right. It meant the shock front had to have this value at all times and could not exist at any other value.

Then another surprise followed. It was next discovered that now a match of energy supply to demand, within generally ±1%, occurred over the entire pressure range computed for the i-ther! At least this applied up to the present era. This simplified everything since it meant the pressure term in the equation that had been the cause of so much difficulty could be ignored altogether. It meant that the early discovery of the Hubble constant being directly proportional to creation rate was no longer questionable. This relation could now be written $C_{SH} = 3H_0$ where C_{SH} is the creation rate at the shock front. All factors could now be represented accurately by a set of simple mathematical equations with Hubble's law accurately representing the motion of all particles from primaries and dust grains up to galaxies.

The new discovery also meant that very high confidence could now be attached to the red-shifts and size of the universe predicted by the computer code that is now available. It means also that the cosmologist has available a new tool for analysing the red-shift data provided by astronomers.

One troubling matter arose from the study of the 'pre-history' of the i-ther – the history up to the point where detailed pressure/radius profiles are computed starting about 15 billion years ago. It was found that at an early time the speed of the shock front would be absurdly slow. Primaries would diffuse out in front to produce collision breeding without annihilation. The rate of breeding, as shown in §7.3, is absolutely huge being 10^{43} times that consistent with observation. Consequently the shock front will advance faster by 'diffusion breeding' that will be proportional to the average random speed of primaries. A proportion E_X of this speed had to be guessed. As shown in Chapter 10 this dramatically reduces the time for the i-ther to grow from one metre to 12 billion light years in radius during the pre-history time. Also shown is that if E_X is not more than about 0.1 it makes little difference to growth in later epochs. However, the need arises for an accurate value for E_X and the potential for its evaluation already resides in the opposed energy dynamics on which the Big Breed theory is based.

Consequently a fourth project has appeared:

The fourth project is for a competent mathematician to build on the dynamics of the Big Breed theory. By considering collision breeding, by primaries diffusing forward from the shock front, an accurate equation giving the value of E_X is to be formulated.

It is unfortunate that the grain size of the i-ther is so small as to prohibit direct experiments for the verification of its existence. This can only be inferred from the way the Big Breed theory, which depends upon it, makes predictions matching observation. However, further validation is promised from computer simulation of the sub-microscopic behaviour of primaries in the act of collision breeding and subsequent annihilation.

Chapter 11 therefore offers the computer specialist a golden opportunity. First a two-dimensional display needs to be developed with the bugs eliminated. Then as further described in Chapter 11 a fully three-dimensional display needs to be achieved. The aim is to prove the spontaneous formation of filament and blob like annihilation cores and demonstrate the self-organising power of i-ther to confirm the predictions of the mathematical theory of chaos.

So the fifth project could demonstrate the power of i-ther to self-organise by chaos and so evolve the creative power needed to allow the quantum world to emerge.

This would be the crowning glory of the theory and provides the most challenging test, of all those mentioned earlier, for the programmer of some super-computer.

We have not finished yet, however. In **Chapter 12** the creation of matter is considered to see how the big bang theory could be rescued. This is permitted as a result of the Big Breed theory solving its major problems by providing the energy requirements needed. The study also considered the bending of light at the stage of photon decoupling from the initial ball of plasma when its radius was something less than 500,000 light years. The astonishing prediction appeared of a radius of curvature for light moving tangentially of less than 1/400,000 or the surface radius. Even at 19% of the surface radius, the radius of curvature turned out to be only about 0.003% of the surface radius. However, this increased as the radial direction was approached becoming infinite when moving purely radial.

This meant, however, that photons would yo-yo back and forth across the origin point with very few reaching the growing surface.

Even in the present era the radii of curvature are generally far smaller than the radius of the universe, being only 15% of it at the shock front. The forces of matter and space gravity were additive.

However, a final problem arose that was resolved on 27/8/2010. Until this point was reached the gravitational force had been considered to be caused by a negative buoyancy effect. The associated theory of quantum gravity, whose derivation is provided in another of our books, showed pressure gradients surrounded massive objects and would produce a force of about the right order of magnitude if sub-atomic particles existed as solid little objects. It then followed that the pressure gradients caused by the accelerating expansion of space would produce similar forces – space gravity. If such forces were real, however, they would prevent galaxies from accelerating away with the i-ther. Now in our book *CREATION SOLVED?* unification of electromagnetism and gravity by Spaniol and Sutton (1994) is described but electromagnetism is shown to be abstract. Furthermore the interpretation of quantum theory given in all our books suggests that matter is also abstract so that sub-atomic particles do not exist as real solid objects. Therefore gravity also needs to be considered abstract and therefore not caused by pressure gradients. Then space gravity would also not exist. The final correction discarded gravity as a force of infinite range: it had to be a range large enough to pull galactic super-clusters together but no greater.

These changes permitted remote galaxies to accelerate with the i-ther but also eliminated the matter and space gravity blue shifts. So now the total red-shifts observed were due to the Doppler shift and that due to light having to speed up due to the opposing speed of recession. It also followed that the Hubble constant assessed from red-shift data also depended only on the sum of these two red shifts.

So the sixth project is another computer simulation starting at the photon decoupling stage following the big bang. A three dimensional sphere is to be represented in which points are chosen at random to represent the emission of photons. Their direction of emission is also chosen at random. Then trajectories are traced, using the equations provided in Chapter 12 but with space gravity deleted. The distance from the origin point of a small volume is to be specified and the concentration of primaries passing through is it to be the output. Accelerating growth of the universe is to be taken into account with the computation terminated at the present era. Most important is the orientation of intensity with respect to the radial direction. This is expected to show increasing anisotropy as distance from the origin point is increased. Comparison with measurements of

the CMB should return an accurate distance of our galaxy from the origin point of the universe.

Chapter 13 followed in which the accumulated experience allowed a study to be attempted for assessing the ages of both the universe of matter and i-ther. The major unknown is the speed difference between the path of primaries of the i-ther and the shock front caused by diffusion of primaries in advance of that front. This excess speed was assumed to be within the range of 0 to 30% of the average random speed of the primaries, i.e. $E_X = 0$ to 0.3.

The age of i-ther could be infinite at $E_X = 0$ but dropped to less than 50 billion years (BY) when $E_X = 0.05$. At 0.3 the big bang had to start before the i-ther had emerged from the void so this was an impossible condition. $E_X = 0.25$ appeared as about the limiting practical value. The age of the universe of matter, however, had to be about 31 BY: not the 13.7 BY conventionally accepted. This was because the new theory had shown, for the first time, that the size of the universe can be estimated from astronomical data was about 24 billion light years in radius. It also located the Earth close to the origin centre of the universe. This meant that the time for matter to travel out from the origin centre to the position at which it released the light we observe had to be added to the travel time of the light.

So what has been provided are opportunities for six highly prestigious research projects. It is my hope these will have inspired some readers to make further efforts to uncover the truth: to show where the driving force behind the universe really lies. Another wish is for this book to open up a new approach for the analysis of astronomical data since existing cosmological analysis is obsolete.

These are all projects for the mathematician or computer specialist but in our companion volume *Quantum Gravity etc.* several experiments in orbit are proposed that could lead to a new form of space exploration – measuring the velocity structure of the quantum vacuum and one laboratory experiment.

Some of these projects are potentially of Nobel Prize winning value. Unfortunately we do not have the correct qualifications to be eligible for consideration. However, we hope others will benefit by continuing this research where we have had to leave it unfinished.

1 The big bang theory predicts a rate of expansion of the universe 10^{120} times greater than observation allows, which invalidates that theory as presently formulated. The Big Breed theory cuts this figure to 10^{43} and this applies only for a brief period of 'inflation'.

2 The Big Breed theory provides the 'inflation switch-off means' the big bang theory lacks. The 'opposed energy dynamics' basis shows cores of annihilation arise that do this.

3 The result was the prediction of a universe in a state of slowly accelerating expansion before the discovery of this speeding up!

Cosmologists invented 'Dark Energy' to make the big bang theory match observation, but admitted this to be totally mysterious, the Big Breed theory has resolved the mystery.

4 Theorists assume a long range force of repulsion is responsible for the acceleration of item 3. The Big Breed theory shows this is wrong. Instead, the acceleration is produced by pressure gradients that develop within the i-ther caused by the high speed collisions of its primaries. Each spherical shell element of i-ther has a pressure inside greater than that opposing from outside. Consequently the pressure and density at the origin point are higher than anywhere else falling to the minimum value at the edge. (This would happen also if repulsive forces were the cause: something all theorists ignore)

5 From study of the CMB cosmologists say Dark Energy accounts for 74% of all the energy making up the universe. This shows a wrong direction is being pursued since i-ther, which is really the Dark Energy being sought, constitutes all there is. It is the ultimate substance on which everything else depends.

6 For the first time Hubble's Law is derived theoretically and shown to be directly proportional to the rate of volume increase per unit of existing volume of i-ther. This net volume creation rate remains unchanged up to the present era.

7 Also the net energy creation rate that is responsible for the accelerating increase of both volume and pressure of i-ther is shown to match the demand needed to provide 6 above up to the present era. This is because the increase of net creation rate required to produce the pressure increase is closely matched by the increase of energy provided due to that pressure increase.

8 The i-ther is shown to exist as a shock-fronted sphere in which the pressure immediately behind the front is 0.15 of the liquidus value where the net creation is zero. This is the condition yielding the fortuitous discovery permitting the validity of item 7. This is no ordinary shock front since it operates on different mechanics.

9 An extra 'diffusion' speed up of the shock front is an extra feature needing further analysis.

10 A major new component causing starlight to be red-shifted has been identified that is additional to the Doppler red-shift and more than twice as great. It is due to the continual expansion of space causing light to speed up in approaching any observer.

Furthermore 'space gravitational' and 'mass gravitational' blue shifts are also derived of much smaller amount. The need to revise astronomical data has been shown necessary since the result drastically reduces the value of the Hubble constant H_0. For example, from the same astronomical data H_0 is reduced from 71 to 45 km/s/Mpc.

11 Also for the first time a theory has been provided that predicts the size of the universe as a function of H_0. It has a radius of about 25 billion light years at the present time, as would be measured by instantaneous information transport.

12 The i-ther is not matter since it consists of a balanced mixture of 'primaries' of two kinds that are opposite and complementary. The i-ther was shown to emerge from the void whilst satisfying the conservation of both energy and momentum.

Again this is new physics.

When opposed primaries collide only in twos the need to satisfy momentum conservation forces collision breeding. This allows creation from the void. The same law forces annihilation when primaries collide in large numbers. This provides the creation switch – off means the big bang theory lacks.

13 Again for the first time a structure is shown to spontaneously form that is fed continuously by energy. This is the condition needed for further self-organisation, as predicted by chaos theory, and so can be expected to occur.

14 These features, 12 and 13, provide the energy required by a big bang type of matter-creation scenario that also matches the weird behaviour existing at the quantum level of reality.

15 Although we do not have any privileged position at the origin point of the universe the analysis has shown from three sources of data that we must be well within 1% of the distance to the edge.

16 That the universe may act like a hall of mirrors is not new since the Astronomer Royal said this is something he suspected and Einstein said light could bend back on itself. However, other factors appear from our study. Certainly at an early time the radius of curvature of light will be so small due to high gravity that light will tend to yo-yo back and forth across the origin centre.

However, in order to permit remote galaxies to keep pace with i-ther during its accelerating expansion gravity had to be considered as an abstract force unified with electromagnetism. It also had to be of very long but finite range.

The possibility of the universe acting as a hall of mirrors, as first suspected, had then to be ruled out.

For the first time the size of the universe can be predicted and so represents a valuable achievement of the Big Breed Theory. A radius of about 28 BLY is returned with the Hubble constant reduced from the presently accepted value of 71 to 33 km/s/M_{PC}.

17 We offer six areas for new research, some of the projects having Nobel Prize winning potential. This we hope will induce others to continue the research at the point where we have had to leave it in the state described in this book.

APPENDIX V PART II

A NUMERICAL APPROACH FOR GROWTH OF THE I-THER

In the previous chapter an analytical solution was finally derived that appears to have eliminated the problems of inconsistency from which earlier attempts suffered. However, a way to check the result and possibly provide greater detail is given in this chapter. In is possible that long range wave effects could produce oscillations having periods measured in millions of years.

V.1 First a Check on Methodology

Pressures have been shown, in previous chapters, to change with both time and distance. An alternative approach can check the validity of the methodology previously adopted.

Checking requires the introduction of partial differential coefficients as illustrated in FIG.V.1. At **A** the representation has all partial pressure increases made positive as required for setting up the equations whilst at **B** the component $(\partial P/\partial r)dr$ is shown with this negative - as in the actual case in order to drive the acceleration. The actual pressure change in the element of volume considered takes place along the dashed line and is dP. This is the result of the change with time $(\partial P/\partial t)dt$ and the change with radius $(\partial P/\partial r)dr$.

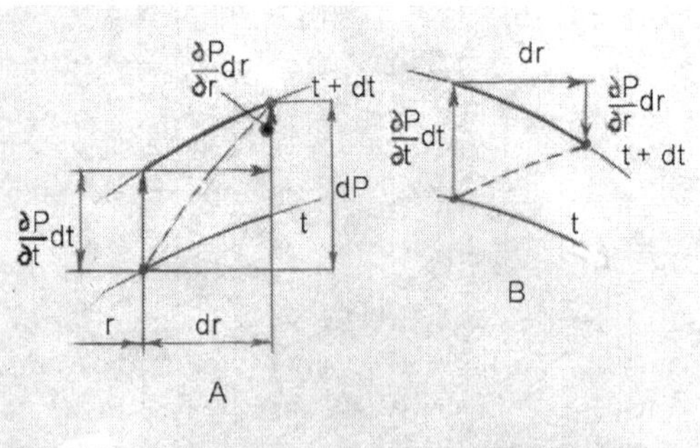

FIG. V.1 SHOWING PARTIAL DIFFERENTIAL REPRESENTATION FOR PRESSURE GAIN IN TIME AND RADIAL POSITION

It follows that:

$$dP = \frac{\partial P}{\partial t}dt + \frac{\partial P}{\partial r}dr \quad \textit{so that} \quad \frac{\partial P}{\partial r}dr = dP - \frac{\partial P}{\partial t}dt \qquad \text{[V.1.1]}$$

This permits a numerical method of integration to be written in which the mechanics is represented by FIG.V.2. The path of a representative element of volume is to be considered. At time t the element is part of a spherical shell of inner radius r and thickness b. It is formed as a small circular disc cut from this shell that has a radius $r\theta$ and is accelerated by pressure differential $(\partial P/\partial r)dr$ acting across thickness b.

In time $t + dt$ the inner radius increases to $r + dr$ and at the same time the thickness increases to $b + db$. The cone half angle θ does not change since the element is constrained by the remainder of the shell. The procedure is to find the radial constraints of distance $b+db$ of the element at time $t + dt$ due to the motion of all other cylindrical shells. This procedure initially has to assume pressure remains constant during interval dt though the pressure will differ from element to element. Due to increase of both r and b each element volume at time $t + dt$ will have increased with its new value determined. Then a second value for the change in volume is to be found from energy creation. The latter is expected to be greater than the former and then the pressure gain of the element will be the ratio of the volumes obtained by these two calculations.

This provides a new pressure from which to start the next step of the calculation. From this a new value for net creation rate C_{NT} is obtained from a curve such as given in FIG.10.1 showing C_{NR} as a function of P/P_L. Here $C_{NR} = C_{NT}/C_{SH}$ where C_{NT} is the creation rate at pressure P and C_{SH} the value at the shock front. P_L is the pressure at the i-theric liquidus condition.

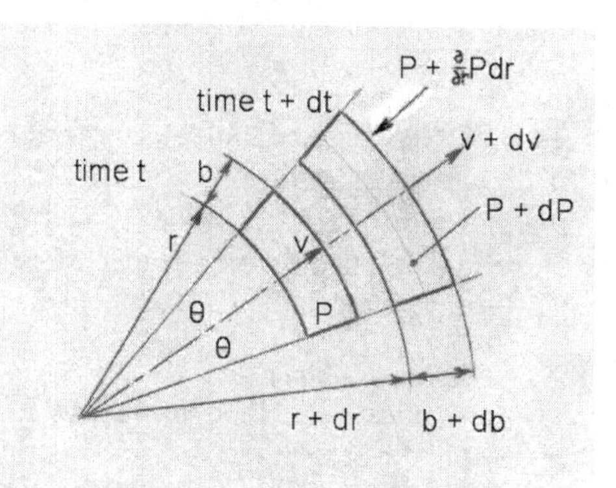

FIG. V.2 SHOWING AN ELEMENT OF VOLUME ACCELERATED AND INCREASING DUE TO CREATION

A pressure gradient $\partial P/\partial r$ can be found for each shell from the array involving all shells. This will establish the acceleration of each shell as required for the determination of dr/dt and consequently yields $r + dr$ for the next step. The routine continues step by step as far as is required

In the next section a method of computation is derived that enables a start to be made from an early epoch, using the pressure/radius profile 15 billion years ago, provided by equation [10.2.4] as the starting point.

V.2 The step by step computation

The way the rate of net creation C_N will vary with pressure is predictable in principle and efforts made to achieve this could be attempted. A plausible relation was argued in the previous chapter and is illustrated in FIG.10.1. This gives the expected shape of the $C_{NR}/(P/P_L)$ curve and allows a history for the evolution of the i-ther to be computed.

More simply the calibrated values from FIG.10.1 can be used to give absolute creation rates. In this case no matching of coefficients is required. First the new radii are obtained from [10.2.4] to provide the array of radius r_n values. The volumes of the elements for the two adjacent P/r profiles follow given by:

$$V_{1n} = \tfrac{4}{3}\pi\left[r_{1(n+1)}^{\ 3} - r_{1n}^{\ 3}\right] \quad \& \quad V_{2n} = \tfrac{4}{3}\pi\left[r_{2(n+1)}^{\ 3} - r_{2n}^{\ 3}\right] \qquad \text{[V.2.1]}$$

But now the change in volume dV_n in time dt has to be calculated from creation rate. The volume change is first assumed to occur at a constant pressure P_{1n}. The value of dV_n then depends on the calibrated C_N rate and the arithmetic mean of V_{1n} and V_{2n} i.e V_{an} so that the new volume becomes V_{C2n} where:

$$V_{C2n} = V_{1n} - V_{an}C_N dt \qquad \text{[V.2.2]}$$

The negative sign assumes going back in time. Now V_{C2n} will not be the same as the value V_{2n} obtained from [10.4.9] and [10.4.11} and the latter is the true value. The ratio of these two volumes can therefore be used to obtain the pressure P_{2n} for the next P/r profile using:

$$P_{2n} = P_{1n}V_{C2n}/V_{1n} \qquad \text{[V.2.3]}$$

The procedure can then be repeated until the i-ther is as small as required. At least that is the intention. In practice the apparently simple procedure, given in GWBASIC code "UnivCre39" has led to unexpected difficulties so that at the time of writing no success has so far been achieved.

V.3 Wave action is expected!

The iterative procedure so far described cannot address any wave action that will be involved. The author has considerable experience of wave effects as a result of designing and developing a 'gas wave turbine'. This machine used pressure waves for compression and expansion within the cells of a rotor. As a consequence of this experience he appreciates that with creation rates rising with time due to pressure increases, that compression waves will start from the central point and spread out. They will reflect back as a train of rarefaction waves and produce what is known as quarter-wave resonance. So pressure oscillations will be superimposed on the steady pressure rise taking place everywhere. Consequently the iterative procedure previously described needs to be used only for providing a starting point from which a numerical finite difference method needs to begin.

The next section is devoted to providing the equations that will enable an accurate solution for evolution of the universe to be made. Ancient Indian psychics have told the world that there is a creator for the universe called 'Brahma' who breaths in and out with a periodic time measured in many thousands of years. What a surprise it would be if the wave action outlined predicted a similar periodic time!

V.4 Derivation of a finite difference solution

The aim now is to start from at least 15 billion years ago with a P/R profile obtained from Chapter 9 and then compute forward in time up to the present era.

For most spherical shell elements of internal radius r and thickness b its volume can be written to sufficient accuracy as:

$$V_b = \pi\theta^2 r^2 b \qquad [V.4.1]$$

Since from the perfect gas equation: $\rho = 3P/v_p^{\,2}$ the i-ther mass in the element at time t becomes: $m_b = \pi\theta^2 r^2 b\, 3P/v_p^{\,2}$ [V.4.2]

Now force is simply mass×acceleration i.e: $m_b \times dv/dt$ in this corrected version and can be equated to the force due to pressure gradient from [7.6.1] using FIG.7.6 and so:

$$F = -\pi\theta^2 r^2 \frac{\partial P}{\partial r} b = \frac{3\pi}{v_p^{\,2}}\theta^2 r^2 b\, P \frac{dv}{dt} \qquad [V.4.3]$$

Then [8.4.3] above can be re-arranged to yield the acceleration:

$$\frac{dv}{dt} = -\left(\frac{\partial P}{\partial r}\right)\frac{v_P^{\,2}}{3P} \qquad [V.4.4]$$

When r is zero or small the approximations involved are inaccurate. To improve accuracy with finite values of b and 8 figure computing it is easier to use the exact equations. The entire sphere is represented as elements of equal thickness b with each having an internal radius r represented in an array so the nth radius is r_n and the next $r_{(n+1)}$. Then the value of element thickness b_n is constrained by being trapped between the two adjacent elements so that: $b_n = r_{(n+1)} - r_{(n-1)}$ and the volume of the element is now written:

$$V_b = \tfrac{4}{3}\pi\left[r_{(n+1)}^{\;3} - r_n^{\;3}\right]\sin^2\theta/4 \qquad [V.4\ .5]$$

Using the same logic as for the derivation of equation [10.4.4] now yields the more accurate form:

$$\frac{dv}{dt} = -\left(\frac{\partial P}{\partial r}\right)\frac{v_P^{\ 2}}{P}\frac{r_n^{\ 2}\left(r_{(n+1)} - r_n\right)}{\left(r_{(n+1)}^{\ 3} - r_n^{\ 3}\right)} \qquad \text{[V.4.6]}$$

Then using the simple equations of motion for this array of elements in which suffix $_1$ means at time t with suffix $_2$ meaning at time $t + dt$ we can write:

$$v_{2n} = v_{1n} + \frac{dv}{dt}dt \quad \& \quad dr_n = v_{1n}\,dt + \frac{1}{2}\frac{dv}{dt}dt^2 \qquad \text{[V.4.7]}$$

It also follows that: $r_{2n} = r_{1n} + dr_n$ [V.4.8]

Then the volume of the entire spherical shell element at time $t + dt$ becomes V_{b2n} where:

$$V_{b2n} = \tfrac{4}{3}\pi\left[r_{2(n+1)}^{\ 3} - r_{2n}^{\ 3}\right] \qquad \text{[V.4.9]}$$

And V_{b1n} is similarly expressed for time t with suffix $_1$ instead of $_2$.

Volume increase due to creation (Relating to Chapter 7)

Now the volume of the same element at time $t + dt$ is to be obtained a different way - from the amount of net creation taking place in the element in time dt and with pressure P assumed temporarily to remain constant at value P_1. Equation [7.3.15] is applicable but V_C will be ignored since it has already been argued that its effective value will be very small. The revised equation is:

$$\frac{dP}{P} + \frac{dV}{V} = C_N\left(1 - \frac{V_B}{V}\right)\frac{V_B}{V}dt \qquad \text{[V.4.10]}$$

And the dP/P term is ignored since $dP = 0$ as the initial assumption. Furthermore V is the average volume of element n from t_1 to t_2 and is: $\left(V_{b1n} + V_{b2n}\right)/2$. It follows that:

$$dV_{bn} = C_N\left(1 - \frac{V_B}{V}\right)\frac{V_B}{V}\frac{\left(V_{b1n} + V_{b2n}\right)}{2}dt \qquad \text{[V.4.11]}$$

Since C_{NA} was shown by equation [7.5.8] to be directly related to the Hubble constant and is the same as $C_N(1 - V_B/V)V_B/V$ the value of this parameter can be obtained from astronomical data for the present era. Then values for other conditions derived from the analysis can be made with respect to FIG.7.1.

C_{NT} =0 at $V_B/V = 0$ and also at $V_B/V = 0.25$ but with positive values in between these limits, varying almost in the manner of a parabolic arc as shown in FIG.7.1 given as the difference between equations [7.4.1] and [7.4.2].

Volume increase due to creation (Relating to Chapter 10)

The revised creation equation relating to FIG.10.1 replaces [V.4.10] by [10.6.1] and is:

$$\frac{\delta(PV)}{PV} = C_{RC}\, C_{SH}\, \Delta t : \quad \textit{Noting that} \quad C_{SH} = 3H_0$$

$$\textit{Where} \quad \frac{\delta(PV)}{PV} = \frac{dP}{P} + \frac{dV}{V} \qquad \text{[V.4.12]}$$

And the dP/P term is ignored since $dP = 0$ as the initial assumption. Furthermore V is the average volume of element n from t_1 to t_2 and is: $(V_{b1n} + V_{b2n})/2$. It now follows that:

$$dV_{bn} = C_{RC}\, C_{SH}\, \frac{(V_{b1n} + V_{b2n})}{2} dt \qquad \text{[V.4.13]}$$

Deriving the end of step pressure P_{2n}

For either case, relating to Chapters 7 or 9, the following steps are equally applicable.

At time $t + dt$ the same element would, if pressure remained constant, occupy volume V_{b2nC}, which in general will differ from V_{b2n} given by [V.4.9] where the new value (with dV_{bn} from either [V.4.11] or [V.4.13] according to choice of creation rate) becomes:

$$V_{b2nC} = V_{b1n} + dV_{bn} \qquad \text{[V.4.14]}$$

Since the volume will not have changed and will remain as given by [V.4.9] the pressure will change instead. So this means the pressure $P_{2\,n}$ cannot be the same as P_{1n} but has to be corrected And the dP/P term is still ignored since $dP = 0$ as the initial assumption. to become:

$$P_{2n} = P_{1n} \frac{V_{b2nC}}{V_{b2n}} \qquad \text{[V.4.15]}$$

Hence:

$$\frac{dP_n}{dt} = \frac{P_{2n} - P_{1n}}{dt} \qquad \text{[V.4.16]}$$

However, this is not required for substitution in equation [V.4.6] since the required partial differential is given directly from the arrays of P_2 and b_2 as:

$$\frac{\partial P}{\partial r} = \left(\frac{P_{2(n+1)} - P_{2n}}{r_{2(n+1)} - r_{2n}} \right) \qquad \text{[V.4.17]}$$

(This result actually resolves the concern regarding what was thought might be conceptual error number two. The original derivation given in Chapter 7 is shown to remain valid in this respect. Only error number one is also now corrected.)

The routine has to make the calculation from [V.4.17] for all values of n in the array to produce a pressure/radius profile at time

$t_2 = t + dt$.

Substitution of [V.4.17] in equation [V.4.6] using the array providing P_{2n} and b_{2n} yields the end of step acceleration. as:

$$\frac{dv_{2n}}{dt} = -\left(\frac{P_{2(n+1)} - P_{2n}}{r_{2(n+1)} - r_{2n}}\right)\frac{v_p^2}{P_{2n}}\frac{r_{2n}^2\left(r_{2(n+1)} - r_{2n}\right)}{\left(r_{2(n+1)}^3 - r_{2n}^3\right)} \qquad \text{[V.4.18]}$$

This equation has to be used for the ends of the array and can be used throughout. It will be slightly more accurate, however, for the remainder to use both elements adjacent to element n to determine $\partial P/\partial r$ so that the above now becomes:

$$\frac{dv_{2n}}{dt} = \left(\frac{\left(P_{2(n+1)} - P_{2(n-1)}\right)2}{r_{2(n+2)} + r_{2(n+1)} - \left(r_{2n} + r_{2(n-1)}\right)}\right)\frac{v_P^2}{P_{2n}}\frac{r_{2n}^2\left(r_{2(n+1)} - r_{2n}\right)}{\left(r_{2(n+1)}^3 - r_{2n}^3\right)}$$

[V.4.19]

For n=1 the central sphere is identified. So equation [V.4.18]is used. With 50 elements in the array values of both P and r for n=51 are required so that a value for n+1 is made available. For all other elements equation [V.4.19] is to be used.

These provide the value of dv/dt for use in equations [V.4.7] and [V.4.8] for the next time step. This is started by replacing suffix $_2$ by suffix $_1$. The process can then be repeated indefinitely to obtain a complete solution to the creation problem.

Actual programming, however, has not been as simple as expected. The solution proceeds for several time steps and then goes violently unstable. Efforts to eliminate the problem are under development.

V.5 More appropriate units

The second is an inconvenient time unit when time is measured in billions of years Therefore the computation carried out uses a billion years – BY as the time unit with distances measured in billions of light years – the BLY: unit.

Conversion Factors 1 year = 31.56×10^6 s: **c** = 2.997925×10^8 m/s by equating numerators to denominators so the original units cancel:

Time 1 BY = $31.56\times10^{6\ +9}$s = 3.156×10^{16} seconds.

Distance 1 BLY = $3.156\times10^{16}\times$c = 9.4615×10^{24} metres

The computer programs use speeds in m/s and accelerations in m/s^2

to obtain distances in metres and then convert using the above. Now v_P is in m/s and dr in billions of light years so first convert

Hubble Constant H_0

Cosmologists quote the rate of expansion of the universe as:

$v = H_O D$ with $Ho = 71$ km/s/M_{PC} (M_{PC} = megaparsec)

Where v is the recession velocity. Then we can put:

$$H_O = \frac{71{,}000\ m}{MPC \times s} \times \frac{1\ MPC}{3.26 \times 10^6\ LY} \times \frac{1\ LY}{31.56 \times 10^6\ s \times c\ m/s}$$

Each numerator and denominator is equivalent and each step is arranged so that the unwanted units cancel out. Hence this converts to: $H_O = 2.3019 \times 10^{-18}\ s^{-1}$. ($c$ as c_D given below) Also

H_0 in $M_{PC} = 977.3\ (v/c_D)/D$ with D in 10^9 light years i.e. BLY

V.6 Data for use with ECM and the Big Breed Theory

Solar System

$G = 6.67259 \times 10^{-11}$ Nm^{2}kg^{-2} is the Newtonian gravitational constant (1998 data).

$c_D = 2.997925 \times 10^8$ m/s is the **speed of light** at Earth orbit.

$m_E = 5.976 \times 10^{24}$ kg is the mass of the **Earth** in kilograms

$r_D = 149.6 \times 10^9$ m (metres) **mean orbital radius of Earth**.

(Apogee radius = 152.0×10^9 m: perigee radius = 147.0×10^9 m)

$r_{ES} = 6.378 \times 10^6$ m is the equatorial radius of the **Earth**. (0.003% greater than measured across the poles)

$v = 29.79 \times 10^3$ m/s the mean **orbital speed of Earth**

$m_S = 332{,}946 \times$Earth mass is the mass of the **sun**

$r_{SS} = 0.696 \times 10^9$ m the equatorial radius of the **sun**

$r_p = 108.2 \times 10^9$ m is the mean orbital radius of **Venus**.

$r_1 = 45.9 \times 10^9$ m perigee or perihelion radius of **Mercury**.

$r_2 = 69.7 \times 10^9$ m apogee or perihelion radius of **Mercury**

Sub-Atomic Particles (A **proton** has 1836.1×electron mass.)

$m_e = 9.1091 \times 10^{-31}$ kg is the mass of an **electron**

$m_p = 1.67252 \times 10^{-27}$ kg is the mass of a **proton**

$m_n = 1.67482 \times 10^{-27}$ kg is the mass of a **neutron**

$a_p = 1.1 \times 10^{-15}$ m is the approximate **radius of the proton**

$a_0 = 5.29167 \times 10^{-11}$ m is the 'Bohr radius' of the **H atom**

$h = 6.6256 \times 10^{-34}$ J-s (Joule seconds) is **Planck's constant**

1 ev = 1.6021×10^{-19} J (joule or Nm) is the energy gain when a charge equal to that of an electron falls through one volt.

ANGULAR SIZE DISTANCE RELATION

On November 1st 2010 it was realised that the equations derived in Chapter 9 concerning red-shift will also predict refraction of light. This will produce a magnification of the size of distant galaxies that will, to some extent, offset the reduction expected from distance. Then a Googol search of the internet showed the effect has already been observed by astronomers.

DERIVATION

Photons do not move like cannon-balls and in Chapter 9 it was argued that the momentum carried by a photon has to be based on its flat space value c: independent of spatial variation in the density of space (i-ther). This is also supported by the theory of quantum gravity associated with the Big Breed theory. This gives the correct value for the perihelion advance of planets that is over-predicted when momentum is assumed to be affected by space density.

It needs to be emphasised that, in the pressure/radius profiles computed, the speeds of recession include the effect of reduction in pressure with radius increase. However, this reduction makes the flat space recession speed, now to be given the symbol u, and the flat space speed of light c, smaller than the observed values. Since these reductions are in the same proportion the effects cancel.

The frame of reference is to be at an origin placed observer

The momentum of a photon, moving with speed c-u, in a radially inward direction, will be defined as $p_u = (c-u)E/c^2$ whilst its momentum component in the tangential direction is $p_v = v \times E/c^2$.

If another photon is moving at an angle α, measured from the radial direction, this angle has to be based on the local frame between R_1 and R_2 representing the element Δr under consideration.

Hence: $v = c \times \sin\alpha$ - near enough $v = c \times \alpha$.

A photon moving forward in time across a radial distance Δr from station 1 to 2 is traced back in time from 2 to 1. Relative to an origin-centred observer equating radial momenta $p_{u2} = p_{u1}$: yields:

$$\frac{E_2}{E_1} = \left(\frac{c_2}{c_1}\right)^2 \frac{(c_1 - u_1)}{(c_2 - u_2)} \qquad \text{[AS.1]}$$

Equating tangential momenta yields:

$$\frac{v_1}{v_2} = \frac{E_2}{E_1}\left(\frac{c_1}{c_2}\right)^2 \qquad \text{[AS.2]}$$

A photon moving at angle α_1 to the radial will be refracted to angle α_2 in moving from 1 to 2 and will be given by:

$$\alpha_1 = v_1/c_1 : \ \& \ \alpha_2 = v_2/c_2 \qquad \text{[AS.3]}$$

Substituting for E_2/E_1 from [AS.1] into [AS.2] and substituting the result for v_1/v_2 in [AS.3] yields the angle ratio from [ASA.3] as:

$$\frac{\alpha_1}{\alpha_2} = \frac{(c_1 - u_1)}{(c_2 - u_2)}\frac{c_2}{c_1} \qquad \text{[AS.4]}$$

If light moves on an exact radial path from one edge of a galaxy of size S, then it will appear to have angle α_0 from the opposite edge. Then the apparent size will be: $S_{AP} = \alpha_0 R$
where R is the radial distance of observer to that galaxy.

Owing to refraction, however, and going back in time, at some arbitrary radius r_2 the angle will have reduced to α_2 and then in moving further distance Δr to r_1 will have reduced further to α_1. Consequently the gain in size as seen from an observer at r_2 will be:

$$\delta S = \Delta r\left(\alpha_2 + \alpha_1\right)/2 \qquad \text{[AS.5]}$$

Hence the apparent size S_{AP} will be related to the true size S by:

$$\frac{S_{AP}}{S} = \frac{R}{\sum_0^R \Delta r\left(\alpha_2 + \alpha_1\right)/\left(2\alpha_0\right)} \qquad \text{[AS.6]}$$

If the average size S of a galaxy is assumed to be 100,000 light years (10^{-4} BLY) then with R measured in BLY the apparent observed angular size α_0 in arcseconds becomes:

$$\alpha_0 = \frac{20.62}{R}\frac{S_{AP}}{S} \qquad \text{[AS.7]}$$

Results of a computation using "Un141110" had the input data:
H_0=36 km/s/M$_{PC}$: P_{SH}/P_L=0.15: E_X=0.1: R_{MO}=20 BLY: ΔT_B=1.3 BY
The predictions for the present era (M=15) yield R_{sh}=44.4 BLY with u_{SH}/c_0=1.5 and the origin pressure becomes $P_0/P_L = 0.619$.

TABLE AS I ANGULAR SIZE (Tracing light path back from R=0)

N_F M	Z	u/c_0	c_T/c_0	R_{BLY}	S_{AP}/S	α_1	α_0
3 14	0.113	0.053	1.053	1.283	1.027	15.04	16.59
5 11	0.455	0.193	1.240	5.250	1.102	13.07	4.33
16 9	0.780	0.300	1.377	8.021	1.158	11.66	2.98
38 7	1.136	0.397	1.506	10.88	1.221	10.42	2.31
54 5	1.407	0.511	1.637	13.82	1.291	8.97	1.93
78 3	1.68	0.619	1.762	16.87	1.370	7.60	1.68
83 2	2.297	0.680	1.828	18.34	1.415	6.82	1.59
94 1	3.514	0.734	1.892	19.87	1.463	6.13	1.52

The above represents the low extreme for the Hubble constant at 36 and the high extreme will be assumed as H_0=71. This had to adopt

R_{MO} = 9.5 BLY and ΔTB = 0.65 resulting in P_0/P_L = 0.546. This pressure is about the maximum allowable and higher values of R_{MO} return still higher pressures – which is why R_{MO} is so low. The pressure limit is set by the need for the i-ther at any neutron star to have $P/P_L<1$ (the black hole value in our theory). In *QUANTUM GRAVITY* i-theric pressure at the surface of a neutron star of 1.4 solar masses is shown to be about 1.5 times the value at Earth.

This computation gave the radius of the shock front for the present era as only 21 BLY moving at $1.43 \times c_0$. A similar α_0 variation resulted with S_{AP}/S = 1.38 at R = 9.85 BLY where u/c_0 = 0.71 and Z = 3.35. Again no minimum of α_0 appeared.

These results seemed disappointing at first since published theories, from an internet search, suggested that much larger magnification factors than given by equation [AS.6] existed. Indeed a peak minimum value of α_0 is reported at a distance of 5.7 BLY where Z=1.6. Then α_0 increases with greater increase of distance.

However, the assumption has been made that the average size of distant galaxies is the same as local ones despite i-theric pressure reducing with radius. Our theory therefore suggests that the true size of galaxies increases as pressure reduces: S increases $\propto 1/P$.

Next it no longer seems possible to predict the maximum recorded value of red shift, which is well over 6, unless some energy absorption by the Aspden effect occurs (See Chapter 9 p.216). The extra shift might enable the Aspden effect to be quantified from observed red-shift and luminosity-distance data.

If F is the flux (energy/unit area) of photons arriving at distance r from a star and $KFdr$ is the loss due to absorption by free electrons then the energy balance can be written:

$$F = (F + dF)(1 + dr/r)^2(1 + Kdr)$$

Expanding and omitting second order terms and rearranging in integral form, with F_S and r_S referring to the star surface, yields:

$$\int_{FS}^{F} \frac{dF}{F} = -2\int_{rs}^{R} \frac{dr}{r} - K\int_{rs}^{R} dr \qquad \text{[AS.8]}$$

$$\therefore \log_e(F/F_S) = \log_e(r_S{}^2/R^2) - K(R - r_S)$$

The observed luminosity distance D would assume K=0. Then: $F/F_S = (r_S/R)^2$. This in [AS.8] yields:

$$KR = \log_e(D^2/R^2) \qquad \text{[AS.9]}$$

The code can now be used to evaluate both R and K by inputting H_0 and R_{MO} values until observed and theoretical red-shifts match.

C_U - the ultimate speed of primaries
C_{SH} - net creation rate at shock front where $P_{SH}/P_L = 0.15$ always.
C_{NT} - net creation rate at pressure P/P_L from [10.4.4] p.222
$C_{NR} = C_{NT} / C_{SH}$
C_{NA} Net creation constant.
C_{RT};C_{BT};C_T see [7.3.10] – [7.3.16], pp. 177-178 creation rate constants for collision breeding with no annihilation.
C_{PF} - phase coupling factor – for motions of +ve & -ve primaries
c speed of light: c_T observed speed of light allowing for non-uniform density of i-ther: c_D datum speed of light c_0 at origin
E - sum energy: E_0 rest energy (of particle): E_K its kinetic energy
$E = E_0 + E_K$
i - impulse: $di = d(mv)$: and I is moment of inertia mk^2
k - radius of gyration
F - force in Newtons
H_0 - Hubble constant in s^{-1} = value in km/s/Mpc×3.242×10^{-20}
P - pressure: P_L i-theric liquidus pressure: P_0 at origin centre: P_{SH} at shock front of i-ther
p - momentum = mv: suffix p or P – positive (mass or energy)
M – profile number: M=0 15dT_B 109 years ago: M=15 now.
m - inertial mass: m_0 rest mass: m_K kinetic mass: $m = m_0 + m_K$ - kg
n - number of primaries per unit volume: suffix n or N - negative
N-Path line number: N=0 @R=0: N=100 outer path line: 101 @ R_{SH}
R - Radius in metres m; or LY light years; or BLY - 10^9 light years
r - radius units as above.
V - volume in m^3
v - velocity: v_p of a primary of +ve mass; v_N or v_n for a primary of -ve mass. Velocities in m/s. Average v_p =1.465×c = 4.389×10^8 m/s
x,y,z - spatial co-ordinates
t - time in seconds: s
ρ - mass density kg/m^3
ε - energy density of i-ther J/m^3: $\varepsilon = \rho C_U{}^2$ (ρ of i-ther not matter)
$\alpha, \beta, \phi, \theta$ - angles also ε sometimes means angle.
σ - diameter of a primary - metres
λ - mean free path of a primary - metres

Units and Definitions

BY means 10^9 years: 10^9 years=3.156×10^{16} seconds (s)
BLY means a distance of 10^9 light years:
R metres (m) = 9.4615×10^{24} × R in BLY.
'i-ther' is a sub-quantum all-pervading medium
'primaries' are the substance of i-ther and of all that exists
'History of i-ther' is specified as starting from 15 BY ago.
'Pre-history of i-ther' starts when its radius was 1 m to 15 BY ago.

NAME INDEX

SUBJECT INDEX

SUBJECT INDEX

REFERENCES

Abbott, F. (1988): *Baby Universes and Making the Cosmological Constant Zero*: Nature,Vol.336, 22/29 Dec.1988, p 711

Adler, Stephen (2004): *Quantum Theory as an Emergent Phenomenon* Cambridge University Press. ISBN 0521831946

Aspect, Alain et al (1982): *Experimental Tests of Bell's Inequalities used in Time-varying Analysers*: Phys.Rev.Letts,Vol.49.No.25: 20 Dec.1982

Aspden, H. (1984): *The Steady -State Free-Electron Population of Free Space*: Letters Al Nuovo Cimento. Vol.41,N7(1984)

Backhousc ct al, (1991): *Essentiul Pure Mathematics* Longmans

Blanchard, C.H. (1969*): Introduction to Modern Physics* Sir Isaac |Pitman & Sons Ltd., London

Bohm, David. (1980) *Wholeness and Implicate Order* Routledge & Kegan Paul, 39 Store St., London WC1E 7DD

Davies, Paul (1989) *The New Physics*: Cambridge University Press

Dingle,Herbert (1972): *SCIENCE at the Crossroads* Martin Brian & O'Keeffe, London

Feynman, Richard P.(1985): *QED, The Strange Theory of Light and Matter*; Princeton University Press, New Jersey (1985)

Greene, Brian, (1999): *The Elegant Universe: Superstrings, Hidden Dimensions, and the Quest for an Ultimate Theory* Jonatan Cape, London 1999

Gribbin, John: (2005) *Deep Simplicity: Chaos, Complexity and the Emergence of Life.* Penguin Books Ltd., 80 Strand, London WC2R ORL

Guth, Alan **&** Steinhardt, Paul (1989)**:** *The Inflationary Universe* The New Physics Ed. Davies, Paul: Cambridge University Press

Hawking, Stephen W. (1988,1996): *A Brief History of Time* Bantam Press, London

Hinton, G.E. (1992)*How Neural Networks Learn from Experience* Scientific American, Mind and Brain issue Sept.1992, pp 105-109

Hulse, R.A. & Taylor, J.H. (1975): Astrophys.J.Lett.195, L51-L53

Jeans,J.H. (1887,1984): *The Dynamical Theory of Gases* 4th Edition Cambridge University Press (1st Edition 1925 Dover Publications Inc.)

Kaku, Michio (2009): *Physics of the Impossible*: Penguin Books

Lerner, Eric J.(1992): *The Big Bang Never Happened* Simon and Schuster, London, West Garden Place, Kendal St. London W2 2AQ

Michelson, Albert A. & Morley, Edward, W. (1887): *On the Relative Motion of the Earth and the Luminiferous Ether*: The American Journal of Science, No.203,Vol.134,P.333-345, Nov.1887

Michelson, A.A. & Gale, H.G. (1925): Nature **115** No 2894 566

Novikov, I.D.(1983): *Evolution of the universe* Cambridge University Press

Pearson, Ronald D. (1990*): Intelligence Behind the Universe!*

ISBN 0 947823 21 2 Headquarters Publishing Co.Ltd., 5 Alexandria Rd., London W134 OPP (out of print)

Pearson, Ronald D.(1991): *Alternative to Relativity including Quantum Gravitation*: Second International Conference on Problems in Space and Time: St. Petersburg, (Sept.1991) pp 278-287. Petrovskaja Academy of Sciences & Arts Chairman Local Organising Committee: Dr. Michael Varin: Pulkovskoye Road 65-9-1 St. Petersburg 196140, Russia. FAX: (812) 291-81-35 Phone:Alexandre Alekseev: office:(7) (812) 291-36-73, Home:(7) (812) 173-55-69 E-Mail: consym@saman.spb.su.

Pearson, Ronald.D (1992): *Origin of Mind*: (the first derivation of a true solution to the problem of the cosmological constant) 70 page booklet available from author

Pearson, Ronald.D.(1994): *Quantum Gravitation and the Structured Ether*: Proceedings of the Sir Isaac Newton Conference. St. Petersburg (March 1993) pp 39-55 : see above for address except that this conference was hosted by the Russian Academy of Sciences.

Pearson, R. D.(1997): *Consciousness as a Sub-Quantum Phenomenon*: Frontier Perspectives, Spring/Summer 1997, Vol.6,No.2 pp70-78 (inadvertently omitted from list of contents)

Pearson, R. D. (2005) *A Paradigm-Shifting Physics Supports Immortality!*: Consciousness Series 7, (2005) Indian Council of Philosophical Research, Darshan Bhawan, 36 Tughlafabad Institutional Area, Mehrauli-Badarpur Road, New Delhi-110062

Pryde, J.A. (1966): *THE LIQUID STATE* Hutchinson & Co. Ltd. London W.1

Schwarzschild, B. (1998): *Very distant supernovae suggest the cosmic expansion is speeding up*: Physics Today, 51(6),17-19,1998

Smolin, Lee (2007): *The Trouble with Physics* Allen Lane/Penguin

Spaniol, Craig (WVSC) and Sutton, John F. (NASA-GSFC) (1994): ***'G'*** Russian Academy of Sciences (1994) pp201 (The Sir Isaac Newton Conference. See Ref. above for address)

Starobinski,A,A & Zel'dovitch,Ya,B, (1988) *Quantum Effects in Cosmology*: Nature. Vol.331,25 Feb.1988

Tryon, E. P. (1984): *What Made the World?* New Scientist,8/3/84, p14-18

Weinberg, Stephen (1989): *The Cosmological Constant Problem* Reviews of Modern Physics, Vol.61 (1) Jan 1989

Whittaker, Edmund T (1951) *History of the theories of aether and electricity*: The philosophical Library ISBN 0-88318-523-7

Will, C.M. (1988): *Was Einstein Right?* Oxford University Press,Walton St., Oxford, OX2 6DP

Wright, E,L (2005): *Supernova Cosmology* http://www.astro.ucla.edu/~sne_cosmology.htmlWright

Wright, Selwyn E. (2008): *Problems with Einstein's Relativity*: Trafford publishing.2008.

OTHER BOOKS IN OUR SERIES

CREATION SOLVED?

244 pages: A valuable Primer for the technical book of the series called *Quantum Gravity via Exact Classical Mechanics*.

Despite containing hardly any maths, this book shows how physics has lost its way. The concept of 'potential energy' seems simple and obvious but is being misused. It is still resulting in false logic: the source of major false prediction. The book also shows why negative pressures cannot be assumed to cancel mass-energy. This is shown to be the major reason why the big bang theory makes a hopelessly wrong prediction. Then by restoring lost expertise from an allied discipline these methodologies are superseded so that many vexed questions are resolved.

The need for a new exact mechanics to replace Einstein's relativity theories is established. Then the reader is enabled to derive a revised version of Newtonian mechanics to yield an exact theory. This is done by leading readers to do the maths themselves.

One appendix covers Newton's mechanics in detail.

Another summarises the differential and integral calculus.

These appendices are designed to enable intelligent readers, who have not yet studied the calculus, to carry out the led derivations for themselves.

(This book was originally CREATION SOLVED? PART I)

The next book of the series is an essential companion volume to the book we hope you have just read. It is:

QUANTUM GRAVITY

via an Exact Classical Mechanics ECM

233 pages

This second book of our series details the maths of a novel approach that provides a revision of Newton's mechanics. This elevates it from an approximate theory to one which is exact. Newtonian mechanics is limited to low speeds and weak gravity but ECM theory is applicable up to the speed of light and the extreme gravity of black holes.

The theory starts off with quantum compatible assumptions and yet matches all the achievements normally considered unique to Einstein's theories of both special and general relativity. Consequently a solution to the vexed question of finding a paradox-free solution to quantum gravity appears naturally.

Since the theory was required to operate equally at both macroscopic and sub-quantum levels of existence the basis had to be the particle nature of light and matter: not its wave nature. This is because electromagnetism does not exist in the sub-quantum domain. The surprising result is that the 'time dilation' effects predicted by Einstein are replaced by electro-mechanical mechanisms that cause clocks to lose time. Equations identical to Einstein's appear but clearly have a very different interpretation.

All the experimental checks are matched. These are the doubling deflection of starlight grazing the sun, the Shapiro time delay, the perihelion advance of planets and the life extension of cosmic rays.

To eliminate the internal contradictions that mar relativity theories speed induced mass increases and magnetic effects of moving charge are measured from 'local space'. The latter is the centre of the Earth up to a radius possibly greater than the distance to the moon. Then the local frame switches to a sun-centred position.

A large section of the book is devoted to design of new experiments most of them aimed at determining the position of the interface between these frames of reference.

Knowledge of the calculus is assumed.

(It was originally sold as CREATION SOLVED PART II)

THE BIG BREED THEORY
of creation of the universe

64 pages. This book is a popular introduction with many illustrations in full colour, showing how breeding and annihilation can account for the existence of a creative medium leading to the formation of our universe.

This is not recommended for people who have read the present volume but might wish to make it available to those having little mathematical understanding.

(Available for internet download as well as purchase. A cheap version having only the covers in colour is also available)

INTELLIGENCE BEHIND THE UNIVERSE II

The grand unification of physics and survival

326 pages

This popularisation of our six book series links survival via the Big Breed theory of creation and is still available. However, the original work has now been revised to improve accessibility. It is now recommended as one of two alternative options:

OPTION 1: is recommended for people who want a brief introduction to the history of science but totally reject the possibility that an immortal mind can exist. It shows how physicists have lost expertise in classical mechanics that has led to the failures they now admit are blocking progress. It then provides solutions described without any maths. This 193 page book is called:

Physics Hits the Buffers: Why?

Read Our Solution!

Written mostly for young non-technical people aspiring to science

OPTION 2: is an easy read with no maths or technical jargon. It is recommended for people who do not totally reject the possibility of there being a mind separate from brain with the potential of immortality. It gives a history of the way survival of death was proved experimentally by a number of scientists starting with Sir William Crookes in 1870 and shows how scientists still resort to corruption to discredit the evidence. The author's own investigations give further support crowned by a theory that unifies physics with survival and the 'paranormal'. This 234 page book is called:

INTELLIGENCE BEHIND THE UNIVERSE III

The grand unification of physics and survival

All available via the websites:

www.lulu.com or www.bigbreed.org and www.darkenergymysterysolved.co.uk. Some are listed on Amazon

Watch out for the DVD that starts with a lecture by Pearson and ends with a computer experiment.